Public Involvement in Energy Facility Planning

About the Book and Editor

Because the power industry is anticipating greatly increased generating capacity requirements in the 1990s, political controversy over electricity demand and supply is likely to return to--and perhaps surpass--the level of rancor experienced during the 1970s. Fortunately, a sizable number of utility companies have come to believe that destructive confrontation can be avoided through public involvement early in facility planning, as well as through broader corporate policy.

If these initiatives might be termed the beginnings of a "revolution," so far it has been a quiet one in the sense that published literature is sparse. This book seeks to improve the situation by bringing together, under joint sponsorship of the American Planning Association and the Edison Electric Institute, an extensive collection of articles, proceedings, case studies, and other relevant accounts of the state of the art. It is the most comprehensive work to date on how the concept of open planning has been put into practice by progressive electric companies.

Dr. Dennis W. Ducsik is a research associate in the Energy Studies Group of the Center for Technology, Environment, and Development at Clark University, Worcester, Massachusetts.

Public Involvement in Energy Facility Planning

The Electric Utility Experience

edited by Dennis W. Ducsik

Routledge
Taylor & Francis Group
LONDON AND NEW YORK

First published 1986 by Westview Press

Published 2019 by Routledge
52 Vanderbilt Avenue, New York, NY 10017
2 Park Square, Milton Park, Abingdon, Oxon OX14 4RN

Routledge is an imprint of the Taylor & Francis Group, an informa business

Library of Congress Cataloging-in-Publication Data
Public involvement in energy facility planning.
(Westview special studies in natural resources and energy management)
"Annotated bibliography on citizen participation in power facility siting": p.
1. Public utilities--Location--Decision making--Citizen participation. I. Ducsik, Dennis W., 1946- II. Series.
HD9685.A2P83 1986 363.6'2'068 86-1620

ISBN 13: 978-0-367-28469-5 (hbk)

Contents

PART THREE
SELECTED PUBLIC INVOLVEMENT PROGRAMS

Tables

Figures

Foreword

For decades, the electric utility industry sought locations for its new power plants in a fairly simple and straightforward manner. In many instances, all that was required was a flat tract of land of suitable size, an adjacent source of cooling water and, hopefully, reasonable proximity to the source of the load to be supplied. With the advent of the 60s and 70s, individual and collective attitudes toward such projects changed as the country experienced a heightened sensitivity to environmental values. At the same time, however, utilities were still taking advantage of economies of scale, particularly with respect to power plants, with the size of these units becoming increasingly large. Under such circumstances, the clash between environmental values and utility construction projects was inevitable. Such conflicts often resulted in, if not outright collapse of the proposed project, spectacular battles on the legal front and in the media. Storm King Mountain (N.Y.), Kaiparowits (Utah), Seabrook (N.H.), Turkey Point (Florida) -- all are reminiscent of major struggles between utilities and environmental interests.

Confrontations such as these led several utilities to realize that decisions which did not adequately take into account the views of affected groups and individuals would be difficult (if not impossible) to implement, perhaps even after regulatory approval was secured. To such companies, the concept of public involvement in planning, having gained legitimacy in transportation-related and urban renewal applications, seemed to be an attractive alternative to the old way of making siting decisions in strict confidence. Citizen participation theory holds, basically, that public consent for a proposed action cannot be developed in the absence of public trust that social values and concerns have been evaluated in parallel with technical considerations. And this is best achieved in the context of an open, visible, and credible decision-making process to which the public has access.

Early utility efforts at open planning were fashioned on little more than the notion that it was somehow "the right thing to do," or that it would "improve the company image." As time went on the practice of public involvement matured to the point where the subject found its way onto the agenda of many trade and professional meetings in the power industry. By the early 80s, the community of utilities with several years of experience had become sufficiently large to warrant the creation of a separate forum, in which practitioners could share information on their respective programs and evaluate the relative merits and future of public participation.

In March of 1982 the Edison Electric Institute took an important first step in that direction -- under the auspices of its Energy and Environment Committee -- by convening a workshop which brought together the key figures in utility participation programs all over the country. Before adjourning, the attendees urged that a permanent entity be established to advance the state-of-the-art of public involvement in electricity planning. Responding affirmatively later in 1982, EEI decided to create a standing Task Force on Public Participation. In its first year of operation the Task Force, in cooperation with the EEI Consumer Affairs Committee, sponsored the preparation of a handbook specifying the details of techniques for utility-citizen interaction. A training session to introduce the manual to utility personnel was subsequently held in Chicago, Illinois, in the summer of 1984.

This volume, the second major outgrowth of the 1982 workshop, represents a companion to the handbook effort in that it explores some of the larger issues associated with the design and implementation of public involvement programs. It encompasses much of the generic literature in the field, including not only reprints of major journal articles but also significant unpublished material such as the proceedings of the 1982 workshop (and of a similar event held in Canada the year before). What's more, it serves as the first formal outlet for reports on some of the more established company programs, together with several case histories of individual projects (three of which are covered in depth). The discussions herein should be of interest to those in municipal as well as investor-owned utilities; to those looking for information on participatory approaches to transmission routing as well as power plant siting; and to those who want an honest appraisal of the risks of public involvement as well as the benefits.

While the concept of public participation within the electric utility industry is firmly established, its future applications in various aspects of industry operation are largely untested. Much practical experience in the area of facility siting has been gained and documented. But what of applications beyond siting? Many utilities, and perhaps the industry itself, are in a state of transition marked by changing technologies, financial conditions, and customer expectations. This has created significant opportunities for the industry to develop new applications of public involvement in such areas as consumer affairs and strategic planning, or in dealing with emerging environmental issues like sulfur emissions and waste disposal. Utility management, in other words, will continue to be challenged to solicit public response in the evaluation of electricity-related problems and their proposed solutions.

Bernard J. Bujnowski,
Chairman, 1982-1984,
Task Force on Public Participation,
Edison Electric Institute

Acknowledgments

For assistance, support, and encouragement in the preparation of this manuscript, my gratitude extends to an array of people with whom, to the person, it has been more than a pleasure to work. First among these are my coauthors, who not only produced text in a cogent and timely way but also graciously tolerated my sometimes considerable editorial intervention. So, too, did my typists--Jane Adler, Terry Reynolds, and Karen Shepardson--prove to be highly diligent and accommodating, to say nothing of totally competent in operating the arcane (as yet to me) technology of word processing. Working behind the scenes, but by no means obscure to me, were Gerald Edgely and (particularly) his successor, Rick Loughery, staff members at the Edison Electric Institute to whom belongs the credit for most capable administration of that organization's sponsorship of the project. Finally, I owe a special debt of thanks to two executives of the Pennsylvania Power and Light Company: Gerry Farber, who gave me my first major opportunity to work with the network of citizen participation practitioners that exists in the electric utility industry; and Bernie Bujnowski, who saw clearly, from his position as first chairman of the EEI Task Force on Public Participation, that documenting the experiences of that practitioners network could go a long way toward stimulating its further growth.

Dennis W. Ducsik

1

Introduction: The Quiet Revolution in Utility-Citizen Interaction

Dennis W. Ducsik

On the pages of Public Utilities Fortnightly in early 1975, Mason Willrich -- then a professor of law and now a vice-president for corporate planning at the Pacific Gas and Electric Company -- challenged utility executives to adapt innovatively to the profound changes in operating climate the power industry had witnessed since the mid-60s.[1] The touchstone of his argument was that an ambivalent, volatile, and increasingly hostile political environment had supplanted the societal values which traditionally favored utilities, i.e., consensus on growth, technological optimism, confidence in government, and deference to business judgement. Under these new conditions, Willrich said, the take-it-or-leave-it attitude toward implementation of facility development plans was no longer appropriate, because it engendered substantial and unnecessary bad will. As an alternative he recommended that broad public participation become a principle of operation in all major decisions, including (among other things) the siting of power plants and transmission lines.

Willrich, of course, was neither the first nor the last to recognize the bankruptcy of the traditional "decide-announce--defend" approach to making electricity decisions and to endorse, in the alternative, the concept of "open planning". As can be seen from the sampling of statements in Appendix A, throughout the 70s a number of

---

This chapter is based in part on "Power Plants and People: A Profile of Electric Utility Initiatives in Cooperative Planning", APA Journal (Spring, 1984), and is reprinted with the permission of the editor.

opinion leaders issued calls for early, informal, and two-way communication between utilities and the public, arguing that constructive dialogue seldom occurs in the distinctly adversarial and highly legalistic arenas of government regulation. The former Carter Administration even attempted to institutionalize the principle of collaboration via its proposed Nuclear Siting and Licensing Act of 1978, which would have directed the Nuclear Regulatory Commission to "encourage or require" utilities to "engage in open and advance planning (for nuclear reactors)...with opportunity for the Commission and interested parties to participate (therein)".[2] Recently echoing the same basic theme, moreover, was a U.S. General Accounting Office report to the House Committee on Energy and Commerce, entitled The Federal Government Should Encourage Early Public, Regulatory, and Industry Cooperation in Siting Energy Facilities.[3]

Although power system expansion programs are now dormant in many regions of the country, it would hardly be prudent to infer that controversy over facility siting is therefore a thing of the past. In this regard it is important to note that the electricity industry and the federal government alike are still projecting substantial new capacity requirements to the year 2000 and, consequently, a need for many new sites. Consider, for example, that:

* the much-publicized DOE report on The Future of Electric Power In America has put the growth figure at 53 to 105 percent of the supply in place as of 1981,[4] and a contemporary study by EPRI has indicated that a similar projection band would require between 205 and 630 new sites just for big nuclear and coal-fired units;[5]
* Secretary of Energy Donald Hodel has pointed out that something in the neighborhood of 20-30 percent of present capacity will become physically or economically obsolete in twenty years, which to him means that between 120 and 180 GWe (150 'typical' nuclear units or 200 to 300 coal-fired units) will be needed just for replacement purposes.[6]

Thus, the prevailing view among utility managers is that, as Secretary Hodel puts it, "if we are going to meet even our minimum additional requirements, the necessary power plants ought to be on the drawing boards today." Further evidence to this effect lies in the results of a 1982 EPRI poll, which revealed that 42 percent of utility executives thought they would order a new nuclear reactor within 10 years.[7]

These scenarios are the subject of rancorous debate with environmental and consumer groups, of course, who point especially to the problematic aspects of nuclear power and who vigorously promote energy efficiency and 'soft' production technology. But one must also realize that the locational problem would still remain even if the large baseload power plant were to become a 'dinosaur' in technical and economic terms, as activists these days hope and claim will occur. This is clear from the aforementioned EPRI study's estimate to the effect that a total of between two and four thousand new sites could be needed by the turn of the century, with the majority in either case being associated with small-scale hydroelectric generation and other renewable resource-based technologies. And this does not include still other categories of facility planning activity in which political resistance has become a fact of life, such as the siting of transmission lines and of rail access to coal-fired stations (especially those recently converted from oil).

With these and related issues (e.g. rate-setting) still controversial in the public eye, it is obvious that the political atmosphere surrounding questions of electricity demand and supply is no less polarized now than it was during the previous decade. Correspondingly, it cannot be said that the need for creative approaches to conflict resolution in the utility sector has diminished in recent years. Very much to the contrary, if anything this need has grown more acute!

That increasing contention can be seen on the horizon does not suggest, however, that citizen participation will necessarily experience rapid penetration in the power industry. In this respect it is important to understand that the historical attitude of most utilities toward interactive modes of planning has always been unreceptive. Although some interest in the idea was in evidence by the late 60s, throughout the 70s the conventional wisdom was that getting together with affected interest groups would constitute an unnecessarily radical departure from established public relations strategies. This view was conditioned in part by the advent of NEPA and a variety of pollution control laws at the Federal level, together with the passage of pre-construction licensing laws by many states -- all of which made citizen "participation" (albeit in the narrow, adjudicatory sense) a formal part of the regulatory process. Such reforms on behalf of environmental protection, the industry hoped, would resolve most questions of plant/site acceptability and obviate the need to develop altogether new strategies for dealing with the public and its concerns.

The industry's wait-and-see policy was also strongly motivated, it should be noted, by acute wariness over the prospects for constructive collaboration. In fact, as this author learned from a series of interviews in the mid-70s,[9] the belief was widespread among utility planners and managers that citizen involvement in siting would invariably turn into a 'can of worms,' both politically and in a technical sense. And it was equally clear that a substantial burden of proof would have to be satisfied if this fear of participation-induced problems was ever to be overcome, if only because so much was at stake in the facility planning function. One respondent put it this way:[10]

> First of all, the issue of siting has become far more complex. It is now no longer a matter of building fossil fuel power plants or hydro-electric power plants which were often in remote areas and at a time when urbanization was not as widespread as it is now. Today, complex electric generating systems are being utilized, requiring a complex analysis for siting... vast investments are involved, and to open a decision-making process to "undefined forces" would make the investment much too risky. A corporate management thus faces a difficult decision: they are involved in a complex issue with vast sums of money committed to generate electricity for the future. In these times when every decision has to be justified to the stockholders as well as to the public and government, a trial and error process 'open' to other parties for decision-making is very difficult to justify. Most corporate decisions are based on 'cost/benefit' analysis and with all the pitfalls involved in public participation, to convince management to undertake a process with potentially very high costs and potentially high but generally invisible benefits is at best a difficult proposition. Thus, acceptance of public participation has been slow.

For similar reasons, the bulk of managerial personnel today remain negatively predisposed toward the concept of public involvement, especially in the more pivotal aspects of corporate decision-making.

But there are also definite signs that risk-averse attitudes are giving way, if gradually, to a conviction that sound business practice requires a dialogue with interested parties throughout the course of planning. Among the several manifestations of this incipient transition is that a vanguard of utility officers has become vocal in warning that strict reliance on representative institutions for expressions of the public interest can result in dangerous political isolation. One company president recently said, for example, that "failure to give weight to public concerns in making business decisions is to put your business at risk."[11] Another executive, in writing specifically about the value of consumer advisory panels, put it this way:[12]

> The idea of asking a group of citizens to advise us on how to run "our" business has been a difficult concept for most utility managers to get used to. And that is precisely why many utilities have decided they need consumer advisory panels--as a constant reminder that we cannot any longer consider this solely "our" business, not when our business has such a major impact on the public and not when energy costs are taking an ever-increasing chunk of the consumer's income.

Rhetoric does not always correspond to reality, of course, but in this case there is reason to believe that a significant number of companies give more than lip service to the cooperative school of thought. One indicator is the noticeable trend toward establishing permanent advisory boards or councils;[13] another is the frequency with which community interaction is now an important element in socioeconomic impact mitigation.[14] What's more, as Chapter Four of this volume reports, there are at least 34 North American utilities (including five in Canada) that have set aside their concerns and engaged in some form of nonadversarial interaction with the public while planning a power plant. There are even a handful of pioneering companies for whom citizen participation apparently has become a matter of routine across-the-board. Thus, while participatory ventures are still very much the exception and not the rule, there is certainly a new constituency in place where one did not exist a decade ago.

Interestingly, if the aforementioned activity can fairly be termed the beginnings of a "revolution," so far it has been a rather quiet one. No trade association or other industry-wide organization has yet developed a clearinghouse to foster cross-talk on public involvement activity; indeed, there has never been a full-scale conference for the new breed of outreach professionals to discuss and evaluate issues in participation programming. No one has even conducted a comprehensive and in-depth survey with an eye toward widespread circulation of the resulting data base.[15] In short, there has been very little information-sharing of the kind that would benefit both current practitioners and those who are giving interactive methods serious thought for the first time.

Also serving to retard the rate at which innovation has diffused through the industry is the fact that written accounts of utility experiences have not often appeared in the trade literature. Indeed, as anyone who has ever attempted to assemble a bibliography on the subject will readily attest, outreach projects are not always well-chronicled in a summary report or other descriptive material. Typical of what turns up in a careful search is shown in Appendix C, which includes:

* only 30 works available through standard publishing outlets, with the majority tucked away in a scattered array of technical journals, workshop transcripts, or other sources of relative obscurity and limited distribution;
* a mere 18 additional unpublished reports, most of which are internal company documents that do not appear on any indexes or computerized lists and are available only upon request;
* only one in-depth case history, also unpublished, which really takes an episode apart and puts it back together again in a richly detailed fashion.[16]

More could be found with more exhaustive effort, no doubt, but the fact would remain that the documentary record is far from voluminous and is certainly disproportionate to the level of program activity achieved to date.

Not only is the available literature sparse; it is also quite primitive when it comes to interpreting the experience gained. What few accounts have appeared in print are largely descriptive and anecdotal, as only a half-dozen or so could fairly be called issue-oriented. What's more, virtually nothing has been written about the bad experiences, or the disappointing aspects of those which were successful on the whole. This is certainly understandable, given that doing so would inevitably touch on points of embarrassment and sensitivity to the parties involved. But the lack of negative feedback is nevertheless unfortunate, because there is just as much to be learned from failure as from success--and possibly even more!

Why have the lessons of the past remained such well-kept secrets? Part of the reason may be that utility managers are not accustomed to writing formally about their experiences, or are not inclined to devote resources in any substantial amount to applied research (especially that which has an academic flavor). It may also be that companies developing interaction programs have simply been too preoccupied with day-to-day operations to keep records good enough to reconstruct the experience accurately, let alone conduct retrospective evaluations. Whatever the cause, however, we know from an informal survey conducted in 1980 that the problem of "organizational amnesia" is a serious one. This poll found that 10 out of 22 respondents had not formally reviewed their participatory experiences; that all but four of the 12 who did conduct assessments relied exclusively on internal sources; and that only three companies were using (or attempting to develop) a well-defined methodology or model for evaluation.[17]

Regrettably, scholars and research-oriented professionals have not made much progress either in developing a body of empirical knowledge in this field. Indeed, it almost seems to have escaped their notice altogether, considering that the number of published analyses of even narrow issues can be counted on one hand.[18] This paucity of substantive reference material stands in marked contrast to that which exists for citizen participation in other areas of project planning (e.g., water resources, highway location), where an enormous amount of applied work has been done.[19] The literature is also miniscule compared to the myriad articles and books the research community has put out on resistance to power plants, in which the anatomy of controversy has been painstakingly dissected many times over.[20] Thus, ironically, we have acquired much more insight on the mechanics of zero-sum confrontation than on the workings of cooperative problem-solving.

In the last few years, fortunately, some initial steps have been taken to collect and disseminate information on participation activity, and these deserve acknowledgement. First to be noted are the efforts of Ontario Hydro in Toronto, a company always among the leaders in the open planning field by virtue of its extensive outreach programs as well as considerable in-house research. In November of 1981 Ontario Hydro staff were instrumental in arranging Canada's first national workshop on utility-sponsored participation, which brought together representatives from nine companies to compare notes on their respective experiences in both plant siting and transmission routing. Much reference was made at that event to regulatory and cultural conditions unique to Canadian provinces, but many of the issues broached were relevant to the U.S. as well. Helpful discussions took place, for example, of various types of public "interface"; of the different locations for the participatory function within companies; of techniques that had proved workable as well as several operational problems; and of areas for further research.

In this country, coincidentally, the first major overview report on early public/regulatory/industry cooperation in siting power plants and other energy facilities was published--by the U.S. General Accounting Office--just one week after the close of the Canadian workshop.[21] In preparing it the GAO staff contacted and/or visited nearly 150 organizations and individuals, and in doing so identified 14 specific instances of industry-initiated open site planning. Four of these cases, all involving generating stations, were selected for discussion in the report because they were judged to be quite representative of the findings and conclusions the study team had reached. Thus a second sample of descriptors is available to convey the flavor of specific episodes, and to serve as the basis for extracting at least some initial hypotheses about the practical workings of an open planning process. (For a digest of the GAO findings and recommendations, see Appendix B.)

Another recent and vigorous attempt to explore participation initiatives among U.S. utilities was undertaken in March, 1982 by the Edison Electric Institute. At its Washington, D.C. offices the EEI convened a "Workshop on Utility Experience with Advance Public Participation in Planning," a two-day gathering chaired by this author and attended by 31 persons from 17 companies across the country. The agenda comprised a chairman's presentation to review industry activity in the aggregate; three subsequent case histories describing the programs and projects of selected companies; and a series of discussion sessions covering topics like benefits and risks of public involvement, strategies and techniques for effective interaction, and internal corporate organization/administration. Here too the focus was on participation in facility planning, although reference was made at several points to applications beyond siting as well.

This brings us to the present volume, which (as noted in the Preface) is an outgrowth of the 1982 EEI workshop.

There it was recognized that information on prior experiences and current programs was sorely deficient, and that a concerted effort should be made to disseminate existing literature (both published and unpublished) and to encourage further authorship among utility personnel and other planning professionals. Toward these ends, publication of a compendium was deemed to be a particularly appropriate measure.

Part One of this book consists of two background articles, reprinted from the Journal of the American Planning Association and Public Utilities Fortnightly respectively, which present most of the arguments (pro and con) often heard in speculative discussions about the workability of citizen participation. The former evaluates six basic concerns typically voiced by utilities about the practical implications of "democratizing" the site selection process, and is particularly helpful in understanding the basis for the industry's historical trepidation in this domain. The latter focuses more narrowly on the relative merits of consumer advisory boards, but the issues raised are of broad interest since committee formation is a logical and increasingly common element in public involvement programs more generally. The piece is especially interesting, moreover, because its cautionary discussion of potential pitfalls evoked a rather extensive and polemical response from several utility executives, whose letters to the editor and direct correspondence with the author are also included as a means of further illustrating the point-counterpoint nature of recent dialogue within the industry.

The implicit message of Part One is that judgments about cooperative planning should not be based on abstract argumentation or sweeping generalizations, and the subsequent papers therefore manifest an experiential approach to the question. In Part Two, the discussions remain at a generic level but center on the views of seasoned practitioners of participation, as reflected in the proceedings from the two major workshops held in 1981-1982. These are not verbatim transcripts, but narrative reports by the respective chairpersons summarizing and integrating the various lines of inquiry pursued at each event. And both incorporate some useful overviews: of company programs and techniques, in the case of the Canadian report; and of patterns in the design and outcome of some 58 specific projects, in the case of the EEI report.

Part Three takes a closer look at individual programs and projects sponsored by four utility companies: Ontario Hydro, Portland General Electric, the Tennessee Valley

Authority, and United Power Association. In each case information is presented on the genesis of the participation management function, the nature of commonly employed methods of interaction, and the company's most noteworthy episodes of public involvement. The closing chapter of Part Three recounts the work of a major consulting firm for public sector clients in the U.S. and Canada, and includes both normative statements of participation principles and operative flow charts to put them into practice. Taken as a whole, these five chapters provide a good sampling of recent and ongoing efforts to advance the state-of-the-art of open planning methodology.

The company-specific chapters of Part Three, by the way, are instructive not only with respect to the process of utility-citizen interaction but also as examples of the diverse issues the process can address. For Ontario Hydro (as well as UPA), the principal context of application has been transmission planning, and the experiences described range from urban siting of a transformer station to developing a locational strategy for an entire regional network -- the first experiment in open system planning known to this author. For Portland General Electric, the decision of whether to go ahead with a proposed hydroelectric plant at a river site of great ecological, aesthetic, and historic importance provided a timely opportunity to experiment with a new approach to community relations. And at TVA, the mobilization of public opinion on matters like the possible deferral of nuclear units, the disposal of low-level radioactive wastes, and the implementation of PURPA (the Public Utilities Regulatory Policy Act) demonstrates the relevance of collaborative techniques in areas other than facility siting per se.

The bulk of utility experience lies in the siting/routing category, however, and for this reason Part Four is devoted almost exclusively to more probing assessments--by non-utility observers--of four additional site selection projects sponsored by investor-owned utilities: Northern States Power (St. Paul, Minnesota), Northeast Utilities (Hartford, Connecticut), and Pennsylvania Power and Light (Allentown, Pennsylvania). Of these, three occured early in the 70s and one toward the end; three were successful and one was deemed a failure; and three involved baseload coal facilities whereas the fourth dealt with a pumped-storage unit. What is perhaps most significant about these cases, on the other hand, is not their points of departure but the fact that they all shared a common feature: the use of a citizen advisory committee as the principal mode of interaction. Publication of these detailed accounts will thus contribute to a fuller understanding of the key process design issues associated with this outreach device, not only for a siting study but also (as in the PP&L case) as a permanent "sounding board" on other matters of company policy.

All in all, it would not be an exaggeration to say that this collection of papers stands as the most comprehensive work to date on how the open planning process has functioned under the auspices of electric companies. This is tangible progress, yet it should also be said that there is still a great deal of research to be done before a truly ample body of knowledge exists on the subject. One should not lose sight of the fact that, for the vast majority of interactive projects, field data has yet to be assembled and subjected to the kind of scrutiny that can illuminate relationships between process structure and outcomes. One wonders, for example, what contextual factors (as opposed to the choice of program strategy and technique) have the greatest influence on the course of events? Similarly, what specific organizational features bear most heavily on the conduct of involvement programs? These and other issues in what it takes to make participation consistently workable have not, given the preliminary nature of this document, been addressed as fully and as carefully as they eventually need to be.

Thus, the direction for further work is readily apparent. If guidelines for future public involvement projects are desired, we need to analyze a much greater number of cases both individually and in comparative terms, with care taken to obtain feedback from all respective parties. In this way a much clearer picture can be developed of why each project turned out the way it did, of where the process is most vulnerable to breakdown, and of what might be done differently when the next generation of outreach initiatives are formulated. By including a longitudinal component in these investigations, moreover, we can identify changes in how both the utilities and affected interest groups have perceived and dealt with each other over time, and in how the prospects for constructive resolution of electricity-environment disputes have been altered as a result.

What cannot be expected of careful post-mortems, of course, is that they will point to ten easy steps or some other formula for guaranteeing the success of a participatory project. The process of interaction can never be entirely predictable because it involves human beings, and because its fortunes are too closely tied to the idiosyncracies of each context of application. But to say that participation is more art than science is not to suggest that its operation cannot be improved through scientific study. The great value of systematic inquiry lies in its ability to separate the real difficulties from the imagined and to distinguish the manageable forces from the uncontrollable. This is far preferable to evaluating the concept of public involvement in the abstract, where conclusions are necessarily based on speculative reasoning and preconceived notions rather than on a concrete body of evidence.

The importance of separating fact from fancy through research on participation experience is further emphasized if one bears in mind that, among the overwhelming majority of "uninitiated" electric companies and citizen groups alike, the notion of getting together in a cooperative forum is still a very intimidating one. Nor will it be otherwise until the process becomes less of an unknown quantity in operational terms, until we know more about where the pitfalls lie and how they might be avoided. The industry in particular seems likely to remain cool to the idea unless the veil of uncertainty and misinformation about it is removed, at least to the point where experimentation is viewed as calculated risk-taking rather than a total stab in the dark. The prospects for more widespread participatory planning in the future, in other words, will depend a great deal on how effectively the events of the past are brought to light and evaluated. A wealth of experience has accumulated; the time has come for serious efforts to collect and reduce the data, so that the learning curve of participation can be articulated even more effectively than it has been in this volume.

Notes

1. M. Willrich, "The Electric Utility and the Energy Crisis," Public Utilities Fortnightly, vol. 5, no. 1, pp. 22-28 (January 2, 1975), and vol. 5, no. 2, pp. 25-34 (January 16, 1975).

2. S. 2995, introduced to the 95th Congress on March 17, 1978.

3. U.S. General Accounting Office, The Federal Government Should Encourage Early Public, Regulatory, and Industry Cooperation in Siting Energy Facilities, Report

EMD-82-18, Washington, D.C. (November 13, 1981). A digest of this report is included herein as Appendix B.

4. U.S. Department of Energy, The Future of Electric Power in America: Economic Supply for Economic Growth, Report of the Electricity Policy Project, Office of Policy, Planning and Analysis, DOE/PE-0045 (June, 1983).

5. See J. Sherman Feher, "Future Siting Requirements for Electric Utility Power Plants," Electric Power Research Institute, Palo Alto, California; presented at the annual conference of the American Association of Geographers, San Antonio, Texas (April, 1982).

6. Remarks presented by Secretary Donald Hodel before the Great Lakes Conference of Public Utility Commissioners, White Sulphur Springs, West Virginia. Reported in U.S. Department of Energy, DOE This Month, Washington, D.C. (August, 1984).

7. See "Industry Plots Nuclear Revival," in Power Line, newsletter of the Environmental Action Foundation (March, 1984).

8. See, e.g., M. Lyons, "Power Politics," in Resources, newsletter of Resources for the Future, Inc., Washington, D.C. (February, 1983).

9. D. Ducsik, "Citizen Participation in Power Plant Siting: Aladdin's Lamp or Pandora's Box?" Journal of the American Planning Association, vol. 47, no. 2, pp. 154-166 (April, 1981). This article is reprinted herein as Chapter Two.

10. Excerpt from a memorandum by S.R. Kaderali, Pacific Gas & Electric Company (September 9, 1975), provided courtesy of W.B. Allen, Director of Environmental Quality, PG&E.

11. R. Short, "Corporate Survival, or the Canary-in-a-Coal-Mine Theory of Management," remarks by the president of Portland General Electric Company at the City Club, Portland, Oregon (January 22, 1982).

12. Letter to the editor from G.T. Martin, Vice-President of Texas Electric Service Company, Public Utilities Fortnightly, pp. 67-68 (December 3, 1981). See Chapter Three for the full text of this letter.

13. Two prominent examples not covered in this compendium are Seattle City Light and Kansas City Power and Light. SCL has established several standing citizen committees to address planning issues such as load forecasting and energy conservation policy. And KCP&L has given an independent advisory board a significant role in its long-range planning efforts.

14. One contemporary report on mitigation activities at some 38 power plants found, for example, that interaction with the community played a role in 20 cases, or just over 50% of the sample. See W. Metz, "The Mitigation of Socioeconomic Impacts by Electric Utilities," Public Utilities Fortnightly, vol. 106, no. 6, pp. 38-42 (September 11, 1980).

15. The only survey data known to this author is that which Ontario Hydro of Toronto, Canada, has collected and printed in a preliminary in-house report entitled North American Electrical Utilities Public Participation in Route and Site Planning, Community Relations Department (November 1980).

16. See Michael Harwood, Getting Together: Can the Electric Utilities and the Environmentalists Learn to Communicate Sensibly, Thus Saving Us All a Lot of Trouble?, unpublished manuscript, Washington, Connecticut (October 1974).

17. Ontario Hydro, op. cit. note 15, supra.

18. Aside from the aforementioned GAO report, the only sizable work in print is a book by David Myhra entitled Energy Plant Sites: Community Planning for Large Projects, Conway Publications, Atlanta, Georgia (1980). As a handbook on socio-economic impact mitigation, this has much to say about how a utility should relate to site-area residents and government officials, and the author makes some pointed comments about how openness has been mishandled on occasion by project managers.

19. See J.L. Creighton, Public Involvement Manual: Involving the Public in Water and Power Resources Decisions, U.S. Department of the Interior, Water and Power Resources Service, Washington, D.C. (1980). See also U.S. Department of Transportation, Federal Highway Administration, Effective Citizen Participation in Transportation Planning, 2 volumes, Washington, D.C. (1976).

20. See, e.g., D. Stever, Seabrook and the Nuclear Regulatory Commission, University Press of New England (1980); S. Ebbin and R. Kasper, Citizen Groups and the Nuclear Power Controversy, MIT Press (1974).

21. U.S. General Accounting Office, op. cit. note 3, supra.

PART ONE

Issues and Concerns

2

Citizen Participation in Power Plant Siting: Aladdin's Lamp or Pandora's Box?

Dennis W. Ducsik

For over a decade now, one of the most intractable problems on the energy-environment scene has been rancorous political opposition to the construction of big power plants. Born in the late 1960s at places like Storm King Mountain, Turkey Point, Bodega Head, and Four Corners, the resistance movement has snowballed to the point where it is now rare for a major project involving an electric generating facility to go unopposed. This lack of public acceptance has even been institutionalized in the form of national organizations such as the Environmental Action Foundation, to which groups all over the country turn regularly for help in fighting utility expansion plans.[1]

Although the reasons for political resistance to power plants are many and complex (especially in the case of nuclear reactors), concern over the choice of site and its environmental and social implications usually plays a key role. Indeed, with the possible exception of "No to Nukes," the most familiar cry to a utility executive's ears is almost certainly "Don't Put It Here." It is a cry, moreover, that seems likely to be heard with increasing frequency throughout the 1980s and beyond. Electric companies are already hard pressed to come up with locations that are technically capable of development as well as

---

Reprinted by permission of the Journal of the American Planning Association, vol. 47, no. 2, pp. 154-166 (April 1981). A condensed version of this article also appears in K. Gilbreath, ed., Business and the Environment: Toward Common Ground, 2nd edition, The Conservation Foundation, Washington, D.C. (1984).

suitable on economic, environmental, and political grounds.[2] As the tradeoffs involved become even more acute in the future, it follows that confrontation between the industry and affected interest groups can be expected to intensify as well.

The ultimate source of difficulty, of course, is that no large-scale power facility can be designed to completely avoid imposition on a landscape, an airshed, or a body of water in its vicinity. The laws of thermodynamics together with engineering and economic realities simply preclude it. Nor is it realistic to think that siting disputes are a thing of the past in light of recent progress in reducing the nation's consumption of electricity. Even load growth at the modest annual rate of two percent translates into a significant need for additional generating capacity, and there will always be older plants to replace irrespective of the rate of increase in demand. Some would argue, in fact, that it is desirable to accelerate the retirement of certain oil-fired units as a means of reducing dependence on foreign petroleum products and improving air quality in cities, where many of these units are located. Thus, a sizable complement of new sites must still be found in the years ahead.

Can anything be done to avoid the grim prospect of more disruptive conflict over the location of power plants? In this author's view, the time has come to give serious consideration to an idea that has actually been around since the early 70s and found useful application in fields such as highway location and water resources development,[3] but not to any significant extent in the electricity sector. This is the concept of 'open' planning, which calls upon electric companies to voluntarily incorporate some form of citizen participation into the site selection process. Many variations are possible, but the key feature of this approach is that interaction with concerned citizens should take place <u>during</u> the process of elimination that is commonly used in the evaluation and choice of alternative locations. Such collaboration would occur, moreover, well in advance of the regulatory process which now provides, in the form of licensing hearings, the principal opportunity for persons affected by a siting decision to influence it directly. Under the participatory scheme, a formal public hearing would mark the end point of dialogue between the company and the public at large, not the beginning.

Over the years this notion of citizen involvement in site planning has gained the endorsement of prominent individuals and organizations outside industry circles,[4] and has even been written into proposed federal legislation.[5] On the whole, however, it has yet to penetrate the thinking of those in the best position to initiate new approaches to the politics of siting, i.e., the utilities themselves. Instead of reconsidering the longstanding policy of keeping sites confidential until license applications are ready to be filed, most companies take the position that legislative reform is needed to ensure that siting disputes will be settled more expeditiously within the framework of federal and state certification procedures. As to suggestions that the company might try to get together in a constructive fashion with environmentalists and other citizen advocates, typically these have either been greeted with the utmost skepticism or dismissed as altogether outlandish.

With the climate of alienation that has long prevailed between utilities and their ardent antagonists, this attitude on the part of the electricity industry is perhaps understandable. The problem, however, is that for several years now the idea of collaborative siting has been so negatively prejudged that very few companies have been willing to experiment with it, even on a limited basis, and there has been practically no careful evaluation of it in the industry literature.[6] As a consequence of this unfortunate lack of attention, very little is really known as to whether citizen participation might be a viable means of stemming controversy over where to put future power plants, or, for that matter, any major energy facility likely to engender serious environmental opposition.

This article is intended as a first step toward better understanding the relative merits of the participatory approach to siting. The principal objective is to develop a realistic and balanced perspective on the matter, one that accounts in particular for issues that are important from the standpoint of the power industry who, after all, must take on the bulk of the risks before any benefits can be realized. After a brief review of the rationale advanced on behalf of collaboration, the discussion concentrates on the major concerns a utility might be expected to have with respect to operational implications. These concerns have been identified by the author in a series of informal (though extensive) communications with thirty-two persons with experience in the siting domain, over half of whom are executives or technical professionals employed by major electric companies. A systematic appraisal of the data generated in this study has been completed,[7] and it is data generated in this study has been completed,[7] and it is hoped that the effort has yielded some insightful propositions that can serve as a guide to future inquiry on the subject.

THE RATIONALE FOR COLLABORATION

Advocacy of citizen participation in power plant siting is, in large measure, an outgrowth of nearly universal dissatisfaction with adjudicatory hearings as the basic forum for communication among those with a stake in the decision process. The source of consternation from the utility standpoint is that intervenor groups, capitalizing on the procedural complexity of the regulatory system, have turned such proceedings into the "marathons among legal track meets."[8] The resulting delay and uncertainty can wreak havoc on facility planning efforts which, because of the technical complexity and vast financial commitments involved, must proceed in an orderly fashion. Equally distressing to the industry is the tendency for disputes to wind up in the courts, before judges who are ill-equipped to engage in technology assessment and who are inclined, therefore, to base decisions on strictly procedural considerations instead of the substantive merits of the project.[9]

Environmentalists, too, are disillusioned with the traditional hearing process, albeit for very different reasons. Coming as it usually does at the end of the licensing process - which itself does not begin until after facility plans are ready to be consummated by the utility - the hearing is widely regarded as an exercise in futility.[10] This breeds frustration, distrust, and anger on the part of concerned citizens, who sense quite rightly that many important options have been foreclosed and that their involvement comes too late to have a meaningful influence on the project (except perhaps to stop it altogether).[11] Contention thus tends to gravitate naturally away from issues of substance and toward questions of procedure.

Such difficulties have precipitated much discussion and debate as to how the rules of practice governing hearings might be modified to make them less vulnerable to dilatory tactics on the one hand,[12] and to facilitate effective contributions from intervenors acting in good faith on the other.[13] There is another school of thought, however, which holds that political turmoil over electricity decisions will never be obviated within the context of agency hearing rooms, where the heavily legalistic and distinctly adversarial climate by itself induces polarization of views and inhibits compromise solutions. More importantly, by the time a hearing convenes it is simply too late for mutual adjustment to occur on any matters except the relatively minor details of plant layout and design. The basic problem, in other words, lies not so much with what goes on during the hearing itself, but rather with what does not transpire in the way of communication during the earlier stages of the planning process.

It is against this background that suggestions arise to the effect that citizens be invited to participate directly in site selection, beginning at a time when options are open and plans can be developed in a way that accounts fully for environmental and social concerns. In theory, this creates an opportunity to avoid a great deal of unnecessary conflict and to take corrective measures well in advance of the eleventh hour, when changing plans can result in a dramatic escalation of overall project costs. Front-end costs, of course, would almost certainly go up with the implementation of a collaborative approach, and some additional time would probaly be needed. However, if the process succeeds in narrowing the range of issues to be hashed out at the licensing stage, the probability of inordinate delay at this crucial point - where the financial penalties of slippage are particularly severe - seems likely to decline.

Even without the potential for smoother scheduling, there is another and probably more important benefit that could accrue to a utility in the wake of a successful interactive endeavor, i.e., an increase in its general credibility in the public's eye. If nothing else, a more open process would seem to be a good way to ameliorate problems of misinformation and confusion that so often in the past have served to erode public confidence.[14] The resulting goodwill might even spill over in the long run into the arena of price regulation, where any relaxation of political tensions would be a most welcome development to beleaguered utility executives attempting to secure timely rate relief.

Underscoring this line of argument is the fact that the political context within which investor-owned utilities operate has undergone profound change in recent years, and even the most friendly critics of the power industry agree that it should adopt new modes of behavior in response to present circumstances. Mason Willrich, for example, has asserted that each power company should take an active role in efforts to form a "new political consensus."[15] Among the specific steps Willrich advocates, interestingly, is that utilities should first involve affected interest groups in the development of methodology and criteria to guide the siting process, and then join with these same groups in an open-ended search with the aim of establishing an inventory of acceptable sites for various types of plants.

At least on the surface, these and other exhortations on behalf of citizen participation appear perfectly logical and reasonable. Why, then, has the concept yet to be embraced by the electric utility industry? One possible explanation, usually espoused by those with a perspective untempered by field experience, is that such reluctance is merely a reflection of hidebound conservatism on the part of utility executives. While there may be a grain of truth in this view, it must also be said that many of the proponents of collaboration are equally guilty of taking its efficacy completely for granted, as if it were an Aladdin's Lamp. Before attempting a judgment on the matter, therefore, it seems advisable to try to appreciate the misgivings of those most familiar with the particulars of how the siting process operates in practice.

THE NATURE OF UTILITY CONCERNS

As any utility planner or consultant will readily attest, finding sites that are capable of development and potentially licensable -- in engineering, economic, and environmental terms -- is an exacting task that takes several years of effort, costs millions of dollars, and often produces documentation which occupies an entire bookcase.[16] Further complicating the situation is the fact that site selection has become almost as much a political process as it is a technical one. Indeed, the continuing mobilization of opposition forces and the increase in hostility toward utilities generally has led the industry to weigh questions of public acceptance heavily against other relevant parameters, beginning even with the evaluation of broad candidate areas during the early stages of reconnaissance.[17]

Against this background, one might reasonably expect that both the political and technical implications of collaborative planning would be of concern to the electric companies. As it turns out, this was precisely the case among the utility respondents with whom this author has communicated. From the political standpoint, the idea actually seemed antithetical to what most considered a rational response for those who come under siege, i.e., to 'pull in the wagons' and do whatever else seems necessary to minimize vulnerability to attack. From the technical standpoint, the idea seemed equally at odds with the conventional notion of what is required in solving vastly complicated planning problems. The kneejerk reaction, in other words, was not what one might call favorable.

These general misgivings on the part of utility respondents can be translated into six distinct (though interrelated) statements of concern. It is appropriate now to review them, one by one.

Concern no. 1: Will early disclosure of planning information put the company at a political disadvantage, and thereby prejudice its ability to purchase and/or license a preferred site?

By its very definition, collaborative planning suggests that information about alternative plant locations and their potentially adverse impacts should be made available in advance to persons outside the utility itself. The desirability of sharing information in this manner, however, was seriously called into question by several respondents. One comment frequently made was that early disclosure of potential sites would stir up the public unnecessarily and wouldn't allow the company enough time to prepare an adequate defense against irresponsible or incorrect allegations. Why risk embarrassment by opening the files before you have all the appropriate answers, knowing that critics will be quick to capitalize on the situation to bolster their own position and credibility, and that the media (which thrives on controversy) will probably blow things way out of proportion?

Another concern about providing ammunition to utility opponents is that it will ultimately be used to complicate the licensing process, or, worse yet, to preclude the purchase of a preferred site altogether. The latter possibility is especially worrisome in states where utilities do not have access to the power of eminent domain, in which case there is nothing to prevent an opposition group (or, for that matter, a wealthy individual opposed in principle to industrial development of any kind) from buying critical parcels of land and then refusing to sell. Even in states where utilities have the power of condemnation, moreover, there is a reluctance to use the process because it is perceived negatively by the public at large, provides another opportunity for procedural delay, and can result in higher acquisition costs due to speculation (which is viewed as a rip-off to customers and stockholders alike).

Concern no. 2: Can environmentalists behave in a rational and constructive manner and be willing to accept compromise solutions?

Underlying the foregoing reservations about early disclosure is a fear that citizen activists would continue to engage in zero-sum political struggle, regardless of any attempt on the part of the company to initiate a constructive dialogue. This in turn seems based on a set of perceptions regarding the motives and behavioral characteristics of environmentalists. They are accused, for example, of having a parochial, hands-off mentality that fails to acknowledge the existence of tradeoffs ("can't they say anything but no?"). Another prevalent image is that advocacy groups constitute a disloyal opposition, i.e., they have no stake in seeing a broad public interest served and do not feel accountable for the consequences of their actions. Many utility respondents also assert that "intervenor types" are protecting vested interests at best, and are power-hungry, ego-tripping, or out to discredit the establishment at worst.

In addition to characterizing the opposition as inflexible, irresponsible, and self-serving, many respondents indicated that citizen activists have highly emotional tendencies and are therefore inclined to overreact in ways that will undermine an open planning process. A final concern was that "manipulative" environmentalists would use the open planning platform as a means of attracting public attention, and might not even participate if they thought their ability to selectively control the flow of information to the media would be diminished.

Concern no. 3: Can laypersons participate effectively in a complicated planning process without having the skills and qualifications normally required of the technical professional?

Foremost in the minds of several respondents was the issue of whether and how participating citizens would be able to cope with a planning problem of enormous scope and complexity. Some felt, to begin with, that the highly technical nature of power plant siting simply precluded direct participation by anyone other than qualified professionals, who have the specialized knowledge and experience necessary for intelligent planning. These respondents seemed to feel that the involvement of unsophisticated laypersons would inevitably lead to chaos, and is therefore inconsistent with the need to maintain an orderly and efficient site evaluation process. Others were more willing to entertain the possibility that concerned citizens could be educated on key technical issues, but saw the task as a monumental one consuming an inordinate amount of time and resources.

Concern no. 4: Will environmental advocates recognize that there are no perfect solutions, given that planning operates in the presence of numerous constraints and ambiguities?

A second technical issue raised by several respondents arises from the fact that site planning is a process fraught with constraints and uncertainties. Two recurring comments were that the range of potential sites might not be sufficiently wide to satisfy concerned citizens,[18] and that the prevailing legal and institutional framework might also limit the planner's range of motion in developing suitable alternatives.[19] Another common worry expressed by utility respondents was how participating citizens might react in the face of environmental information that is qualitative, diffuse, incomplete, or otherwise intangible. Such information, it was noted, is not only hard to get a handle on but also resists comparison to more easily quantified engineering and economic considerations (the so-called 'apples and oranges' problem).

Being keenly aware of these limitations, the experienced planner realizes that the search for ideal solutions can be unending. Will environmentalists be indecisive in the face of this reality, or unwilling to balance the need for more and better information against resource and time constraints? What some respondents seemed to fear most is "paralysis by analysis," where "pipe-smoking alternative-posers" argue interminably for keeping options open and for studying each and every site in great detail.

Concern no. 5: Is it possible to keep the focus on site selection when many environmentalists are more interested in larger, generic issues such as need for power and the future of nuclear technology?

A fact often pointed out by respondents is that environmental groups these days are concerned not only (and perhaps not even primarily) with specific impacts, but also with a number of larger questions surrounding the future expansion of power supply systems (e.g., need for power, use of nuclear technology, etc.). Is it possible, however, for such matters to be dealt with in the context of an open planning forum, established for the purpose of resolving specific site-related conflicts? Utility representatives see little hope for this, in part because some of the most important issues are generic ones and beyond the company's control altogether. They argue, consequently, that some means must be found to keep the focus on site selection per se, or else the collaborative process will fail. At the same time, they fear this may constitute a wholly unacceptable pre-condition for the environmentalist concerned with the more general implications of long-range power system plans.

Concern no. 6: Will the company get burned in the end anyway, due to the impossibility of achieving universal happiness with the outcome of the planning process?

Quite apart from the issue of what should be on the agenda for discussion is whether, in any event, a cooperative effort would be fruitless because of the difficulty utilities and concerned citizens seem to have in agreeing on anything at all. Several respondents pointed out that the basic difference between the industry and even its most "reasonable" opponents is at the level of values and perspectives, a fact which all but guarantees that everyone cannot be made happy with a final decision. To some this is particularly troublesome because, with the present legal context, there may be little to prevent a tyranny of the minority in the form of protracted court battle, regardless of the outcome of the collaborative planning exercise. With this in mind, a few respondents suggested that there be a quid pro quo for open planning, perhaps in the form of limits on the right of participants to challenge decisions after a certain point.

All in all, if the above reactions to the concept of public involvement in siting were to be characterized in a single word, that word would have to be "dubious." To most respondents it seems palpably irrational to attempt to build a collaborative relationship when there is little hope of arriving at accommodation but a very real possibility of creating chaos and intensifying conflict. Would this not, in effect, be playing right into the hands of one's enemies? Thus, it is the prospect that open planning could well turn into a Pandora's Box that stands as the major disincentive to its use by the electric utility industry, which simply views citizen participation as too much of a gamble.

Considering the high stakes involved in making power plant investment decisions, one can hardly call into question this desire to avoid risk. However, it does seem reasonable to ask if the utilities, who have grown accustomed to operating in an intensely adversarial mode, have an accurate perception of the risks to begin with. The opinion here is that probably they do not. In contemplating the difficulties they envisage, in fact, a number of potential mitigating factors have come to mind, along with some thoughts as to how specific problems might be avoided through careful process design. The concatenation of misgivings put forth by utility respondents is thus deserving of a closer look.

POLITICAL OBJECTIONS EVALUATED

Of the six basic concerns articulated by utility respondents, the first two are closely related and reflect a basic fear that going public at an early stage would somehow backfire in a purely political sense. Without confidentiality, it is argued, opposition to potential sites will be stimulated before the company is ready to meet it, and intransigent activist groups will simply have additional opportunities for disruption. Would not an open process, then, make siting even more contentious and vulnerable to intervenor tactics than it already is?

Consider first the assertion that premature disclosure will give rise to unnecessary public concerns and embarrassment on the part of an ill-prepared utility. While this is certainly a plausible scenario under business-as-usual conditions (where the utility announces and defends a single, preferred site), one wonders if such difficulty would really be encountered within the open planning context. Since the whole idea is to begin interaction with affected interests at a time when a range of options exists and adjustments can still be made in response to expressed concerns, it seems unlikely that anyone would expect the utility's data and analyses to be anything other than preliminary. This is not to say that advance notice of conceivable facility locations would be received by the public with total equanimity. Nor can one guarantee that the free flow of information associated with collaborative planning will not give rise to misconceptions about the alternatives available and their impacts. Nevertheless, at least there is an opportunity to deal constructively with the public's fears and mistaken impressions. It is difficult to imagine how such an effort could give rise to anything like the anxiety engendered by the conventional "decide-announce-defend" process, in which dialogue tends to polarize shortly after affected persons learn of the project and begin to sense that most key decisions, for all intents and purposes, have already been made.

A second argument that seems less than persuasive is that site planning information must be kept secret lest it eventually be used against the company in some manner. First of all, it is hard to see how this might pertain to the case of environmental data, considering that extensive impact statements are required in connection with the regulatory process st both state and federal levels. Also seemingly overdone is the fear of speculation, insofar as even a two or three million dollar increase in land cost -- which is often occasioned by rumor in any event -- is quite small relative to overall project expenditures. While this is by no means an insignificant amount, neither is that which the utility loses for each day or week that a new power plant is delayed as a result of public opposition that might have been avoided.

One issue that cannot be dismissed so easily is land availability, which is crucial and, in some states, must be handled very delicately. No utility will disclose a preferred site if there is a chance that acquisition will be precluded by a landowner who refuses to sell, or by a zoning change made at the behest of opposition forces. It helps, of course, that most electric companies have access to the power of eminent domain and can request a state override of local zoning ordinances by the public utilities commission. While neither of these legal devices have ever been politically palatable to the industry, they both would seem to be important prerequisites for effective open planning. In fact, using them to implement a siting decision arrived at through collaborative means, and thus more easily identifiable with collective interests, may even help remove the traditional stigma of 'big brotherism' that has made the public uncomfortable with such proceedings in the past.

This line of reasoning, to be sure, offers little comfort to the utility professional who subscribes to the second major type of political concern identified in this study, that having to do with the alleged motives and behavior of "extremists." With environmentalists expected to exploit every available opportunity for subversion, it is understandable that the industry would see little to be gained (and much to be lost) in attempting a collaborative siting process. The obvious question to be raised here, however, is whether the problem of extremism is as widespread or a threatening to the viability of a cooperative effort as the statements of utility respondents make it seem. Without denying that there is a darker side to human nature which creates some difficulty for open planning, this author is inclined to doubt that such a highly pessimistic view is warranted.

It is important to note at the outset that some respondents seemed to view the environmental community in terms just as monolithic as environmentalists have always viewed the power supply establishment. Indeed, many of the adjectives chosen to characterize intervenors (intransigent, self-serving, irresponsible, and so on) seem borrowed from the very lexicon activists have long employed with reference to utility actions. Drawing this parallel is useful because it serves to remind us not only that reasonableness is a very subjective thing, but also that neither side in a polarized debate is likely to have a very accurate perception of the other. Such perceptions, after all, are based on observations of behavior that may be idiosyncratic to an adversarial context. Any sociologist will confirm that even the most docile citizen can be transformed into an activist if the circumstances are right; witness, for example, how parents angry over the busing of school children and farmers aroused by governmental food policies have taken to the streets in recent years. In this light one might argue that much of the extremism decried by the utilities is really a natural response on the part of concerned citizens to a perception that, unless they resort to unorthodox tactics, their strongly held values will be excluded altogether from the decision process.

These observations suggest that the key issue is not whether environmentalists are capable of acting rationally, but whether they have come to view cooperation as a viable alternative to confrontation. Utility respondents often asserted that activists possess a vested interest in the adversarial process, but other factors seem to belie this notion. Thoughtful environmentalists recognize, for example, that prolonged radical behavior can diminish their credibility and be self-defeating in the long run, and that the best way to influence decisions is usually to participate in them directly on a constructive basis. Citizen groups have even been known to seek interaction with utilities on their own initiative,[20] and some recent reports show that traditional adversaries in the business and environmental communities generally are making concerted attempts to identify common ground.[21] Thus, while the environmental movement certainly has a permanent stake in advocacy, there is no reason to think this necessarily stands in the way of a collaborative relationship with those responsible for the siting process.

Still, it would be naive to suggest that the problem of extremism will disappear altogether. Some will undoubtedly enter the participatory arena bent on opposition, and there will be others for whom no amount of interaction in the spirit of accommodation would make a difference. What is to prevent such persons from disrupting an otherwise useful process of interaction among parties willing to seek compromise? Is it desirable, as has been suggested, to simply exclude them and/or put constraints of some kind on subsequent legal challenges?

The inclination here is to reject such limitations on a number of grounds, not the least of which is that environmentalists would probably decline to participate at all, and for good reason. First of all, as has been noted above, it is by no means clear how one distinguishes the extremists from the responsible parties, given the relative nature of such characterizations. Moreover, a so-called open process that is available to some and not others is a contradiction in terms, and would probably give rise to legitimate skepticism about the utility's motives (e.g., "this must be a clever attempt to divide and conquer"). This would discourage participation, as would the pressure those excluded from the process seem likely to put on those inclined to stay (via charges of selling out, etc.).

Likewise, fairness will always require an adjudicatory procedure for those who are adversely affected (even after open planning) because one cannot assume that the requirements of due process will automatically be met in the course of such an informal undertaking. In the alternative, if elaborate procedures were devised to police the process, flexibility would be reduced in a manner reminiscent of the highly structured adversarial approach that collaborative planning is designed to supplant in the first place. Thus, once again it seems likely that a restrictive participatory policy would only serve to aggravate the existing climate of distrust and thereby reduce the chances for success.

At first glance this appears to be a Catch-22 situation: without certain parties participation won't materialize, and with them it can only degenerate into confrontation. What is being overlooked, however, is the possibility that the open planning process itself might create an unfavorable climate for extremism. One should remember that charges of secrecy, double-dealing, and other deceitful practices rely heavily on emotional appeal and are most effective when the utility is perceived as a villain. Extremists also benefit when there is confusion as to the facts of the case, and often rely on outright distortion and obfuscation. With open planning, however, insofar as it seeks to dispel alienation and clarify the issues of choice, much of the basis for radical political action might well be removed.

This concept of reduced vulnerability to extremist action seems equally applicable in the case where dissatisfied participants undertake legal action after open planning has concluded. Both regulatory officials and members of the judiciary, who are not unaware of the prevailing political climate, may be much less sympathetic to continuing challenges in the face of an obvious good faith effort by the utility to incorporate all views and suggestions into the siting process. This is particularly true if a negotiated settlement has been reached, in which case the company would have the support of a constituency much broader than it had ever enjoyed in the past in connection with its siting decisions.[22] Such political capital, in fact, would in all likelihood prove valuable in almost any area of utility endeavor that is much in the public eye.

TECHNICAL OBJECTIONS EVALUATED

Proceeding in a manner similar to that of the foregoing section, there are a number of points that can be made in response to the third and fourth concerns utilities seem to have about citizen participation. Here the prevailing view is that the process could just as easily break down in the technical sense, either because of an inability on the part of concerned citizens to grasp the analytical complexities of the subject matter, or because of an unwillingness to accept the groundrules by which site selection normally operates. Is it not impossible, respondents ask, to "democratize" an endeavor that is complicated and needs to be conducted in a very structured and orderly fashion? Clearly there is a special paradigm of planning underlying this concern, and its basic premises need to be scrutinized.

To begin with, consider the issue of disparity in expertise. Respondents have argued that because the siting process is highly technical, it should be reserved for technical professionals who have acquired the necessary experience and expertise. This is an argument subject to criticism, however, on grounds that it underestimates the capabilities of laypersons and overestimates those of experts. The fact is that many planning issues can be dealt with only through careful reasoning and common sense -- not expert analysis -- and in this regard the engineer is no less prone to stupid mistakes than anyone else. Further, on certain matters it is the average citizen who is the true expert, and in such cases direct interaction with the public at large could be of real value to the planning team and improve the overall quality of the siting effort.

One topic a utility staff might not be completely familiar with, for example, is that of socio-cultural conditions related to land use. Impacts in this domain often have significance only for those who live in a particular area and possess knowledge and understanding that no outsider has. Such information is rarely obtainable by surveys, polls, and other consensus-oriented techniques, which are not sufficiently disaggregated to reliably foresee public acceptance issues, judge their severity, and direct planner attention to local problems and opportunities.[23]

This is not to say, on the other hand, that one should expect citizens to participate directly in the sort of detailed ecological or meteorological studies that utility specialists carry out. Indeed, this is far from the real objective of participation, which is to incorporate a broader set of values and perspectives into the process of making the difficult tradeoffs that have been confronted heretofore in a vacuum by the utility, only to be challenged at some later (and much more inconvenient) time.

Most planners would readily admit that the siting process, however complex, is at root a matter of judgment and even intuition; witness, for example, the 'apples and oranges' problem often mentioned as a key difficulty in making site comparisons. Here it is the purely technical methods of analysis that tend to break down, and it is also where collaboration with affected interests -- a dialectic inherently capable of operating on subjective information -- could be most advantageous.

Establishing that concerned citizens have a legitimate role to play in siting, of course, says nothing about whether or not the process can be made to go smoothly. Achieving effective two-way communication between technical professionals and laypersons is rarely a straightforward matter, particularly when there is a good deal of anticipatory mistrust present (as there would be in the power plant siting case). Also, it is clearly preposterous to suggest there should be an environmentalist looking over the shoulder of every siting engineer, an image that quite understandably makes utility people shudder ("don't ask the people to help you plan, you'll not know whether you're coming or going."). How, then, should participation be structured to be workable?

No obvious answers can be given, although one point which needs to be brought out is that collaboration does not necessarily connote a truly joint effort. What seems most important is that the public not be totally insulated from the technical work, because a lack of understanding of the thought processes followed and judgments made by planners can easily give rise to a perception that the weak or questionable aspects of various alternatives are being concealed in technical jargon. Thus, it is essential that lay participants at least be informed of what is being done, and be encouraged to make whatever observations or suggestions they feel are appropriate (e.g., as to new lines of inquiry, the criteria to be applied in making tradeoffs, and so on). Beyond this baseline the degree of interaction can vary considerably as a function of what is expected from the participants, what sort of capabilities they exhibit, or where their major interests lie.

Another problem to be faced is that of educating participants on the idiosyncracies of site selection, for there can be no constructive dialogue unless all parties to the process have achieved a modicum of literacy on its substantive aspects. This could be a prodigious task requiring considerable investment of time and effort on the part of the utility. Here again, though, one must not overlook the fact that many concerned citizens are intelligent, serious students of the issues who could rather quickly comprehend at least the broad technical dimensions of the situation (if not the specific engineering details). Even if this were not the case, what is wrong with an endeavor that helps those outside the utility gain deeper appreciation of the nature of the siting problem? Rather than bemoan it as too costly an investment, it seems the industry should welcome this educational opportunity as one that is entirely consistent with its long-standing public relations goals.

The final issue at the interface between participation and planning involves the manner in which imperfections in the real world of planning are to be dealt with in the collaborative context. The hope, of course, is that citizens working on a cooperative basis with the utility would come to realize that decisions are never made under ideal circumstances, and that it is necessary to bound the problem and even muddle through on occasion in order to maintain a timely and efficient evaluation process. Realizing this goal is still not easy, however, and much will depend on the extent to which a spirit of give and take develops in the course of interaction. Can anything be done to foster flexibility and otherwise loosen up the planning process? There are not set formulas to prescribe, but certain steps and gestures do come to mind as being necessary in this regard.

It would be important, for example, to ensure that the project schedule allows sufficient time for planner-citizen interaction, recognizing that environmentalists might be inclined to approach the impact assessment process in a more cautious manner than experienced site planners. Planning lead times already span a couple of years or more, and as long as participation begins at an early point this would seem to allow the necessary breathing room. If not, perhaps the objective should be to find potential sites for generic plant types, rather than a specific unit for which a target construction date has been established. In either event, one must take care that the time available is not used in an undisciplined way, perhaps by adopting a mutually agreeable timetable together with procedures by which it can be modified in the event of unanticipated circumstances.

Aside from arranging an appropriate time horizon, it is important that the utility be sensitive to the fact that if the alternatives presented are not sufficiently broad to present real choices, participation will be suspected of being a thinly-disguised effort to market a decision that has already been made (even if it has not!). If such a perception is to be avoided, attempts must be made to broaden the range of options to the point where there is room for compromise. It might be desirable, for example, to begin with a reconnaissance of broad regions of interest instead of narrowing the focus immediately to evaluation of a small number of final candidate sites.

Yet, it will not be so easy to relax other types of constraints or to remove certain ambiguities from the impact assessment process. What then? Here it would seem that the utility's attitude toward such difficulties, and not the difficulties themselves, is the most important factor in determining the viability of the collaborative approach. It may well be, for example, that questions involving uncertainty will not become bottlenecks if the company acknowledges that there are good and bad ways to deal with incomplete information and employs methodologies that are designed to explicitly support a judgmental approach. Likewise, having a wide range of options available may not be so crucial if the company shows it will actively seek to develop new alternatives and to ease particularly burdensome constraints. It might even be the case, ultimately, that all parties could work together to effect legal and institutional changes that would ease the siting dilemma, changes that might not otherwise be possible without the support of a bipartisan constituency.

What is being argued here, really, is that in most cases the substance of planning need not pose major difficulties for a collaborative effort if the style of planning is adjusted in appropriate ways. What seems essential is that participating citizens come to feel that the company is keeping an open mind and is trying, in the face of imperfect circumstances, to be responsive to environmental and social concerns. If this can be achieved, the notion of blending citizens into a highly technocratic planning process seems not so disquieting after all.

ON THE SEARCH FOR COMMON GROUND

Several arguments have now been made to the effect that constructive interaction between utilities and concerned citizens is not a wholly implausible scenario. Nothing has been said, however, about whether the process can actually converge on a mutually acceptable course of action, an issue quite apart from political and technical considerations per se. Respondents noted in their fifth and sixth concerns that discussions could founder at the very beginning -- over non-siting questions like need for power or choice of plant technology -- as well as subsequently when disagreements emerge as to the relative merits of specific locations. Thus, if the parties involved will never see eye to eye and will reach an impasse sooner or later, why bother with collaboration at all?

Consider, first, the dilemma posed by the fact that advocacy groups will probably raise a number of larger, generic issues unrelated to the siting problem itself. On the one hand, to broaden the scope of inquiry to include such matters is to create additional potential for contention, especially where nuclear plants are concerned. On the other hand, insofar as the specification of capacity requirements and plant types predetermines to a large extent the options and impacts in siting, it is understandable that environmental groups would vigorously object to any attempts to decouple these underlying decisions from the process.

In light of strongly held sentiments to this effect, it seems clear that questions of power demand and alternative means of supplying it must be dealt with somehow if a collaborative siting process is to proceed. Ideally, such questions should be debated and resolved in a separate forum and at a more appropriate level of policy-making, and it is perhaps encouraging to note in this regard that state and federal regulatory agencies have begun to implement new procedures for review of utility demand forecasts and supply plans.[24] It is highly problematical, though, as to whether the environmental community will have confidence in these new mechanisms to the point where plant-related issues would not prove distracting to a participatory siting process. For one thing, not all states have adopted legislative provisions enabling evaluation of long-range system plans. For another, there is evidence that few states are truly capable of effecting a comprehensive overview of utility planning and its relationship to other programs and policies having to do with environmental resource management.[25]

It is conceivable, therefore, that existing forums for consideration of long-range plans will prove unacceptable to citizen advocates in some respects. What then? One possible strategy might be to obviate the need to consider this aspect of decision-making altogether by involving citizens in a broad reconnaissance effort aimed, not at siting a specific facility, but at identifying suitable locations for a variety of generic plant types. The attractive feature here is that, because no final decisions have been made as to when or even whether a plant will be built, the situation might be less threatening than it would be if the utility were committed to bringing a particular facility on line at a preordained time. Most utilities maintain an inventory of potential sites anyway, and seeking public input to modify or expand that inventory could provide public acceptance information useful to system planners and project designers alike.

Not to be overlooked, however, are several potential disadvantages to this site inventory approach. One of these stems from the fact that a site may look good at one time but be unacceptable later as technology, environmental understanding, or regulatory requirements change. Unless the open planning effort is a continuous one (perhaps it should be!), changing conditions could render moot any tentative agreements reached. Another, probably more severe obstacle is that land banking proposals are often disdained because they evoke the image of a self-fulfilling prophecy. Some people, for example, might think twice about endorsing a site for a large generating facility, even tentatively, for fear of prejudging the utility's attitude toward mitigating demand or exploring alternative energy sources. While it is most unlikely that a utility would build a plant merely because a suitable site is available, the symbolic and perpetual drawback to the environmentalist of being associated with a utility site inventory is, nonetheless, not to be discounted.

Thus, a utility may well have no choice but to direct a participatory process toward siting an actual facility. In this case, perhaps the only way to keep the focus on the issue of location is to start with a plant that seems least likely to raise the hackles of the environmental constituency. It certainly seems unwise, for example, to apply a collaborative approach for the first time to the siting of a nuclear reactor, considering the ongoing debate over radioactive waste disposal and the change in public attitudes toward accidents in the wake of Three Mile Island. One need not be so pessimistic, however, over the prospects for effective cooperation in the case of other major facilities such as coal-fired or hydroelectric plants, which at least do not suffer from a deep-seated mistrust in the basic technology that is employed.

An even better way to minimize attention might be to choose a station of intermediate size (e.g., a 500 megawatt cycling coal plant) for the initial outreach effort. To be sure, this is not the type of plant the industry usually has great difficulty siting; but the real objective is to break the ice of mutual distrust and begin to develop a cooperative working relationship, and to achieve this it seems best to start with a rather limited participatory experiment that has a reasonable chance of success. Thereafter it might be possible to tackle the issues associated with larger facilities, where the stakes are higher and the the conflicts more difficult to resolve. Indeed, the best course of action would probably be to follow up the initial siting endeavor with joint deliberations on how a future collaborative siting process might be structured, in light of the experience gained.

With this basic strategy of working up to the more controversial projects, one begins to wonder if it might not be feasible after all to address, as an adjunct to the site selection process, the questions of need for power and alternative supply strategies. Certainly there is no a priori reason why some issues like, say, the validity of load forecasts or the prospects for purchasing power cannot be scrutinized in an open forum, at least to some extent. Without the opportunity to engage in such deliberations, there may be little inclination on the part of the citizens to attend to the question of siting at all. Is there any hope of resolving anything? This author, for one, is not about to underestimate what can be accomplished if only the present atmosphere of hostility and suspicion can be dispelled.

This is not to say that compromise solutions which enjoy substantial bipartisan support will be easy to come by, and this brings us to the final impediment to effective open planning, i.e., the problem of honest disagreement. Utility respondents argued in the end that, if nothing else, severe polarization of views would ultimately preclude closure on whatever issues are brought under consideration. In their opinion, then, the mere fact that reasonable minds can be expected to differ is a compelling argument against interaction. Is it really so? Only, it would seem, if one believes that either (1) it is necessary for all conflict to be resolved as a result of the process, or (2) it is unlikely that any significant amount of conflict can be resolved. To subscribe to these propositions, however, would be to lose sight of some of the most basic concepts upon which all attempts at compromise are based.

One important premise is that it is often possible to achieve substantial agreement on a course of action without reaching consensus on underlying goals and values. The history of decision-making in our pluralistic society is fraught with examples where those with seemingly irreconcilable differences at the level of broad principles have come to terms within a context of specific proposals. One must take care to remember, in addition, that universal happiness is not the criterion upon which any process of conflict resolution should be judged. The ultimate goal, rather, should be to avoid unnecessary conflict and seek accommodation whenever possible, and to have all participants feel comfortable with the way the process was conducted. This view of what is to be expected of collaboration readily admits that dissatisfaction cannot be totally forestalled. The hope is that those whose interests suffer will not only be fewer in number, but also more equitably treated and less inclined, therefore, to turn their disagreement into overt resistance. Is this not, after all, the classically conceived essence of democratic methods for managing the clash of opposing views?

To be sure, avoiding impasse within an atmosphere of divergent interests and long-standing mistrust will still be a formidable task, requiring most careful attention to a number of behavioral realities. One cannot assume, for example, that a utility company would abandon the position of advocacy which grows out of the very nature of its business; nor is it reasonable to expect that concerned citizens will be responsive to all goals and interests, or even have confidence in their ability to make balanced judgments. For environmentalists, moreover, participation in a so-called balancing process may not ring true because they fear co-optation and a resulting loss of constituency. One can easily appreciate the awkward position of a citizen advocate who is asked to make a transition virtually overnight from watchdog to collaborator, especially when there are extremists about who would be quick to make the charge of selling out.

These observations make it clear that one cannot rely on a simplistic 'goody-two-shoes' approach which says that, if people are just nice to each other and lay their cards on the table without playing games, agreements will eventually be reached. Given human nature, it seems more pragmatic to conceive of interaction in terms of hard-nosed bargaining, in which participants are asked only to seek a compromise consistent with their own interests. In this way accomodation can often be reached even if the respective parties don't accept each other's facts, figures, or line of reasoning, because each side is motivated at least initially by the realities of what can be gained from compromise. If a sense of collegiality develops in the course of rubbing elbows (as it often does), all the better, but a relationship of grudging respect can do almost as well.

To argue that partisanship is important even in collaborative planning, of course, says nothing about how conflict should be handled at the operational level. Those familiar with the route to compromise know that it is tortuous even under the best of circumstances, in which all participants feel pressure to reach a decision and have clear incentives to give ground. Still, there is every reason to believe that traditional adversaries can find courses of action that each can live with if the circumstances are right, and it is heartening that more is being learned all the time about how to make this process work. Consider, for example, the work that is being done with techniques such as compensation[26] and environmental mediation,[27] with the latter being applied to a variety of disputes around the country. As these efforts toward improving the state of the art of conflict resolution multiply, it seems reasonable to be optimistic about the prospects for convergence in an open planning endeavor.

CONCLUDING REMARKS

In a sense, this discussion of the relative merits of citizen participation in power plant siting has now come full circle. The hopeful note sounded at the outset soon gave way to a formidable litany of objections which, on their face at least, made collaboration seem a classic example of an idea that sounds good in principle but is patently unworkable in practice. However, further examination of the 'can of worms' scenario put forth by utility respondents makes one realize that some concerns are probably overstated or based on questionable premises, and that others might be obviated through appropriate groundrules to guide the interaction process. There is reason to believe, in other words, that the participatory approach has more to commend it than the electricity industry has heretofore been able to envision.

This author has noted, in fact, a rather distressing tendency on the part of utility respondents to view the concept of collaboration almost exclusively through the jaundiced eye of the political game theorist, and to evaluate it in terms of an unreasonably stringent criterion, i.e., whether or not it constitutes a panacea that could eliminate all political conflict from the siting process.

Clearly it cannot. What it does offer is a means of approaching these inherent difficulties in a manner far more constructive than is possible in an adversarial context where there is every incentive to polarize debate and exploit uncertainty. The industry also seems inclined, unfortunately, to view proponents of participation as wild-eyed idealists who harbor the naive expectation that consensus is attainable through gentle persuasion. Yet this, too, is far from the truth, as most professionals in the field of collaborative problem-solving recognize full well the realities of value conflict and the need, therefore, to select modes of interaction specifically designed to operate under such circumstances.

This is not to say that issues of concern to the electric companies should be taken lightly, or that the lines of reasoning advanced on behalf of open planning constitute a mandate for immediate adoption. Many of the counterarguments presented in response to industry misgivings are admittedly laced with "ifs," "mights," and "coulds," and while plausible they hardly comprise a definitive rejoinder to the utility position. It seems clear, moreover, that in some ways collaboration is a delicate balancing act that requires a great deal of practice before it can be performed with regularity and precision. Thus, a substantial burden of proof remains to be satisfied before one can conclude that citizen involvement should become a principle of operation in all siting endeavors.

The central thrust of this article, however, has not been to demonstrate conclusively the wisdom of opening up the site selection process. Such a determination can never be made with confidence on the basis of a priori arguments, which are intrinsically speculative on both sides as to what sort of scenario might unfold in practice. The only answerable question, then, is whether collaboration should be taken more seriously as a legitimate alternative to the conventional means of dealing with public acceptance problems. In this author's mind the response should most definitely be in the affirmative, for the position is obviously defensible and even persuasive in many respects.

Where does one go from here? Since the basic theoretical issues are now well drawn, it seems best to proceed with inquiries of a more empirical nature. A body of experience is needed to serve as the testing ground for the various hypotheses that have been articulated so far, both pro and con, and to provide insight on a host of issues in the category of process design. How open is open enough? Who should (or will) participate, and how does one deal with the issue of representation? Which participatory techniques are more useful than others? What are the sources of funding and other resources necessary to support the effort? What should be the role of government officials with ultimate responsibility for granting licenses and permits? Clearly, the particulars of mobilizing a pluralistic planning process must be carefully attended to; indeed, the extent to which the technical and political issues addressed herein become significant will probably depend more than anything else on operational details such as these.

As it turns out, an evidentiary base of sorts already exists because of a handful of innovative utilities have experimented in the last ten years with various forms of citizens participation in siting. Some have simply allowed early access to documents prepared in connection with a preferred site,[28] while others have actively solicited the opinion of environmentalists on the relative merits of candidate sites.[29] One company -- Pennsylvania Power and Light -- has even gone so far as to establish a permanent siting advisory task force whose deliberations began, in 1977, with a critique of the general methodology PP&L and its consultants were using in the evaluation of potential sites.[30]

These cases, while limited in number, present a golden opportunity to learn more about the intricacies of open planning. The problem, however, is that the literature at the moment is woefully inadequate in this area. Accounts of utility experiments are typically sketchy and anecdotal, peparing for the most part in a scattered array of trade journals, workship proceedings, and company reports of limited distribution. Thus, although considerable data has been generated, the bulk of it remains little studied and poorly understood. Until this information is brought to light more effectively, collaborative siting will remain an unknown (and, to the utilities, fearful) quantity whose true risks and benefits can only be the subject of speculation.

What this suggests is that it is no longer worth trying to decide in the abstract whether citizen participation is a Pandora's Box or an Aladdin's Lamp, for in fact it is probably neither. What it may be is an idea whose time is coming, because the institutional reforms of the '70s have failed to quell electricity-environment controversies, and because other potential solutions have been too few and far between. The participatory approach, in sum, deserves a lot more systematic analysis in the decade of the '80s than it has received in the decades past.

Notes

1. See R. Morgan and S. Jerabek, "How to Challenge Your Local Electric Utility: A Citizen's Guide to the Power Industry," Environmental Action Foundation, Washington, D.C. (March, 1974).

2. See, e.g., G.A. Brown et al. Power Plant Site Considerations at Charlestown, Rhode Island, University of Rhode Island, Marine Technical Report No. 23, Kingston, Rhode Island (1974); see also R.H. Ball, et al., California's Electricity Quandary: II. Planning for Power Plant Siting, Rand Corporation Report R-1115-RF CSA, Santa Monica, California (September, 1972).

3. See, e.g., A.K. Sloan, Citizen Participation in Transportation Planning: The Boston Experience, Ballinger Publishing Co., Cambridge, Massachusetts (1974); J. O'Riordan, "The Public Involvement Program in the Okanagan Basin Study," 16 Natural Resources Journal 177-196 (January, 1976).

4. See, e.g., Committee on Power Plant Siting, Engineering for Resolution of the Energy-Environment Dilemma, at 12, National Academy of Engineering, Washington, D.C. (1972); Special Committee on Electric Power and the Environment, Electricity and the Environment: The Reform of Legal Institutions, at 8-9, Association of the Bar of New York City, New York, N.Y. (1972).

5. S. 2995, introduced to the 95th Congress on March 17, 1978, included a provision directing the Nuclear Regulatory Commission to encourage or require utilities to engage in open and advance planning for the addition of generating capacity as well as the evaluation of potential sites.

6. One of the better pieces on the subject that has appeared in recent years is an interview with Robert I. Hanfling of the Department of Energy, "Utility Decisions: What Voice for the Public?," Electrical World (December 1 and 15, 1977).

7. D. Ducsik and T. Austin, Citizen Participation in Power Plant Siting: An Assessment, Energy Studies Group Report No. 2, Center for Technology, Environment, and Development, Clark University, Worcester, Massachusetts (May, 1979). These communications took place during 1974 and 1975, with information collected in the form of notes taken during face-to-face interviews and telephone conversations, letters, tape recordings, and marginal notations received in response to concept papers describing the open planning process.

8. T.J. Dignan, "Licensing Hearings: A Modest Proposal," Nuclear News, at 45 (May, 1974).

9. For a good discussion on this point, see Barton, "Behind the Legal Explosion," Stanford Law Review 567 (February, 1975).

10. See, e.g., Ebbin and Kasper, Citizen Groups and the Nuclear Power Controversy, MIT Press, Cambridge, Massachusetts (1974).

11. The necessity of committing to studies and making equipment selections for a given site prior to knowing whether that site would be acceptable is described in J.L. Leporati, Envirosphere Company, "Expanding Role of States in Power Plant Regulations," presented at the 47th Annual Executive Conference, New York, N.Y. (October, 1976).

12. One measure that is often proposed, for example, is the use of a res judicata standard to preclude the rehearing of issues for which there was a prior opportunity for hearing.

13. See, e.g., Boasberg, Hewes, Klores, and Kass, Policy Issues Raised by Intervenor Requests for Financial Assistance in NRC Proceedings, NUREG 75/071, Nuclear Regulatory Commission, Washington, D.C. (July 18, 1975).

14. The inverse relationship between public confusion and public confidence has been documented by Clark and Brownell in Electric Power Plants in the Coastal Zone: Environmental Issues, American Littoral Society Special Publication No. 7, Highlands, New Jersey (October, 1973).

15. M. Willrich, "The Electric Utilities and the Energy Crisis," Public Utilities Fortnightly, vol. 95, no. 1, pp. 22-28 (January 2, 1975); vol. 95, no. 2, pp. 25-34 (January 16, 1975). Professor Willrich is now vice-president for corporate planning of Pacific Gas and Electric Company.

16. See Nagel and Vasseld, "Power Plant Siting Requirements," Planning, vol. 39, no. 2, at 24 (February, 1973); Brown, "Power Plants - Picking the Sites," Consulting Engineer, vol. 40, no. 3, at 133 (March, 1973).

17. Commonwealth Associates, Inc., Nuclear Power Plant Siting - A Generalized Process, prepared for the Atomic Industrial Forum, Inc., Washington, D.C. (August, 1974).

18. Nuclear plants, for example, are subject to stringent regulatory criteria with respect to seismology, distance from population centers, and so on.

19. Among the limitations known to this author are the lack of compensatory mechanisms to mitigate adverse

effects (e.g., liability problems that preclude recreational access to sites; tax laws that confine payments to the host town) and legal constraints on the region of interest for siting purposes (e.g., statutes that disallow out-of-state investment in the rate base; charter provisions and/or pooling agreements that restrict siting to the service territory only).

20. Sierra Club Press Release, "Environmental Groups Form Wisconsin Utilities Advisory Coalition," Office of the Midwest Representative, Madison, Wisconsin (August 22, 1972).

21. See H. Leonard, J. Davies III, Gordon Binder, eds., Business and the Environment: Toward Common Ground, The Conservation Foundation, Washington, D.C. (1977).

22. As it was put nearly a decade ago, "a political compromise with expert and responsible ecology groups capable of representing the public's interest would be a large step toward securing a solid mandate upon which to rest long range construction plans." D. Jopling and S. Gage, "The Pattern of Public Political Resistance," Nuclear News, at 32 (March, 1971).

23. The advantage of citizen participation in providing more disaggregated information on public acceptance has been noted in Calvert and Heilman, "New Approach to Power Plant Siting," Journal of the Power Division, Proceedings of the American Society of Civil Engineers (June, 1972).

24. For a detailed compilation of siting legislation in the fifty states, see Energy Facility Siting in the United States, Southern States Energy Board, Atlanta, Georgia (February, 1978). See also Stevens, "State Perspectives on Energy Facility Siting: Current State Practices," report for the National Governors' Association, Washington, D.C. (December, 1978).

25. See "Survey of Current State Practices for Evaluating Utility Power Supply/Demand Planning," in Electricity Planning - Today's Improvements Can Alter Tomorrow's Decisions, EMD-80-112, Appendix I, U.S. General Accounting Office, Washington, D.C. (September 30, 1980).

26. See, e.g., M. O'Hare, "Not on My Block You Don't: Facility Siting and the Strategic Importance of Compensation," 25 Public Policy 407-458 (Fall, 1977).

27. See, e.g., D. O'Connor, "Environmental Mediation: The State-of-the-Art", 2 EIA Review 9, MIT Laboratory of Architecture and Planning (October, 1978).

28. Examples are Florida Power & Light Co., Potomac Electric Co., and Washington Public Power Co.

29. Examples are Northeast Utilities Co., Northern States Power Co., San Diego Gas & Electric Co., Ontario Hydro, and Duke Power Co.

30. See Pennsylvania Environmental Research Foundation, *Public Involvement in Corporate Policy Decisions: A Report on the Experience of the Public Advisory Committee of Pennsylvania Power & Light Company*, Philadelphia, Pa. (March, 1979). See also Chapter Thirteen, *infra*.

3

Consumer Advisory Boards and Investor-owned Utilities: Rhetoric and Reality

Geraldine Alpert

EDITOR'S NOTE

This third chapter represents a digression in content and format from the remainder of the volume, and thus warrants some explanatory comment. To begin with, the discussions here revolve around citizen participation in relation to consumer issues, whereas the focus elsewhere is on the distinctly different (though by no means unrelated) process of facility planning. Secondly, we have taken the unusual step of securing permission from a journal -- Public Utilities Fortnightly -- to reprint not only a feature article by Dr. Geraldine Alpert, but also several letters to the editor written in response to that article (together with the author's follow-up remarks). Accordingly, unlike the other chapters this one is divided into two parts of separate authorship.

Despite the apparent topical inconsistency, we believe the Alpert article has generic significance by virtue of its focus on advisory boards, a vehicle for participation which has seen considerable application in various aspects of the utility business (not just customer relations). The piece has relevance, also, because it voiced true-to-life concerns with which many practitioners can easily relate. As such it complements the previous article by Ducsik, by specifying an additional set of pitfalls to which public involvement endeavors might fall prey.

Dr. Alpert is a senior consultant at National Economic Research Associates, Inc., White Plains, New York.

In the sum of the Fortnightly material there lies, moreover, a unique opportunity to display the contrasting perspectives on open planning which currently exist in the utility industry. The fact is, Dr. Alpert's critical views evoked a rather polemical reply from several readers, which prompted the editor to write that "rarely has an article published in Public Utilities Fortnightly generated so much written comment...". Uniformly supportive of participation in general and citizen committees in particular, this included letters from:

* vice-Presidents of Pennsylvania Power and Light Company and Texas Electric Service Company, an assistant vice-president of AT&T, and a consultant to Wisconsin Public Service Corporation;
* the chairpersons of the Consumer Advisory Council of National Fuel (Buffalo), and of the Consumer Advisory Board of the Peoples Natural Gas Company (Pittsburgh);
* officials of the National Council of the Churches of Christ and the American Association of Retired Persons.

Also received, interestingly, were three longer statements proposed as rejoiner articles -- one by a power company manager, one by a university consultant to a power company, and one by a representative of the national Conference of Consumer Organizations. This brought to eleven the total of express reactions -- from diverse and clearly bipartisan quarters -- to the provocative Alpert presentation.

Our only real dilemma lay in determining how much of this commentary could be incorporated without suffering the diseconomy of an unduly long manuscript. The question was resolved by excluding all of the longer pieces (which are available from the respective authors*), as well as any

---

* The papers not included here for reasons of length are "Consumer Advisory Boards: Another View," by Peggy J. Scribner, Supervisor, Consumer Research and Liaison, Georgia Power Company, P.O. Box 4545, Atlanta, Georgia; "Consumer Advisory Boards: What They Can Do For A Utility," by Dr. Maureen J. Fleming, Department of Management, University of Montana, Missoula, Montana 59812; and "Consumer Advisory Panels and Investor-Owned Utilities: The Reality of the Issue -- a Rejoinder," by Dr. Louis S. Meyer, Steering Committee Chairperson, Conference of Consumer Organizations, Box 158, Newton Center, Massachusetts 02159.

letter which did not speak from a basis of direct experience with utility-sponsored participation. The latter criterion was derived from the fact that the Alpert article makes empirical reference only to public involvement in the education, health care, and social welfare sectors, which several respondents felt to be a shortcoming in light of the corresponding activity that has occurred in electric utility circles. We hope this complaint is now effectively redressed, not only in the second part of this chapter but also by all that follows in the compendium as a whole.

PART A: THE ORIGINAL ARTICLE*

In the last ten years, citizen participation, consumer involvement, community control, and related ideas have not only been much discussed but have found their way into federal legislation, public policy, and even utility boardrooms. The electric utility industry, faced with mounting pressures from customers, consumer activists, and regulatory commissions to keep rates low when financing and construction costs are rising rapidly, has sought relief through strategies directly involving customers in utility activities.

The establishment of consumer advisory boards or consumer councils, the use of public opinion polling, and the holding of public hearings have become epidemic in the industry. But before we see a wholesale conversion to the principle of citizen participation, it may well be worth examining the implications.

Citizen participation is hardly a new concept of American thinking. Our political process was founded on the cornerstone of public involvement -- through the electoral process, legislative hearings, and a variety of other mechanisms for individuals or groups to press their interests. At the earliest stages of our political development, de Tocqueville observed that the outstanding characteristic of American democracy was its profusion of voluntary associations and interest groups.

Why, then, have we experienced in recent decades an increasing demand for greater levels and different types of citizen participation? The concept of citizen participation received its greatest impetus from sociological theory

---

*Reprinted with the permission of Public Utilities Fortnightly, vol. 108, no. 5, pp. 19-22 (August 27, 1981).

in the 1960s, which suggested that the problem of inner-city ghettos was one of alienation, of the absence of a "sense of community" among the poor and the minorities. The legislative embodiment of the war on poverty, Title II of the Economic Opportunity Act of 1964, was explicitly aimed at developing a sense of community through community action programs to be "developed, conducted, and administered with the maximum feasible participation of residents of the areas and members of the groups served."

"Maximum feasible participation" was also a response to a growing concern among civil libertarians about the rights of the poor and the minorities. The latter perceived themselves and their leaders to be excluded from the political decision-making process and, as a consequence, subject to a societal distribution of benefits and rewards over which they had little influence. Demands for citizen participation by the poor and the minorities were a call for greater control over decisions, with the ultimate goal of influencing the distribution of society's goods or services.

But government policymakers saw "maximum feasible participation" as a way of bringing the poor and the minorities into the mainstream of American life, reducing conflict, defusing opposition, and ultimately winning the "war on poverty." The less radical minority group leadership, for its part, saw increased participation as a mechanism for organizing the poor, developing and training leadership, and eventually increasing the influence of Blacks and the poor on political decision-making.

THE GROWTH OF A MOVEMENT

In the 1960s the demand for increased citizen participation came mainly from the poor seeking a voice in political and economic decisions. In the 1970s, environmentalists and consumer activists, reflecting a new disenchantment with government and altered American consciousness and values, joined the cry.

After Vietnam and Watergate, increasing bureaucratization and regulation made government appear more remote than ever from the intended accountability of the electoral process. Regulatory agencies and their staffs, once viewed as guardians of the public weal, came more and more to be seen as captives of the industries they were to oversee. Accompanying this disenchantment with government was a growing awareness of the social costs of technology and economic development. This concern for the "quality of life" stimulated groups now called "environmentalists" to demand an ever louder say in government and industry decisions affecting the "environment," which has been loosely defined as anything from the siting of industrial plants to the nesting habits of the water grouse.

The consumer movement, easily confused with the environmentalists because the principal actors in both are often the same, is a more recent addition to the political arena. In the ten years since Ralph Nader's "Unsafe at Any Speed" was published, the consumer movement has evolved from the cause of a handful of dissidents to a full-scale social movement that touches virtually every segment of American life.

These social movements have made distinct, but sometimes conflicting, demands on the political system. The poor ask for a "fairer" share of society's goods and services. The environmentalists want an end to the perceived degradation of the environment associated with industrialization and technological development. And the consumerists seek protection from inferior or dangerous products, "excessive" prices, and other abuses in the marketplace.

One demand they have shared, however, has been for a more active participation of citizens in decision-making based on the presumption that access to the process means influence over decisions emerging from the process. In the last ten years, we have seen hundreds of diverse groups join forces on one or all of these issues and demand with a single voice greater participation by "citizens" (read also the poor, minorities, consumers, or the public) in decision-making.

The theoretical value of citizen participation is hardly in dispute, at least among democrats with a small d. But its value, as presently demonstrated in our educational, health care, and social welfare systems and in virtually every part of contemporary life, is less apparent.

DISENCHANTMENT SETS IN

There is, indeed, growing disenchantment with consumer participation among government policymakers as well as the public. But utilities have, nevertheless, begun to experiment with consumer advisory boards and councils as a means of keeping in touch with customers, finding out what they want, softening their attitude toward "the electric company," providing the participation organized consumerism claims it wants, and, perhaps, warding off the specter of "public power" that seems lurking in many corners.

The experimentation carries considerable risks while the benefits are questionable. There appear to be a number of reasons for a utility to organize a consumer panel. To begin with, the prevalent anti-utility feelings certainly seem to call for a response from company managers. Also, a utility may have grown estranged from its customers and their needs. In times like these, moreover, when critical decisions must be made, it may be particularly useful to confer with customers and share with them the choices confronting utilities, not to mention the risk of making the wrong one. Then, too, a voluntary advisory council may prove a preventive measure, forestalling a possible legislative mandate to put consumers on utility boards. Finally, and perhaps fancifully, consumer councils might even turn out to be allies supporting utility rate requests at commission hearings.

In reality, however, utilities invite two sets of problems when they establish a consumer advisory board. The first relates to the organization of the board itself; the second grows out of the board's likely role in relation to other consumer groups, the regulatory agencies, and the utility.

A community advisory board requires substantial amounts of "care and feeding." There are built-in sources of conflict and frustration for its members from inception. For example, the members, often lacking even rudimentary understanding of utility operations, can become discouraged by reading material or discussions that are too technical. Members are not likely to have the special knowledge or formal training to make informed judgements. The frustration this engenders may lead to a more generalized dissatisfaction with the activities of the utility.

Most consumer organizations experience high turnover rates among members. This is often a reflection of high levels of dissatisfaction as well as the voluntary character of membership which entails much time and effort but offers no pecuniary rewards. As a result, a utility's investment in "educating" members, establishing good staff rapport, working out the administrative mechanics, and getting things to run smoothly may have to be made again and again. Most professionals in community organization will agree that resources required to staff and administer a voluntary organization of this nature are substantial. Nor are there assurances that making that investment will bring an adequate return to the investor.

Most of the frustration, conflict, and confusion that has beset such boards in other fields is related to their unwillingness to accept the limited character of their advisory role. Once an organization has been given life and legitimacy, has established rules of operation, formed committees, and hired staff, it requires objectives to focus on. But when the function is purely advisory and there is no assurance of an effort to accomplish those objectives, dissatisfaction is inevitable.

The consequences of such frustration and the absence of power or control over outcomes may lead to conflict in and out of the organization. Advisory boards can become targets of other consumer organizations and be scorned, along with the utilities, as unresponsive to the needs of the community, the poor, the minorities, or other disaffected groups. The odds are considerable that a utility's goals in establishing a consumer panel will not be met.

Creation of such a panel may not be the best way to keep abreast of customer attitudes. Consumers are a heterogeneous group: it would be difficult to capture the range of customer opinion through a "representative" body. Who, after all, does represent the utility customer? Must we include landlords and tenants, rich and poor, the self-employed, and workers in industry? Members of consumer boards are most likely to "represent" no one but themselves.

Does it matter if a board is not "representative" of customer viewpoints? If information gathering is paramount, then it is important for utilities to be able to assess the extent to which a viewpoint is held by its customers and the degree of opposition or support for utility policy. If a consumer board is intended merely to provide a forum for the airing of viewpoints, its representativeness is less significant. But there are better ways to assess, and to provide for an airing of, consumer opinions -- sampling public opinion, for example, or periodic public meetings. A consumer board does provide an ongoing method of testing the waters, but the composition of the board may influence the outcome and lead to misinterpretation of the "test" results.

Also, establishment of an advisory board may generate more conflict than it eliminates, as member selection, openness of meetings, access to information, and other issues intensify anti-utility feelings. Should leaders of organized consumer groups be excluded from the advisory board, for example, the board could be viewed with suspicion by the public and become a target of the groups' attacks. If, on the other hand, representatives of activist groups are included on the board, the utilities could find it more difficult to influence their advisory boards or keep members' activities within limits set by utility managers.

There is, in addition, a substantial risk that providing an organizational base and a focus of activity might increase rather than reduce militancy. Is it realistic to think that advisory councils and boards will be content, in the long run, with merely giving advice? Or that a group of such highly motivated people will confine their advice to issues put on the table by utility management? The experience of hospitals and school boards with advisory councils suggests the answer is no. It is ironic that, as investor-owned utilities consider increasing customer involvement, rural cooperatives that have traditionally involved members in decision-making have been experiencing internal turmoil and even revolts. No amount of participation seems to assuage customer dissatisfaction when rates are going up.

THE REAL ISSUES

The real issues of consumer participation relate to content not process. Participation will be effective from the consumers' perspective only when some transfer of power over decisions takes place. The very nature of an advisory board is a major obstacle to the long-run satisfaction of its members. Yet, for industry, a transfer of control over decisions raises legal and managerial questions. Having a consumer board make decisions that could jeopardize a utility's financial viability or ability to provide service is hardly a prospect stockholders or managers would care to contemplate.

There is a further risk: once the consumer boards are in existence, regulators or legislators may be encouraged to mandate their perpetuation and give them more than an advisory role. The health field provides a telling example of such an evolution. Regional planning was voluntary through the 1940s and 1950s, supported by hospitals and philanthropies. Then, in 1964, Congress passed the Comprehensive Health Planning Act, which provided for consumer advisory boards to plan for health care.

They had no regulatory authority, however, and were welcomed by the hospital industry. But by 1974, the boards had become potent enough politically to push for the passage of another act (PL 93-641) that broadened their influence and created 200 health systems agencies (HSAs) nationwide. The HSAs advise state regulatory authorities on the issuance of certificates of need (required for new construction and equipment purchases, replacement and modernization, mergers, and changes of ownership) as well as on the distribution of federal funds and the "appropriateness" of medical facilities. They also wield considerable influence on other regulatory and legislative matters.

For the hospital industry, this has meant two-to three-year waits for construction approvals (or disapprovals, as is often the case) and the loss of control over significant managerial prerogatives, not to mention the inevitable exposure to the vagaries of HSA politics.

But hospitals have learned to cope. Their staffs attend HSA meetings, sit on local community boards, providing meeting rooms, and generally ingratiate themselves with local planning boards. Some hospitals have even established their own consumer boards as countervailing forces. Hospital managers have learned to explain hospital plans and policies in terms of community needs, write certificate of need applications that win HSA approval, and, in some cases, use the HSA boards as allies in seeking state approval for their projects. The HSAs remain a bureaucratic obstacle but perhaps not the insurmountable barrier hospital management once feared they might become.

Should utilities face similar legislative mandates, they no doubt will be able to duplicate the hospital managers' successful approach. But it is certainly not in the interest of the utilities to hasten such legislation. The health care industry is beginning to witness government disenchantment with consumer boards. Congress lately has been unwilling to allocate funds for HSA implementation activities and one bill proposed in the House would eliminate the HSAs altogether. Dissatisfaction with citizen participation now extends beyond policymakers to communities and even to some consumer groups.

It may be that the lessons of school decentralization, the poverty programs, and the HSAs will not be lost on the regulators and the legislators. Conceivably, the propensity of government to legislate consumer involvement will diminish. The recent development of consumer advocacy or ombudsman programs at the state and federal levels suggests disenchantment with, and eventual replacement for, the advisory boards once favored in legislative efforts to meet consumer demands.

Managers will be hard pressed to find appropriate responses to their consumer concerns -- hostility toward the companies and opposition to rate relief -- that do not threaten their managerial autonomy. But they would be better off resisting the temptation to form consumer boards or councils. In the final analysis, it seems to me, consumers are more likely to respond to utility strategies that affect their pocketbooks than those aimed at gathering their opinions.

PART B: WHAT OTHERS THINK*

"Consumer Advisory Panels Do Work" (Stanley J. Boulier, Assistant Vice President, American Telephone and Telegraph Company)

It was reasonable to have someone examine consumer advisory panels from a negative position as Geraldine Alpert did. No doubt some of the problems she describes can arise. But despite her rhetoric, she does not make her case against panels.

They do work, in the Bell system and elsewhere, to the mutual benefit of consumers and company. Obviously they have to be well conceived and well run, mistakes can be made, and there can be some tense sessions, particularly in the beginning. But none of Dr. Alpert's downside risk fears have materialized in any of the several dozen continuing consumer advisory panels that meet with Bell system managers. And several panels have been in existence since the mid-seventies.

Why do they work? Here are a few things that help:

* Have an "outsider" respected by consumer leaders organize the panel and run the meetings. Sometimes joint leadership works well, too, with a consumer leader and company manager co-chairing the meeting.
* Let the consumers set the agendas. The company can always suggest issues which the consumers may not know about, such as a proposed rate case.

---

* Reprinted with the permission of Public Utilities Fortnightly. With the exception of that by Helenbrook, all letters appeared in vol. 108, no. 12, pp. 65-68 (December 3, 1981).

* Wherever possible, bring subjects before the panel early enough so members can suggest meaningful changes.
* Be willing to listen and seriously consider consumer input. Your views just might change, for the better.
* It's all right to explain, but avoid being defensive. You don't have to win all the time.
* Always have officers or other policy-level executives participate. They need to know how consumers view the issues.
* Have discussions of issues, not one-way presentations. Avoid industry jargon.
* Work so that both sides gain. The consumers make their views known and respected, the managers improve their thinking, and the policies that result should therefore benefit both.
* Make changes. Consumers have to know they are having an effect.

In our experience, working with consumer advisory panels is good business. Consumers become less hostile -- over time -- when they are consulted, and even become supporters on some issues. And the companies gain valued viewpoints and find they can accomodate many consumer needs and desires without much problem.

"Mutual Understanding Is Likely to Result" (Grayson E. McNair, Vice President for Consumer and Community Services, Pennsylvania Power and Light Company)

Geraldine Alpert wrote a thoughtful article which illustrated some of her concerns about consumer advisory boards. While her points are well taken, I disagree with her conclusions about the value of such efforts to utilities. I would also like to point out that her concerns address a very specific form of public participation, i.e., consumer advisory boards. Overlooked are applications of citizen-involvement techniques to other utility matters such as facility siting and the mitigation of power plant construction (and operation) impacts.

Despite the obstacles to successful public participation programs, Pennsylvania Power and Light Company's experience would suggest that they are not insurmountable, nor would we agree that the establishment of such programs "generate more conflict than it eliminates." PP&L began implementing public participation efforts in 1975 when several pilot projects involving ad hoc citizens' groups were organized to assist in the siting of several transmission line and substation projects. Encouraged by early successes, an advisory board was established in 1977 to participate in a site-selection study for a new coal-fired generation station. Having finished its work on the study a year later, this committee is still in operation after four-and-one-half years.

In addition to power plant siting, PP&L's public advisory committee has worked meaningfully with our company on conservation projects and environmental programs. In 1978, a corporate consumer panel of the type described by Dr. Alpert was organized to address specific consumer concerns, i.e., rates, billing formats, provision of services to low income and elderly, etc. This panel has counterparts in each of PP&L's six operating divisions. Citizen committees have also been established at three of our five generating stations to deal with the impacts on nearby communities associated with modifications to existing facilities. A fourth plant committee is working with mitigation programs associated with a nuclear plant currently under construction.

Following the events in March, 1979, at Three Mile Island, PP&L organized a group of local community leaders as a working committee to review the development of the emergency plan for our own facility. Employee consumer panels have also been organized to solicit input to improve consumer practices. In addition to these specific activities, PP&L is also conducting an internal workshop for managers on the administration of public participation programs later this fall.

Based on this type of experience, some of the potential problem areas raised by Dr. Alpert were recognizable. Our interpretation and treatment, however, differ. Concerning the "care and feeding" of such groups, the author notes that members often lack "even rudimentary understanding of utility operations" and "...are not likely to have the special knowledge or formal training to make informed judgements." It is apparent to anyone in the utility industry that we are faced with complex technical and economic issues which need to be continually addressed. Part of the commitment to public participation involves the challenge of effectively communicating the complexities of utility decision-making to the people outside the industry in a manner which allows them to make meaningful input. Projects intended for public input must be examined closely to define mutually with participants a useful role. Once defined, the necessary resources must also be committed. A case in point is the review and comment role of our public advisory committee on the power plant siting study referred to earlier. Of the year spent on the study, half of staff time was spent on background presentations dealing with related issues -- fuel supply, unit size, water supply requirements, etc.

Dr. Alpert questions this investment of staff time because of the "high turnover rates among members" of the advisory boards. I believe, however, that our early commitment of staff resources was met by a like response from public participants in the form of active participation not only in the siting study but in subsequent assignments as well. In fact, terms of office were instituted to assure an orderly turnover of public participants. I do concede, however, that standing advisory committees have a tendency to become stale. There are, however, alternatives to advisory committees which can also be explored.

The author also takes aim at a relatively easy mark with respect to the "cost-effectiveness" of such groups. Despite PP&L's experimentation with various forms of public participation, we have not discovered a way in which to measure the effectiveness or benefits of such activities. But we continue to evaluate citizen involvement programs in an effort to improve these activities, which I believe have contributed to improved decisions in facility siting and consumer programs. Specific areas of contribution have included improved wording of documents intended for public use, communication of local values in siting projects, and a provision of a nonadversarial public forum which allows the staff to test its position and ideas. A review and comment role on a power plant siting study by our public advisory committee resulted in its recommendation of a model siting process which we subsequently adopted.

Of more fundamental concern is Dr. Alpert's assertion that participants will not accept a purely advisory role. Declaring that participation will only be effective when some transfer of power takes place, she cites examples in the health field to demonstrate her point. I am not sure that the analogy is a valid one, given the significant differences between the utility industry and the health services field. Again, our experience has shown that participants outside of the company have willingly accepted an advisory role. Particularly in the area of facility siting, outside participants have clearly recognized that final siting decisions must ultimately be the responsibility of the utility. Citizen groups, in such instances, have been more comfortable in commenting on the process used to select sites, rather than having the responsibility to support a specific siting decision. It has also been my observation that utilities sponsoring citizen participation programs must be willing to document and explain areas of disagreement and reasons for rejecting advice. Serious disagreement may still remain, of course, but a degree of mutual understanding is also likely to result. Being responsive to citizen advice does not necessarily mean accepting all of it.

My purpose is not to overstate the benefits of public participation programs but to counter Dr. Alpert's suggestions that the cost of such efforts are not worth the benefits. The obstacles as she points out are real. But the challenges which face the utility industry, it seems to me, must be met through some consensus of affected publics. The state of the art of participatory planning is too new to conclude that citizen involvement will not work. As recently suggested by one author (Dennis Ducsik, "Citizen Participation in Power Plant Siting: Aladdin's Lamp or Pandora's Box?" Journal of the American Planning Association, April, 1981), it is difficult to see how public participation efforts could give rise to anything like the controversy and problems associated with the conventional "decide-announce-defend" process -- a process which suggests to many that most of the decisions have already been made.

"The Risks Are Overstated" (John W. Wandrey, Wisconsin Public Service Corporation)

"Consumer Advisory Boards and Investor-Owned Utilities: Rhetoric and Reality" causes me some concern. The article implies that having consumer advisory panels is risky. Our company, Wisconsin Public Service Corporation, has had an advisory panel for nearly four years, and many of the risks listed by Dr. Alpert are overstated in my opinion. As an example, she states: "Participation will be effective from the consumer's perspective only when some transfer of power over decisions takes place." This has not been the case in our company.

At the inception of our three panels, panel members were informed that their suggestions would be taken to management, but they would not automatically be adopted. To date, this philosophy has worked very well, and we have not had the reactions she suggests.

She also states that panels are a "peace offering" to the consumer movement. Our attitude is that the panels are a means of obtaining some consumer reactions to our utility policies prior to their becoming problem areas. We do listen to their concerns and do not take them lightly. If a company is not sincere about its commitment to listen to the panels and consumers, then I agree a negative reaction could occur.

It disturbs me that Dr. Alpert chose not to research her opinions and statements with companies who have had years of experience with panels. I feel her statements and opinions would have been sounder if chief executive officers of utilities having panels would have corroborated her feelings. Interviewing panel members that share her opinions would have also strengthened her conclusions. Also, the American Gas Association and the Edison Electric Institute have devoted considerable effort to initiating consumer roundtables and advisory panels in member companies and would have been excellent sources for her to contact and could have strengthened the article.

In conclusion, I am suggesting that whenever future articles are directed to utility executives on subjects of utility matters, getting more perspectives would be beneficial.

"These Groups Provide a Vital Service" (R.T. Martin, Vice President, Texas Electric Company)

Just about every conceivable excuse a utility might find to decide against forming a consumer advisory panel seems to be listed by Dr. Geraldine Alpert in her article. I think I can say with some confidence that most utilities -- including my own -- thought of all or most of these excuses during the process of deciding to establish a panel.

The idea of asking a group of citizens to advise us on how to run "our" business has been a difficult concept for most utility managers to get used to. And that is precisely why many utilities have decided they need consumer advisory panels -- as a constant reminder that we cannot any longer consider this solely "our" business, not when our business has such a major impact on the public and not when energy costs are taking an ever-increasing chunk of the consumer's income.

Those companies -- 24 as of 1980 -- that have overcome the doubts so ably expressed by Dr. Alpert have not experienced the negative results she lists. To the contrary, reports we have received tell us that consumer panels are proving to be valuable assets to utility managements.

Coordinators of a number of existing consumer panels recently surveyed said that to date, for example: no panel has become dissatisfied with its advisory role; none has been attacked by outside consumer organizations; and none has prompted legislation mandating formation of consumer advisory panels. On the other hand, they have provided urgently needed insights into consumer viewpoints for the all too many of us who may let ourselves become somewhat insulated from the public we serve.

I can speak from personal experience when I say these groups provide a vital service. This would be true, if for no other reason, because they do not allow us to become insulated. These groups are especially valuable in the area of communications. Dr. Alpert says panel members can become frustrated by lack of understanding of the technology and terminology of our business. It has not been our experience that our panelists become discouraged or frustrated. These are intelligent people who are able to grasp the gist of an issue pretty rapidly. It has been our experience that when the panel has a hard time understanding our explanation of an issue, a practice, or a technology, this is an excellent clue that we are not explaining it well to the general public, either. And it also may be an excellent clue that there might be some things we could be doing better. In these cases, the panelists very frequently have been able to advise us on how we can improve services to our customers and on how we can communicate better with our customers.

This, I think, gets to what Dr. Alpert calls her "final analysis," in which she says "consumers are more likely to respond to utility strategies that affect their pocketbooks than those aimed at gathering their opinions." The simple fact is that consumer opinion can have a very serious effect on the consumer pocketbook. This is true especially if consumers do not understand the need to raise rates to build new power plants that use cheaper fuels, to replace worn-out units, and to keep up with rising demand. This lack of understanding not only can raise electricity prices higher than necessary -- because of the need for continued reliance on higher priced fuels and inefficient plants -- but also can prevent construction of enough plants to ensure reliability.

So this is why it is so important that we have help in obtaining consumer opinion and so important that we have consumer help in learning how to communicate effectively. Therefore, I would suggest that utilities that may be considering setting up a consumer advisory panel contact the Consumer Affairs Committee staff representative at Edison Electric Institute for more information on advisory panels. The staff also can provide the names of member companies that already have such panels and can share their actual experiences.

"Possible Pitfalls Are Valid, But Need Not Occur" (Mary Helenbrook, Chairperson, Consumer Advisory Council of National Fuel, Buffalo, NY)

Many utilities recently have taken a forward step in a positive direction in the area of consumer/utility relations by the formation of Consumer Advisory Boards. Then, an unresearched article by Dr. Geraldine Alpert appears in the Public Utilities Fortnightly, and it threatens to seriously hamper this forward thinking with a one-sided view of the utility consumer advisory board issue.

While we may not concur with Dr. Alpert's theories on the history of consumer involvement, we wish here to take issue only with her hypothesis regarding the usefulness and functioning of utility advisory boards. As members of the Consumer Advisory Council for National Fuel, we speak from experience. While it is necessary to note that many points Dr. Alpert makes regarding the possible pitfalls of such consumer panels are valid, it is also important to establish that such problems need not occur.

In establishing a board, the intent of the utility management plays a major role. If their intent is to appease public opinion, they will choose a certain type of consumer and maintain control over agenda and future memberships. If, on the other hand, there is a sincere effort on management's part to understand and communicate with their consumers, they will choose a board of forthright individuals representing many segments of the rate-paying population who have something to say and are willing to say it. They will deal honestly and openly with the group and allow it to be self-perpetuating. In a group such as this, there will be an exchange of information -- the consumers learning some of the technical workings and problems of the utility, and the utility management listening to concerned consumers and honestly attempting to find feasible solutions to the problems aired.

There need not be any confusion by the board of their role as advisors since the title of their panel clearly states that they are, nor need there be any frustration because they are not the policy-makers. The satisfaction of having the opportunity to talk with the policy-makers and encourage change is sufficient challenge for any person, even a highly motivated one. And, when dealing with a sincere management, an advisory council will be able to see its impact on the thinking and action of these policy-makers.

Perhaps consumer boards should be thought of as "sensitizors." They are a key ingredient in helping utility management become sensitive to the needs and problems of the people it serves.

There are other ways of assessing and providing for an airing of consumer opinions, such as opinion surveys and periodic public meetings, and they should be used along with an advisory panel as an additional means of consumer communication.

Unquestionably, consumer advisory boards are an important and useful cog in the wheel of consumer/utility relations. With this in mind, should they be mandatory? Absolutely not. The value of such panels lies in the willingness of management to cooperate and learn as well as the sincerity with which they deal with the boards. They should not be legislated or regulated. But, they should continue to exist, as valuable tools, voluntarily utilized.

Geraldine Alpert Replies

I found the responses to my recent article both interesting and informative. And I am not surprised that a number of utilities have succeeded in avoiding at least some of the pitfalls I described. It seems worthwhile, however, to make some additional comments on the issue of consumer advisory boards in light of the extensive responses to my article.

First, in any organization it often takes time for problems to arise. In the start-up period, where interest and enthusiasm of both utilities and members are likely to be high, future success may seem assured. But it most certainly is not. Utilities that have avoided problems at initial stages of council formation may face new problems as board members' (or utility staff) enthusiasm and interest diminish.

Second, it is not at all clear how utility managements can evaluate a successful consumer council. Without a clear set of goals or tasks, without some notion of what can be accomplished, it may be very difficult to tell if a consumer council is "successful." (Making council members or utility managers feel successful is only half the battle.)

Third, from my own discussions with utilities, I have identified certain key ingredients that seem to make some consumer councils work better than others. Those councils that (1) are given well-defined tasks; (2) have a limited term (just long enough to accomplish those tasks); and (3) experience a specific and concrete outcome of their efforts (even if that outcome is only a written response from the utility) have a better chance of running smoothly than groups with ill-defined tasks and open-ended terms of office.

Fourth, and finally, there is no question that improved consumer-utility communication is necessary to make management more responsive and consumers more understanding. There are a number of ways of accomplishing such a goal -- consumer councils, outreach programs, ad hoc task groups, public opinion polls, town hall meetings, consumer round tables, etc. Given the potential risks and problems associated with consumer councils, utilities may want to invest their efforts in alternative ways of improving their customer-utility dialogue.

PART TWO

Views of Participation Practitioners

4

Chairman's Report on the 1982 Edison Electric Institute Workshop on Public Participation

Dennis W. Ducsik

PART A: SUMMARY OF PROCEEDINGS

Most energy planners are keenly aware that political acceptability is seldom easily forthcoming for large-scale facilities, due to concern over environmental and social impacts. This is particularly true in the case of major power plants and transmission lines, which since the mid-60s have been the subject of acrimonious debate and drawn-out court battles between utility companies and activist groups. Rare is the state that has not witnessed this continuing confrontation, a trend which many consider ominous as it becomes ever more difficult to ensure an electricity supply that is adequate, reasonably priced, and provided in a manner consistent with society's other relevant goals.

Few would deny that, ideally, the best way to avoid debilitating conflict in the future is to channel the energies historically devoted to controversy into efforts aimed at cooperation. Toward this end, a concept that has attracted considerable support is that of "open planning," which can be defined broadly as some form of utility-citizen interaction other than that which normally takes place in the heavily adversarial and highly legalistic proceedings of government regulatory bodies. Over the years this concept has been endorsed by a variety of prominent individuals and national organizations. Even more significantly, attempts have been made by a number of pioneering utilities to put the idea into practice, by involving the public to some degree in various aspects of the facility planning process.

Mindful that among power companies both interest and experience in participatory planning have been growing, on

March 4 and 5, 1982, the Edison Electric Institute convened (in Washington, D.C.) the industry's first major workshop on the subject. In attendance, as shown in Table 4.1, were some 31 mid-level managers and executives representing 17 companies from all over the country, about half of whom were invited specifically on the basis of their involvement in facility-related participation programs. The general purpose of the workshop was to provide a forum in which these veteran practitioners and other interested attendees could:

* establish useful professional contacts;
* share information and views on the workings of open planning;
* evaluate their collective experiences with citizen interaction projects;
* consider issues affecting the future of public involvement within the industry.

The workshop was conducted with an eye towards maximizing informal dialogue among the participants, and was structured in such a way as to maintain a balance between "show-and-tell" and free-flowing discussion.

Of the ten basic sessions scheduled for the two-day agenda, four were invited talks designed to provide the group with something of a 'window' on participatory activities across the industry. All were held on the first day, which began with a chairman's presentation that identified generic patterns in some 58 cases of participatory site planning by utilities, according to six basic parameters of interest.* With this statistical profile as background, the group was then afforded a closer look at individual company experiences through:

* a case study of Ontario Hydro's program of citizen participation in route and site selection, with details provided of its application in two specific projects involving transmission facilities;
* a case study of open planning at Northern States Power Company of Minnesota, with reference made to two coal-fired projects and to recent developments regarding siting under the auspices of the state government.

---

* The full text of this overview is included as Part B of this chapter.

Table 4.1
EEI workshop participants

Speakers/Discussion Leaders

Bernie Bujnowski, Pennsylvania Power & Light Company
Rollie Comstock, Northern States Power Company
Joseph Dietz, San Diego Gas & Electric Company
Dennis Ducsik, Clark University (workshop chairman)
Gerald Edgley, Edison Electric Institute
Thomas Fair, Florida Power & Light Company
Gerald Farber, Pennsylvania Power & Light Company
Dawn Ford, Tennessee Valley Authority
Robert Henderson, Ontario Hydro
Howell Thomas, Center for Negotiation and Public Policy, Inc.
H. Jack Young, Edison Electric Institute

Other Attendees

Maryanne Bernard, Edison Electric Institute
Laurie C. Chisholm, Public Service Company of New Mexico
Christine Coffman, Tennessee Valley Authority
Roger Davis, Salt River Project
Jerry Dawson, Tampa Electric Company
Raul Debrigard, Northeast Utilities
Lewis R. Dixon, Wisconsin Electric Power Company
Keith Drescher, Florida Power & Light Company
Don Fraser, Ontario Hydro
Donna Lou Geiger, Bonneville Power Administration
Thomas J. Hadwin, New York State Electric & Gas Corporation
M.P. Killeen, New York State Electric & Gas Corporation
James R. King, Pennsylvania Electric Company
Ruthann Mogen, Portland General Electric
Morgan T. Morris III, Atlantic City Electric Company
Dennis M.O. Regan, GPU Service Corporation
R.E. Rude, New York State Electric & Gas Corporation
Al Qoyawayma, Salt River Project
Barry Rubin, Clark University
Janie Shaheen, U.S. Department of Energy
Sandra K. Shaw, Allegheny Power System
June Small, Florida Power & Light Company
Myles Smith, Georgia Power Company
Norman J. Stevens, Detroit Edison Company
Vern S. Williams, Bonneville Power Administration

A fitting wrap-up for these sessions consisted of a dinner address that evening by Jack Young, senior vice-president of EEI, who reflected in broad terms on changing industry attitudes and circumstances germane to the subject of participation.

The remainder of the event was devoted exclusively to discussion sessions, which covered a broad spectrum of topics. These included, respectively:

* the meaning and relevance of participation;
* benefits and risks associated with increased openness;
* strategies and techniques of interaction;
* internal corporate organization and administration;
* citizen involvement in issues beyond siting;
* actions EEI might take in the future.

A separate discussion leader was designated for each of these six sessions, and a professional facilitator was on hand to help catalyze the discussions.

How can the workshop's accomplishments best be articulated? To summarize the deliberations as they took place chronologically is hardly a satisfying approach, because in true eclectic fashion the focus of attention often ebbed and flowed without regard to topical boundaries. Coherence of exposition, therefore, requires a synthesis of kindred material according to recurring themes the group pursued. Basically, these were threefold:

* first, they examined fundamental questions about the need and value of public input, and in doing so helped define an underlying rationale for participatory ventures;
* second, they described the genesis, scope, nature, and role of the participatory function in their respective companies, thus providing useful information as to the concept's implementation within the industry;
* third, they engaged in considerable shop-talk on the process of establishing an effective interface with the public, from which emerged several management guidelines and a variety of specific "dos and don'ts."

What follows is a summary view of the discussions associated with each of these three major lines of inquiry.

ON THE RATIONALE FOR PUBLIC INVOLVEMENT

Considering the rather substantial departure from past practice the notion of open planning represents, the group felt it necessary to deliberate at some length on the most original of questions: why participation? The issue surfaced at the initial session, where several members (especially those without prior experience) said they hoped the workshop would help clarify the meaning and significance of this type of public outreach for their companies. The rationale for involvement was also on the minds of every speaker and came up spontaneously in open discussions, including those held during the wrap-up sessions at the end of each day. What emerged from all this was not consensus on a bill of particulars that all chief executives would find compelling. But the group did unanimously subscribe to the proposition that a more constructive role in utility planning endeavors needs to be defined for concerned citizens.

The touchstone for this common ethos was a strong sense that public interaction has become, and will continue to be, a fact-of-life for the electric utility industry. The group recognized that persons affected by power supply decisions are increasingly striving for (and getting) a voice therein, and no one really challenged the legitimacy of this claim to meaningful influence, based as it is on long-standing democratic traditions and values. Rather, comments were made to the effect that the industry must adapt positively and innovatively to the reality that decision-making in the future will be as much political as it is technical and economic. As one member put it: "public involvement in our business is no longer a matter of choice but a question of what form it should take and how it should be managed ... whether this involvement is destructive or constructive may well be up to us."

That participation can be destructive refers, of course, to the many times utilities have suffered delay at the hands of intervenor groups who use legal procedures to obstruct the granting of construction licenses. But the group's concern extended also to larger issues, i.e., the very survival of the industry and its capacity for self-determination. Several members noted that a utility is now so much in the public's (often critical) eye that it can no longer afford to operate without extensive input from those it both serves and impinges upon. To do so is to invite suspicion, animosity, and ultimately a greater degree of interposition by outside forces; the media, for example, will become more entrenched as the primary educator and informer, and legislative/regulatory bodies will increasingly abrogate or infringe on private decision-making prerogatives ("if we don't react, the political process will"). Thus, the general feeling was that to stay in control of their own destinies, power companies "must do more than go along grudgingly and to a minimum degree" with externally imposed requirements; instead, they must take the initiative to shape their relationship with the outside world.

An important corrollary to this proposition, also put forward with substantial group support, is that an adversarial attitude toward concerned citizens is unwarranted ("the industry won't benefit from an us-against-them mentality"). Several in the group argued that the public should be viewed instead as a potential resource to the company, capable not only of providing input and feedback on issues, alternatives, and impacts, but also of helping the company keep in touch with societal values and perspectives ("the utility hasn't cornered the market for knowledge"; "we've trusted more consequential issues to pluralistic processes -- why not this?"). It was pointed out, in fact, that an increasing number of utilities have already recognized the value of consultative models, having moved internally toward the use of group processes in high-level decision-making.

The third apparently key element to the rationale for participation was an equally pragmatic one, having to do with an expectation that early interaction can help reduce (albeit not eliminate) conflict and contention once final decisions are reached. The reasons cited in support of this position were basically two-fold. First, compromise solutions are easier to negotiate at an advance stage, because plans are still flexible and issues have not yet turned symbolic (and can thus be dealt with strictly on their merits). Second, where people realize the process has been open and the utility has made a good faith effort to be responsive, they tend to accept the final outcome even if they don't completely agree with it. There was also some feeling among the group that the company's credibility can be improved in the long run by the use of open planning methods.

With regard to support for such claims, the workshop produced considerable evidence of both a general and specific nature. In the analysis of 58 power plant projects with some form of citizen involvement, for example, it was found that the reported incidence of utility satisfaction with the process and its outcome was an encouragingly high 70% (in fact, only 8 cases could be termed an outright "failure"). What's more, individual group members cited some 13 examples from their own experience in which either confrontation had been avoided or goodwill toward the company had accrued to some degree. These included one case in which company staff estimated the information acquired from affected interest groups prevented at least a year and possibly two of delay.

None of this should be taken, by the way, as an indication that the workshop atmosphere was permeated with starry-eyed idealism as to the workability and effectiveness of open planning. Much to the contrary, the group recognized full well that the process is subject to numerous weaknesses, pitfalls, and unanswered questions, enough so that success is by no means always reasonably assured. Foremost in their minds, it seemed, were the difficulties of overcoming mistrust and the vagaries of human nature -- either or both of which can render the course of interaction most unpredictable. And even where the interaction process is reasonably smooth and constructive, it is not always an unqualified success in terms of altering the relationship between the company and affected constituencies. It was pointed out, for example, that sometimes credibility improves only for the company personnel with whom the interaction took place; that credit may be extended off the record only; and that activists will usually reserve the right to oppose other projects the utility may plan in the future.

Another source of ambiguity the group identified had to do with the question of to what extent deliberations outside the realm of facility siting might usefully be conducted in an open forum. Some members reported they were under pressure or desired to involve the public in such matters as rate-setting and long-range system planning, but were at a loss to respond in a meaningful way ("how do you run a participation program that embraces the entire Pacific Northwest and addresses the issues of power demand 20 years in the future?"). And others noted recent trade literature to the effect that companies like Kansas City Power and Light, Seattle City Light, and Wisconsin Public Service have enjoyed a modicum of success with participation in the broader aspects of corporate planning. But there were also conflicting signals in evidence, enough to breed caution in the mind of any prudent executive ("our consumer advisory panel isn't working very well"; "the committee that performed so admirably on siting issues is almost like a cannon loose on a ship's deck now").

Such observations served to remind the group that, whatever basis in evidence might exist for a favorable prognosis, it is most unwarranted to conceive of participation in glowing, panacea-like terms ("those marshmallow swamp words, you know, drive a lot of people up the wall"). The fact is, one simply cannot presume the process will either be smooth or successful; about the only thing one can really expect to encounter is considerable risk and anxiety in trying to make it so, at least initially.

Does this tempering of optimism with realism dilute the rationale for open planning? It does for some, perhaps, especially if they come to believe that positive outcomes depend on lucky combinations of circumstances more than anything else. But it is significant that those with the most first-hand experience with interaction processes tend to reject this fatalistic point of view. One such person in attendance at he workshop even claimed that "you can't lose," to which the rejoinder came: "you can, but you're still better off for trying." Waxing philosophical, the respondent went on to say:

> ... the name of the game is to go through some adaptive change -- its healthy. The natural response for corporations and indeed for persons is to resist change, because it's scary and it's frightening. We try to neutralize it and make it go away, but that may be the most counter-productive behavior, institutionally, that we can be undertaking. Somehow we need to integrate that stress, to regard pain as something positive. Unfortunately this can be very hard to understand and implement.

ON THE INSTITUTIONAL STATUS OF PARTICIPATION

Regardless of how compelling each member of the group may have felt was the present case for open planning, there was certainly no disagreement that a participation movement of sorts was "alive and going" in the industry, and that considerable institutional change had already been effected in this field. As one person put it at the very outset of the workshop, the departmental affiliations and titles held by attendees "would have been mystery words ten years ago, and in some places as short as five years ago." And by the close of the proceedings it was clear that the discussions had provided a useful snapshot of what has been going on around the country in the way of innovative outreach programming. To this the group can point as a second major accomplishment, for they took home a wealth of descriptive information as to the scope and nature of various public involvement efforts.

As to scope, the group was first informed that the trade literature indicates there are at least forty companies (of which 28 are investor-owned) who have used citizen participation to some extent in connection with major generation and/or transmission projects since the mid-60s. While the total number of cases tallied (67) might suggest that experience on the average is spread thin over the industry, in fact the group concluded that there's much more going on than meets the eye. To illustrate this, a representative from Pennsylvania Power and Light Company circulated a chart showing that PP&L has made extensive use of a committee approach in areas like substation siting, plant operation (including emergency planning), consumer issues, and conservation education. Others in the group added that less formal activity is even more abundant ("our president has regular roundtable discussions with various constituency groups, and so do all the senior VPs"). Thus, while the incidence of highly structured or truly collaborative programs may be limited, the group came to realize that outreach in a broader sense was actually a fairly pervasive phenomenon (albeit a low-profile one, even to the point of making it hard to keep track of everything that takes place in one's own company).

As regards the nature of citizen participation programs utilities have implemented, the opening presentation included broad survey information on numerous power plant projects in which participation had played a role to some degree. This indicated, to begin with, that experience with public involvement in this domain has been quite uniformly distributed when broken down according to three salient features: plant type (i.e., nuclear, fossil, hydro); issue agenda (i.e., environmental vs. socio-economic); and interaction mode (i.e., committees vs. individualized contact). Distinguishing the cases on the basis of whether interaction commenced before or after the identification of a preferred site/plant combination, however, revealed some interesting patterns. For instance:

* three out of four cases fall into the "late" category in that interaction focuses on impact assessment and mitigation for a specified project, with

limited (if any) attention given to alternatives considered earlier in the utility's planning process;
* where advance interaction occurs it is typically dominated by environmental issues, uses committees somewhat more often than decentralized communication with the public, and deals rarely with nuclear or hydro facilities;
* the more common "late" form of interaction, by contrast, seems ambivalent toward plant type, is somewhat less inclined to rely on committees exclusively, and is decidedly oriented toward socio-economic problems affecting host communities.

From this broad "map" of where experience has accumulated to date it is clear that, in the aggregate though not in equal proportion, the industry has experimented with a variety of distinctly different strategies.

Later in the workshop the group went on to receive more specific input on how the two largest utilities in North America -- Ontario Hydro and the Tennessee Valley Authority -- have organized themselves to carry out a participation function. The Ontario Hydro speaker, for example, devoted a portion of his talk to the question of the company's internal structure and the genesis of his department's role within it. In response to growing political and governmental pressures (and some bad experiences with the large public meeting format for participation), the company in 1975 created a major new division with overall responsibility for technical studies as well as for obtaining local, regional and provincial approvals for project activity. Explicitly built into the mandate of this division was the objective of maintaining a cooperative working relationship with the public throughout the planning process, even after the necessary approvals are obtained. Housed within the division, moreover, is a separate Community Relations Department whose staff works to integrate citizen participation with the division's more conventional functions like project engineering and environmental planning. The principal interaction techniques employed are information centers, citizen committees, and workshops.

In the TVA program, as described by its director, the group saw a sharply contrasting alternative approach to participation from that of Ontario Hydro. More specifically, there are three major points of departure:

* first, at TVA citizen interaction is not limited to certain types of projects, but is applied across-the-board to all corporate issues of significant public concern or consequence. In recent years, for example, the agency has "opened up" deliberations on its statement of corporate goals and directions, on alternative strategies for disposal of low-level radioactive wastes, and on the possible deferral of nuclear construction plans;
* second, the organization maintains an ongoing dialogue with a broad audience, so that when public input needs to be mobilized the relevant interest groups and concerned individuals are "up to speed." Toll-free lines, extensive computerized mailing lists, periodic workshops and small meetings, and a variety of other educational and opinion gathering techniques (all with careful attention to follow-up) play an important role in this "constituency building" process;
* third, the responsible department (the Citizen Action Office) is part of the staff of the General Manager (comparable to the chief executive officer in a private utility) and operates, therefore, as a kind of in-house service organization with the direct and forceful support of top management. As the circumstances require, the Citizen Action Office develops and helps implement outreach programs for all agency managers, who are even subject to a veto in the event their planned efforts do not meet with CAO approval.

TVA, in short, has put into place a rather pervasive system for "issues management" which, unlike traditional public relations programs, considers detailed feedback from the public to be as essential to decision-making as is information on technical and economic factors.

No other detailed accounts of administrative arrangements were heard during the workshop, but much of the texture or organizational life for the participation professional was brought to light in open discussion. One recurring theme was that participation efforts often suffered from the lack of a unified approach and/or guiding set of policies for the company as a whole, and sometimes develop therefore in a disjointed and incoherent way. TVA seemed to have avoided this difficulty with its highly centralized, almost codified program. But companies with a less extensive participation function were finding, for example, that corporate support for public involvement must be reconfirmed from project to project, or is more readily forthcoming for transmission facilities than for generating stations. Another problem noted was that participation programming often evolves independently in a number of separate departments, without any common point of liaison or uniform standards governing either the interaction process itself or how acquired information is to be used by management. Proposals for greater coordination, in fact, are often met with overt resistance.

A second motif within this general area of group discussion concerned the status and influence those responsible for participation have vis-a-vis other organs of the company. One person noted that a strong corporate commitment to participation (as articulated by a chief executive officer) doesn't always translate into strong staff support down the line. Others seconded this notion with examples ranging from simple reluctance to cooperate ("some staff members seemed to hide their data when they saw me coming") to one case in which an outright attempt was made by a recalcitrant department to alienate a citizens task force. The latter was acknowledged to be an extreme and rare event, but it still served to dramatize the fact that participation is a controversial issue in many companies and that its advocates are likely to develop some "warts and scars." But examples of bonuses and promotions and other forms of professional reinforcement were also mentioned to the group, suggesting that positive effects for change agents can be forthcoming as well.

It should be noted that the central message in all of this was not one of disillusionment, but that -- as with any first generation of organizational development -- there have been some "growing pains" encountered by those on the leading edge of the field. The group fully anticipated that appropriate adjustments and coping mechanisms would evolve through trial and error over time; indeed, some useful suggestions emerged from the workshop itself. Three persons testified, for example, how it had helped in building credibility within their own organizations to consult with affected managers in the early stages of developing a participation plan (which is consistent, of course, with the "first principle" that involvement usually breeds identification with the process). Another strategy that was highlighted as being important for the future was that of internal evaluation, to determine what each specific public involvement activity accomplishes. Even the veteran practitioners admitted techniques now employed for this purpose are primitive and tend to focus on the bottom line (i.e., the fate of the project or issue in question).

Beyond this the group had no special insights to convey on organizational matters, except to assert its belief that further institutionalization of the open planning concept is probably not well served by legislation. It is widely felt, for example, that effecting participatory requirements through agency rulemaking would counterproductively formalize interaction and dilute its effectiveness ("there's no single way to do it ... we need to encourage a variety of techniques"). And one group member whose company originally had supported a state-sponsored siting process ventured the opinion that it was not likely to succeed, because government doesn't have a strong enough incentive to make the process work ("it's just one more regulatory hoop"). This can only happen, he said, if the utility people who run the program "are willing to put their necks on the line to make it work." This, too, was an admonition to which most in the group could easily relate!

ON THE LOGISTICS OF OPEN PLANNING

The third and final area of concern the group traversed at various points along the workshop path was: how should a process of interaction be designed and operated? Few questions are as open-ended and lacking in singular answers, of course, but the group did have instructive things to say about a sub-set of topics on which their attention implicitly converged. These were:

* how is the public to be defined for purposes of outreach and involvement?
* what are the prerequisites of an effective communications program?
* how are the issues uniquely associated with a task force or committee effort best resolved?

Again, what follows is an attempt to capture the highlights of these primary discussion themes.

In the matter of identifying those with whom the utilities should interact, most in the group seemed to believe that participation should be available to all who are (or feel) affected by the project in question -- including those with an indirect stake based on generic values or concerns. This might include, in addition to those individuals and groups with the potential to become the company's adversaries, constituencies like employees and customers, local trade associations, and even the media. One person, in fact, noted that 31 distinct "publics" have been identified within his company's service territory. Others hastened to point out that this complex collection of interests could be expected to neither speak with one voice nor act in anything approaching unison ("it's a mistake to treat the public like a monolith"). The clear message here was that before attempting to deal with such a diverse clientele one needs a thorough understanding of the social landscape, as can be obtained, for example, through the preparation of an extensive "community profile" (several members said they collect data on local demographics, values, and political traditions as a matter of routine).

A related conclusion the group reached was that it is just as dangerous to be naive about the power structures which govern how decisions are really made, especially at the local level, as it is to be simplistic about the array of interests present in society. Some practitioners noted that public involvement is not always accompanied by constructive behavior on the part of public officials, who may find themselves in awkward positions or be tempted to engage in political opportunism (the "political football" effect). Equally worrisome to others was the possibility of manipulation by private individuals operating behind-the-scenes (the informal part of the power structure). Out of this discussion came a sense of need to identify and at least establish a dialogue with all such "influentials," and a few suggestions were made as to how they might be factored into a participation program.

Having a keen sense of audience, of course, is a necessary but not sufficient condition for effective interaction; one must also establish appropriate communication channels. The Ontario Hydro representative described some of the requirements as follows:

> ...a good communication plan is an essential to the public participation program. This involves the early preparation of information material such as background papers and pamphlets, written in layman's language. The communications plan also outlines how to utilize the news media as a communications instrument. This involves not only advertisements concerning meetings and information centers, but news stories and features

> of the proposed project and citizen involvement in the project. It can also involve a public speaking panel, graphic and audio-visual presentations, face-to-face and telephone contact with affected residents, and consultation with local elected and appointed officials. The list goes on and on.

Subsequently during the workshop others picked up on these ideas, stressing, for example, the importance of providing some instruction of a technical nature to participating citizens, and of ensuring that a two-way flow of information is maintained ("we call our group a listeners bureau to emphasize that their function is to listen and learn, not tell and sell"). Interfacing carefully with the media was also emphasized, in an anecdote recalling how an informal understanding with the press to "go easy" in reporting on the early, often stormy phases of a collaborative experiment played an important part in its eventual success.

Obviously implicit in the foregoing is the fact that citizen participation programs require a great deal of time and hard work, and thus impose a rather severe burden of staffing on the company. As one attendee put it, "this sort of effort demands fairly complete support from the rest of the corporation--no one department can do it all." A related point made subsequently concerned skill in interpersonal communication with a lay audience, which some in the company may exhibit (by virtue of having participated in a speakers bureau, for example) but many others will need to develop. Toward this end, two group members indicated their programs included formal in-house training (on how to manage a public meeting, how to diffuse hostility, etc.) and another recounted how staff members went through "dry runs" prior to making presentations before a citizen committee (to eliminate buzz-words, prepare for especially tough questions, etc.).

Apparently, it's difficult to attach a price tag to all of this because participation program costs have not been very well documented in the past--even in ballpark figures. The general feeling was, however, that while the expense is certainly not trivial there is no reason to believe it must be astronomical either. What's more, most seemed to agree with the assertion that public involvement projects are almost always cost-effective if one takes a properly holistic view of the planning process and the stakes involved ("the question is, do you want to pay a little now or pay more later?"; "what does it cost you to have a plant several months late, when you've really got to have it?").

To what else, in the eyes of the workshop group, must the architects of a communication program attend to ensure effective interaction? Perhaps the toughest issue the group wrestled with had to do with the potential conflict between the public's desire to influence the planning process on the one hand, and the utility's right to exercise final decision-making authority on the other. The latter was considered non-negotiable by nearly everyone; to do otherwise (e.g., by committing in advance to responding in a particular way to public input) was seen as sheer irresponsibility to stockholders and customers alike. At the same time, the group acknowledged that participation will never get off the ground if those involved don't see it as potentially worthwhile, and such credibility may be difficult to establish where there is considerable anticipatory mistrust. Certainly, people said, the process will be considered a charade if a range of options are not available at the outset and the company seeks merely to get the public to accept or concur with its determinations ("the process will fail, in my judgement, if the corporation's objective is to secure community endorsement for decisions already made").

The dilemma posed here is obvious: what if the company tries to be flexible and accommodating throughout the process but finds, in the end, that it can't live with a course of action favored by its citizen advisors? The sense of the meeting was that the risk of this happening could never be eliminated totally, although it could be held to an acceptable minimum through careful program design. Among the guidelines to be followed in this respect were, for example:

* be explicit about the fact that the company can't promise too much, but will make a good faith effort to be open to new ideas/alternatives and be responsive to inquiries/suggestions;
* don't ask for information you know you can't do anything with--the opportunity to comment is an empty and counter-productive gesture if it misleads people into thinking something is possible if it really isn't;
* consult with relevant managers to test their willingness to make reasonable concessions and accept less than optimal outcomes; ideally, the company

should feel comfortable accepting any of the courses of action discussed with the public, but if this isn't possible it's essential to know where there's room to maneuver;

* try to bring key decision-makers (even the CEO) in direct contact with concerned citizens, rather than working through intermediaries or a multi-level chain of command; talking to the person in charge is psychologically important to the public and increases the chances their input will be well-received.

The latter two recommendations, by the way, serve as a reminder that the ability of an open planning activity to influence management might well depend on forces internal to the organization, of which some are structural in origin. One group member rhetorically asked, for example, if the staff function in charge of participation must be relatively "high up" to ensure good communication with top executives and duly affect their decisions. But the question was never really answered--a further indication, perhaps, of the element of uncertainty that comes with the participation professional's territory.

If the workshop left certain conceptual issues unresolved, it certainly proved capable on the other hand of producing definitive statements on some of the operational aspects of participation. Some fell into the category of miscellaneous "tips" such as:

* participation must be easy, convenient, and non-intimidating;
* avoid being glib and cavalier;
* keep the options under consideration to a manageable number;
* do away with large public meetings, the kind that resemble a debating society with utility representatives in pin-stripe suits sitting in front of a theatre full of people.

What engaged the group's interest most at this level of discourse, however, were the functional aspects of one particular vehicle for interaction with the public: the citizens task force or advisory committee. In retrospect this was hardly a surprising development, insofar as all the case studies described by the speakers involved such groups, and because several other attendees had either witnessed or been responsible for their operation.

One item several workshop members spoke of was the process of assembling a working committee, in relation to which the delicate question was raised: should "hyper-environmentalists" and other die-hard opponents be offered representation? To some this seemed a Catch-22 situation: if excluded such persons might denounce the process as a sham, but if included they might constitute a disruptive force. But it turns out that experience seems to argue rather unequivocally in favor of the all-inclusive model, on two grounds. First, those who refuse to enter the open planning arena stand to lose credibility for passing up the chance to influence the process (this happened to an agricultural group in one of the Ontario Hydro cases). Second, those who join a committee but use it as a platform to incite controversy usually meet, eventually, with intolerance from the rest of the group (evidence to this effect was found in the Northern States Power experience). Thus, it seems, the non-adversarial approach has its own self-protecting mechanisms that can dampen extremist behavior and foster a climate of cooperation.

Another topic that elicited considerable discussion was how the role of the utility vis-a-vis the citizens group should be defined ("should the company chair the committee and provide all the staff, or stay in the background and let it be completely independent?"). One school of thought was that a hands-off policy was best, because it allowed the group to fashion its own identity and develop a learning curve of its own ("the company shouldn't be the answer-man--let the group struggle to discover what you already know"). Others in the group felt uncomfortable with this view, however, because it seemingly left the process too unstructured and "out of reach" compared to the alternative of having some form of joint effort. Also, it was argued that elbow contact is essential for productive give-and-take, and is more likely to lead to a relationship of mutual trust and respect ("one of the wisest decisions we made was not to turn the process over to an intermediary; they just get in the way"). But keep in mind that both approaches have met with success in effort. Also, it was argued that elbow contact is essential for productive give-and-take, and is more likely to lead to a relationship of mutual trust and respect ("one of the wisest decisions we made was not to turn the process over to an intermediary; they just get in the way"). But keep in mind that both approaches have met with success in the past (as well as some disappointment), so the arguments can hardly be considered conclusive on behalf of either one.

One point on which all protagonists found they could agree, however, was that it's difficult if not impossible to prescribe a narrow issue agenda for such a committee. At best one can expect to hear a lot of general griping, especially on rate-related issues, but it is also highly likely the group will want to delve into a variety of substantive matters (examples cited by workshop participants included load forecasting, conservation policies, the kind of pesticides used on transmission corridors, and executive salaries). Needless to say, the worrisome prospect here is that the situation might get out of hand, with the group beginning to feel they have a right to demand information on any subject, no matter how extraneous ("what happens if you open the door and find you can't close it?"). This can create considerable anxiety within the company ("one of our other departments was most unhappy to see that dead issue resurrected") and even within the committee itself if the point at issue is of concern only to a small minority.

Does it follow, then, that the company should take steps to limit or control the scope of discussions in the interest of expediting the process? Perhaps surprisingly, the weight of sentiment in the group came down against this proposition, at least as a matter of explicit policy. The basis was threefold:

* first, an unrestricted agenda was thought desirable as a means of allowing people to "get things off their chests," and to pose questions that may be tangential to the decision at hand but bear heavily on the company's overall credibility;
* second, excluded or suppressed issues will probably get raised anyway at some future point and be much harder to lay to rest ("if you can withstand scrutiny, it's better to do it early and in an informed setting than down the line in response to a skilled lawyer's interrogations");
* third, the point was made that the danger is probably overstated to begin with; one group member said, in fact, that sometimes their citizen's committees made progress faster than the company itself.

The group was not entirely cavalier about the issue of restraints, however. They did agree that getting agreement on a timetable was an important objective, in that it would encourage a committee to set priorities and find other ways to operate efficiently (e.g., by forming sub-committees to investigate subsidiary issues).

As a capstone thought it is perhaps fitting to note that, if the group arrived in the end at one abiding insight on the question of limits, it was that there is a limit to how tightly one can hope to plan an interaction process in advance. While useful handbooks are available on many operational aspects of participation, the group saw that it also involves an element of trial and error which can never be reduced to step-by-step instructions. To look for this is only to deny the imperative of staying flexible, and, ultimately, of being willing to confront risk. As one workshop member put it:

> You can study these things 'till you're blue in the face, and you can borrow from everybody's experience, and you certainly can be a lot smarter and more knowledgeable than we were at the time. But there is a unique quality to each case, and there are always surprises ... the notion that you can have a nice game plan, is just not realistic.

FUTURE DIRECTIONS FOR EEI

Last on the workshop agenda, appropriately, was the question of future directions. To address it initially the group was presented with a graphical landscape of participation and asked to describe what was on it. Several images appeared, but the dominant one was that of uncertainty -- "a winding road with fog and potholes." Thus, it is hardly surprising that the closing discussions centered on the need to lessen ambiguities and make the terrain ahead somewhat easier to negotiate.

The best way to do this, they felt, would be to continue and strengthen the process of information sharing and issue exploration that had been initiated by the workshop (and became, indeed, its hallmark). Regarding means to this end, the overriding group conclusion was that EEI -- as the national association of investor-owned electric companies -- should consider developing a network among participation practitioners and supporting it with a variety of information collection and dissemination services. Several preliminary ideas were put forth as to what this clearinghouse function might entail, and the group was assured all would enjoy careful consideration by the EEI staff.

The group also recognized that a case study effort could be an important aspect of an EEI networking program, insofar as information on prior utility experiences was acknowledged to be sorely deficient ("we seem too busy moving from one project to another to document what happened and what was learned from each"). This would be particularly helpful, one person noted, in identifying and comparing alternative models for public involvement in facility planning. Another suggested that it might be useful to look at the track record in other industries that have experimented with open planning, to see how they go about the process. In any event, it was clear that much empirical data will be required if the lessons of the past are to be fully brought to light and understood.

While no attempt was made to assign priorities among the aforementioned steps, the group was unambiguous in its endorsement of the need for initiative-taking in the near future. The prevailing view, moreover, was that the Environment and Energy Committee of EEI should assume the lead role in this regard. Happily, as announced in the Preface to this compendium, the Committee has responded to this suggestion and authorized the formation of a standing Task Force on Public Participation, to develop and implement a work plan of its own. Because of this follow-up action, the 1982 workshop seems assured of an enduring impact above and beyond that which it clearly had on those who were able to attend.

PART B: PROFILE OF UTILITY PARTICIPATION PROJECTS*

As the invited chair of the EEI Workshop, this author was charged with opening the proceedings with an overview of electric utility efforts to involve the public in various aspects of facility planning. Since there had yet to appear in the literature any comprehensive report on industry-wide patterns and trends in this regard, it was necessary to conduct a "reconnaissance-level" study

---

* An expanded version of this report, entitled "Power Plants and People: A Profile of Electric Utility Initiatives in Cooperative Planning," appears in the Journal of the American Planning Association (Spring, 1984). The excerpt published herein is done so with permission of the Journal editors. For help in collecting material for this study the author wishes to thank Barry Rubin and Joseph Mannarino, who served as graduate research assistants for the project.

comprising review of available trade literature (both published and unpublished) as well as personal communication with numerous individual practitioners. This generated a sufficient volume of material to identify, in a preliminary way at least, the most salient features of the innovation occurring in the industry as a whole. Such a profile of utility experience should prove useful as a backdrop for the issue-oriented commentary and case history accounts appearing elsewhere in this volume.

The available information on facility-related participation activity, it turns out, can be reduced to provide some insight on three basic questions. First, how many investor-owned and public utilities have involved the public in developing generation projects, and what specific type of facilities and impacts have been involved? Second, how has citizen access been effected, in terms of how early in the planning sequence it has occurred and of what format for interaction has been used? Third, and perhaps most important, how workable has public involvement proven to be, as manifest in the incidence of positive and negative experiences reported by industry personnel? Such questions are a logical starting point for inquiry, insofar as virtually nothing has been published yet on the scope, nature, or results of utility participation programs in the aggregate.

Before proceeding, it must be emphasized that this analysis is not the product of original field investigation or survey research, and is further limited by the fact that the work of several companies has not been documented or is reported only in a sketchy and anecdotal fashion. With the quality of the data base somewhat uneven, it was clearly inappropriate to conduct formal statistical tests of the sort normally applied to questionnaire-derived data, where there is greater assurance of completeness and less need to resort to inference. Even the use of simple chi-square analysis, for example, would impute a degree of rigor to the study that was not attainable given the resources it commanded.

The shortage of first-hand reportage also imposed a severe constraint on the extent to which public involvement initiatives could be assessed in a qualitative manner. For example, although wide variation was noticeable in the nature of the interaction between utilities and affected constituencies, it was not always possible to distinguish the ventures that ought to be counted as truly participatory (whatever the definition of the word) from those which are more aptly characterized as advanced forms of public relations. All that can be done in this regard is to extend the benefit of the doubt where ambiguity exists, on the theory that it would be patently unfair to make such judgements in the absence of reliable empirical research. This may risk overstating the incidence of progressive outreach efforts, but at the moment that seems better than to err on the side of not giving enough credit where credit is due.

Thus, with only a patchwork quilt of information available, there can be no detailed exposition on the quantitative structure of participation activity in the utility industry, nor any serious attempt to gauge the penetration of a collaborative ethic in subjective terms. Nevertheless, instructive things can be said in a less formal vein about the collection of pilot projects undertaken to date. Such is the modest goal of this report.

CLASSIFICATION OF PARTICIPATORY EPISODES

By this author's count and liberal criteria, 34 North American utilities (including five in Canada) have employed some form of citizen participation in some aspect of planning a power plant. The respective companies and the salient features of their interactive projects, which number 58 in all, are listed in Table 4.2. Each case has been classified according to a series of nine descriptors, beginning with the name and ownership of the sponsoring utility, the project name and type of facility (by fuel), and the year public involvement began. Then information is provided on three key aspects of the participation program itself -- point of public entry, interaction format and agenda, and overall outcome. Each is elaborated below.

Point of Public Entry. Generally speaking, power plant siting comprises four hierarchical activities occurring more or less in sequence,* and interaction with the public could be associated with any or all of them. Each can be described as a separate "level" of planning, as follows:

* Level 1 - reconnaissance/screening: this begins with the specification of a study methodology and

---

* For a detailed characterization of the power plant siting process, see Atomic Industrial Forum, Nuclear Power Plant Siting: A Generalized Process, Washington, D.C. (1974).

Table 4.2
Power plant projects with utility-citizen interaction (1965 - 1981)

Utility	Ownership	Project name	Type of fuel	Year interaction began	Public entry level[1]	Impacts of concern[2]	Format of interaction[3]	Overall outcome[4]
Altantic City Electric	Investor	Cumberland	coal	1979	2	M	D	P
Basin Electric Power Co-op. (N. Dakota)	Public	Laramie River	coal	1978	4	M	C	P
		Antelope Valley	coal	1980	4	S/E	C	P
Central Vermont Public Service	Investor	Burlington Area	oil	1969	2	E	D	N
Colorado Public Service	Investor	Pawnee	coal	1976	4	S/E	C	M/I
Duke Power (N. Carolina)	Investor	Bad Creek	hydro	1969	3	E	D	P
		Perkins	nuclear	1972	3	E	D	M/I
		Catawba	nuclear	1972	3	E	D	P
Eastern Kentucky Power Co-op.	Public	Smith	coal	1978	4	M	D	P
Florida Power & Light	Investor	Manatee	oil	1972	3	E	D	P
		Martin County	oil	1972	3	E	D	P
		St. Lucie	nuclear	1974	3	E	D	M/I
Georgia Power	Investor	Vogel	nuclear	1981	4	S/E	C	P
Los Angeles Dept. Water & Power	Public	San Joaquin	nuclear	1978	3	M	D	N
Montana Power	Investor	Colstrip	coal	1972	4	S/E	C	M/I
		Salem	coal	1980	2	M	D	P

New Brunswick Power Commission (Canada)	Public	Point Lepreau	nuclear	1975	3	M	C	M/I
New England Power (Rhode Island)	Investor	Charlestown	nuclear	1978	3	S/E	H	N
Newfoundland and Labrador Hydro (Canada)	Public	Lower Churchill	hydro	1978	4	S/E	H	M/I
		Upper Salmon	hydro	1979	4	M	H	P
		Cat Arm	hydro	1980	4	E	H	P
New York State Electric & Gas	Investor	New Haven-Stuyvesant	nuclear	1978	4	M	C	M/I
		Somerset	coal	1981	4	M	D	M/I
Northeast Utilities (Connecticut)	Investor	Sheffield/Canaan Mt.	hydro	1970	2	E	C	N
		Montague	nuclear	1974	4	S/E	D	P
Northern States Power (Minnesota)	Investor	Sherbourne	coal	1972	2	E	C	P
		Sherbourne	coal	1972	2	E	C	P
Ontario Hydro (Canada)	Public	North Channel	nuclear	1975	1	E	H	P
		St. Lawrence Seaway	nuclear	1977	1	E	H	N
Otter Tail Power (N. Dakota)	Investor	Coyote	coal	1979	4	S/E	D	P
Pacific Gas & Electric (California)	Investor	Diablo Canyon	nuclear	1965	3	E	C	N
Pacific Power & Light (Oregon)	Investor	Jim Bridger	coal	1975	4	S/E	C	P
Pennsylvania Electric	Investor	Coho	coal	1978	4	E	D	P

(Continued)

Table 4.2 (Cont.)

Utility	Ownership	Project name	Type of fuel	Year interaction began	Public entry level[1]	Impacts of concern[2]	Format of interaction[3]	Overall outcome[4]
Pennsylvania Power & Light	Investor	Post-Susquehanna unit	coal	1977	1	E	C	P
Pennsylvania Energy Park Development Group	Investor	Energy Park	nuclear	1975	1	E	D	N
Portland General Electric (Oregon)	Investor	Boardman	coal	1976	4	S/E	D	P
		Willamette Falls	hydro	1980	3	M	C	P
		Marmott Dam	hydro	1981	4	E	D	P
Public Service of New Mexico	Investor	San Juan	coal	1976	3	E	C	P
		New Mexico Generating Station	coal	1981	4	M	D	P
Puget Sound Power & Light (Washington)	Investor	Skagit	nuclear	1973	4	S/E	D	P
Salt River Project Agricultural Improvement & Power District (Arizona)	Public	Coronado	coal	1975	1	M	H	P
San Diego Gas & Electric	Investor	Encina	oil	1972	3	E	D	P
		Combined Cycle unit	oil	1973	2	E	C	P
		Sundesert	nuclear	1977	4	S/E	H	P

Saskatchewan Power (Canada)	Public	Nipawin	hydro	1977	3	M	C	P
		Greenland	coal	1980	2	M	C	P
		Gravelbourg	coal	1980	2	M	H	P
Seattle City Light	Public	Copper Creek	hydro	1980	3	M	C	P
Tennessee Valley Authority	Public	Phipps Bend	nuclear	1975	4	S/E	D	P
		Yellow Creek	nuclear	1975	4	S/E	D	P
		Hartsville	nuclear	1981	4	S/E	D	P
TransAlta Utilities (Canada)	Investor	Keephills	coal	1976	4	S/E	C	M/I
Virginia Electric Power	Public	Bath County	hydro	1976	4	S/E	D	P
Washington Public Power Supply System	Public	Hanaford	nuclear	1972	4	S/E	H	P
		WNP-3	nuclear	1974	4	S/E	H	P
		WNP-5	nuclear	1978	4	S/E	H	N
Washington Water & Power	Investor	Creston	coal	1980	4	S/E	C	P

Key

1. 1 = reconnaissance/screening
   2 = comparative evaluation
   3 = detailed assessment
   4 = review/implementation
2. E = environmental only
   S/E = socioeconomic only
   M = mixed (environmental and socioeconomic)
3. C = committee only
   D = dispersed only
   H = hybrid (committee and dispersed)
4. P = mostly positive
   N = mostly negative
   M/I = mixed or indeterminate
5. Energy Park Development Group was a consortium formed by Metropolian Edison Co., Pennsylvania Power & Light Co., and Philadelphia Electric Co.

criteria, which are applied to a broad region of interest to identify a (fairly large) number of potential sites and then to narrow the field to a (relatively small) number of candidate sites;

* Level 2 - comparative evaluation: this entails systematic study of the advantages and disadvantages of locations identified at level 1 as being capable of development and potentially licensable, culminating in the identification of a preferred site/plant combination;
* Level 3 - detailed assessment: this involves development of site-specific plant layout and design in conjunction with in-depth environmental and socio-economic studies, culminating usually in preparation of a draft impact report;
* Level 4 - review/implementation: this involves careful scrutiny of the proposed project by government licensing agencies and other officials (and the public via agency hearings), often giving rise to mitigation agreements and final adjustments in engineering design.

A related though less finely drawn distinction can be made between the early and late stages of planning. The former refers most appropriately to levels 1 and 2, not only because of their positions in time but also because options are still open to a considerable degree. Levels 3 and (particularly) 4 are considered late because once a preferred site is identified land options are usually acquired and some equipment orders are placed, since the plant design is nearly final. It should be noted, however, that at level 3 there is still considerable opportunity for mid-course correction since formal licensing applications are not yet filed.

Interaction Format and Agenda. These descriptors refer to how participation was organized and what impacts of concern the utility-citizen dialogue addressed. Regarding format, the basic alternatives are to assemble a committee or task force to represent affected constituencies, or to communicate more broadly with various individuals and groups via informal meetings, information centers, and other dispersed modes of interaction. In the case of impacts, the most elementary question is: did the interaction process focus mainly on environmental matters (broadly defined to include public health/safety as well as natural resource impacts), or was it mainly concerned with socio-economic effects on host communities (i.e., with respect to housing, education, transportation, or general infrastructure)? It is possible of course for committees and dispersed interaction to be used in combination, and for both environmental and socio-economic issues to be on the agenda, so a hybrid answer is also provided in each case.

Overall Outcome. Of all the parameters employed, that of outcome is at once the most intriguing and yet, inherently, the least amenable to objective indexing. To make matters worse, the available documentation on participatory projects leaves a great deal to be desired when it comes to the subject of evaluation. Still, most of the reports and statements collected for this study did include some indication of the degree of satisfaction felt by various involved parties. Depending on the nature of such preliminary information, each participation experience has been characterized as being, on balance, mostly positive, mostly negative, or mixed/indeterminate. Before proceeding, two aspects of this method for filling in the "bottom line" warrant further explanation.

First, it is important to note that the distinction between positive and negative experience is not based solely on whether a project comes on line as scheduled, essentially unaffected by public political resistance. While public acceptance is not an unimportant criterion, it does not account for other participation goals such as dispelling the atmosphere of alienation that has long prevailed between utilities and citizen activists, or narrowing the range of conflict that must be dealt with by adjudicatory means. It is entirely conceivable, moreover, that advance interaction which fails to preclude opposition could benefit the company in the long run, insofar as it can look elsewhere for a licensable project rather than proceed with one destined to meet stubborn resistance at every turn. Thus, included among the cases listed as "positive" in Table 4.2 are several that either were not entirely free of controversy or involved relocation (or deferral) of the facility in question.

One must also bear in mind, in contemplating the judgements made herein as to outcome, that nearly all of the reportage to this effect has been generated by utility representatives or other personnel with a direct connection to the industry. Not that this is undesirable per se; in fact, it is essential to learn how seasoned practitioners perceive the viability of public involvement, for it is to them primarily that others in the electricity business will turn for realistic assessments of the concept. Nevertheless, it cannot be denied that where there is uniformity of source there is also potential for distortion. In discussions with those outside the company, for example, utility officials may tend to exaggerate the positive and be reluctant to expose 'dirty laundry.' And, it could well be that citizen participants or disinterested observers would characterize the course of events in a manner quite different from the view espoused by utility managers. Thus the true incidence of constructive interaction may be somewhat at variance with that indicated, on face value, by the data on satisfaction level reported in Table 4.2.

FOUR BASIC PATTERNS IN EVIDENCE

Now one can consider how the various episodes are distributed across each of these major categories, and what (if any) relationships exist among them. Here, four major patterns are in evidence, not including those related to outcomes (discussed separately in the next section). For illustrative purposes it is convenient to arrange the data in matrix form, as shown in Table 4.3 (which indicates how participation experience has evolved over time) and Table 4.4 (which relates point of public entry to the three other descriptor categories of plant type, impact of concern, and format for interaction). Displayed in this manner the numbers can be compared in proportional terms, a straightforward task as long as one pitfall is avoided: because the sample size in each matrix element will tend to be small, and because there can be no guarantee that the inventory of projects is exhaustive, the finer distinctions that appear could easily be spurious and therefore must be ignored.

A summary and discussion of the patterns noticeable in Tables 4.2 through 4.4 follows:

Pattern No. 1: Interaction experience is spread thinly over a heterogeneous group of companies, and has grown at a steady (albeit modest) pace throughout the 1970s.

From the information in Table 4.2 it is clear, first of all, that public involvement has not been the exclusive province of any one class of utilities, and that, for the average company, participation programming is in the incipient stage of development. The companies listed comprise a fairly good cross-section of the industry in terms of region of the country and ownership mix represented; twenty-

Table 4.3
Incidence of interaction episodes over time

TIME PERIOD:	pre-1970	1970-72	1973-75	1976-78	1979-81	Total
TYPE:						
nuclear	1	3	9	6	2	21
coal	0	3	2	8	9	22
oil	1	3	1	0	0	5
hydro	1	1	0	3	5	10
IMPACT:						
environment	3	8	4	4	2	21
socio-economic	0	2	6	8	5	21
mixed	0	0	2	5	9	16
FORMAT:						
committee	1	4	3	7	6	21
dispersed	2	5	6	5	7	25
hybrid	0	1	3	5	3	12
LEVEL:						
reconnaissance/screening(1)	0	0	3	2	0	5
comparative evaluation(2)	1	3	1	0	4	9
detailed assessment(3)	2	5	2	4	2	15
review/implementation(4)	0	2	6	11	10	29
AGGREGATE NUMBER OF EPISODES	3	10	12	17	16	58

Table 4.4
Incidence of interaction episodes, by entry level

LEVEL:	Reconnaissance/ Screening (1)	Comparative Evaluation (2)	Detailed Assessment (3)	Review/Imple-mentation (4)	All Levels
TYPE:					
nuclear	3	0	7	11	21
coal	2	6	1	13	22
oil	0	2	3	0	5
hydro	0	1	4	5	10
IMPACT:					
environmental	4	5	9	3	21
socio-economic	0	0	1	20	21
mixed	1	4	5	6	16
FORMAT:					
committee	1	5	6	9	21
dispersed	1	3	8	13	25
hybrid	3	1	1	7	12
AGGREGATE NUMBER OF EPISODES	5	9	15	29	58

two, for example, are investor-owned and a dozen are public corporations. At the same time, just over half have had but a single episode, with the remainder split between having two or three project experiences. Some companies can legitimately claim to be veterans because they have also used citizen interaction for transmission line routing and other purposes,* but their number is relatively small and a corresponding few apparently have ceased to employ a participatory style of work.** The point is that what has been identified is not a community of the converted, but a collection of utilities with an obvious pioneering bent.

That a continuing commitment remains to be demonstrated at the level of individual companies, however, does not alter the fact that the frequency of interaction has increased in absolute terms since the early 1970s, from three or four projects per year to a level nearly fifty percent higher. Whether similar growth has occurred in proportional terms is unknown, though it would be reasonable to presume so given the general slackening of facility planning activity in the latter half of the decade (and especially since 1978). What's more, after 1975 the pace quickened in terms of the number of companies initiating their first participatory projects, which climbed to 14 during 1976-1978 (compared to six in each of the previous three year periods).*** At any rate, there is no denying that the interaction concept has enjoyed at least sustained and broad-based application over the years; it can hardly

---

* Northern States Power, Pennsylvania Power and Light, and Ontario Hydro, for example, have all had substantial experience with participation in the context of transmission line and/or substation siting.

** In New England, for example, Central Vermont Public Service and New England Power Company do not appear to have utilized citizen participation in subsequent planning endeavors.

*** The rate of company entry into the participation field, it should be noted, is probably influenced to some extent by the ebb and flow of project activity in the industry overall. The spurt in 1976-1978 may thus be attributable in part to the fact that many utilities were reactivating projects which had been on hold since just after the oil embargo. Likewise, with the downward revision in load forecasts and deferral of projects common in 1979-1981, it is not surprising that only five new companies began participation efforts in that period.

be considered a recent fad among a handful of maverick companies.

Pattern No. 2: Participation experience has been evenly distributed with respect to the various types of plant, impact, and format, although the former two have experienced a shifting emphasis over time.

The far right column of Table 4.3 makes clear that public involvement has not been uniquely associated with one specific technology, set of issues, or strategy of communication. The mix of episodes in the facility category (i.e., with the emphasis on coal and nuclear stations, in nearly equal proportion) is what might be expected given the fuel preference and availability trends of the 1970s. Also uniform in the aggregate has been the attention given to environmental versus socio-economic impacts (a 50/50 split), and the use of committees versus more dispersed forms of interaction. There is, however, an interesting dichotomous character to these latter distributions relative to the incidence of the hybrid alternative. That is, three-fourths of all projects deal exclusively with either the environment or socio-economics, not both, and there is an even more marked tendency not to combine the use of a committee with dispersed forms of interaction. The explanation for this one-or-the-other pattern is perhaps more apparent in the former case than the latter, where one can only presume that the circumstances utilities face give rise to a perceived exclusive choice as to the way participation is best mobilized.

The other pattern to be noted here is that, of the three parameters being discussed, only format of interaction has shown no appreciable variation in mix from one time period to another. Notice, for example, how a fairly even distribution across facilities prior to 1973 evolved first into a situation where nuclear plants were involved in 75 percent of all cases during 1973-1975, after which coal and hydro plants took over a similarly dominant share (76% of all cases from 1976 to 1981). A similar flip-flop is evident in the impact-year disaggregation of Table 4.3, although it is somewhat less dramatic. Environmental matters have always been on the agenda in a solid majority of cases, although there has been a noticeable drop in this regard (from 85% prior to 1973 to about 60% since then).* The incidence of socio-economic issues, on the other hand, has risen steadily and dramatically since 1973, to a peak of nearly 90% involvement in the cases occurring in the 1979-1981 period. This is attributable no doubt to the upsurge in concern over the "boomtown" problems surrounding the development of mine-mouth coal plants in the west and elsewhere (of which several are listed in Table 4.2).

Pattern No. 3: Most participation experience has accrued at points relatively late in the facility planning process, in part due to a spurt in the latter six years in eleventh-hour mitigation of socio-economic impacts.

When it comes to planning level, the incidence of interaction has been far from evenly distributed, as can be seen toward the bottom of Table 4.3. The aggregate data in the far right column show a steady progression in frequency of citizen access from the early to the later points of the process; indeed, one out of every two projects has occurred at the review/implementation level. Adding to that the episodes in the detailed assessment category (level 3), one finds that the late/early mix more generally is about three to one. This pattern is most pronounced since 1975, when a full two-thirds of all participatory activity was concentrated at level 4 alone, a clear reversal of the prior situation in which an equally large majority was associated with the pre-licensing levels (1 through 3).

The recent propensity toward level 4 projects goes hand-in-hand, of course, with the growth in concern for socio-economic effects noted earlier as occurring in the same time frame. One can also see this link in the data of Table 4.4. Notice there that at level 4 socio-economics were on the agenda in 90% of all cases (versus 31% for environmental impacts), and that two-thirds of all level 4 episodes focus exclusively on socio-economics. The reverse is true at levels 1 through 3, however, where matters of

---

* The reader should note here and in subsequent discussions that hybrid cases are taken into account in the numerator of these ratios, so the percentages cited will not as a rule sum to 100.

the environment were on the agenda 97 percent of the time (compared to 38% for socio-economics). Such results are not unexpected, because as a general rule environmental issues are foremost in a utility official's mind in the earlier phases of siting, when appropriate project changes can realistically be envisioned, whereas socio-economics normally come to the fore as a site plan nears completion.

This recent spurt of activity at level 4, it should be noted, does not suggest a corresponding erosion of interest in advance participation. As it turns out, the sum of episodes initiated at the other three levels remained basically constant for every three year period since 1973; and during 1979-1981 there was a noticeable renewal of interaction activity vis-a-vis the evaluation of candidate sites (level 2). On the other hand, there also has been a zeroing out of projects initiated at the reconnaissance level, but that could be a transient phenomenon associated more with the paucity of new siting efforts than with anything else.

Pattern No. 4: Beyond the impact-level association, there appears to be only one other clear-cut relationship within the four basic descriptor categories: citizens have rarely been afforded early access to the planning of nuclear and hydroelectric facilities.

A second parameter which seems to vary disproportionately as a function of entry level, as per the data in Table 4.4, is that of facility type. For reasons primarily technical, having to do with geographic constraints on siting, hydro-related episodes are almost totally confined to levels 3 and 4. And for reasons perhaps more political than anything else, there is a marked differential incidence across levels as between nuclear and fossil plants. Note that:

* while the number of nuclear projects at levels 3 and 4 is approximately the same as for fossil plants, the latter outnumber the former by a factor of three at the two early levels;
* there has not been a single instance where a utility sought outside assistance in choosing from a slate of final candidate sites for a nuclear station (compared to nine such instances for non-nuclear units);
* 70 percent of the nuclear outreach efforts occurring prior to the licensing process (level 4) were con-

fined to level 3, whereas the corresponding percentage for fossil episodes was only 30 percent.

Clearly, the industry has opened up the planning process for nuclear facilities in a much more cautious and incremental manner than it has in the case of fossil stations.

By constructing a series of additional matrices (not shown here), one can easily check for disporportionate frequencies in the incidence of other combinations of project features. But such an analysis yields only minor variations which, for reasons given earlier, may be spurious and therefore incorrectly suggestive of preferences in program design.* For the moment, then, it is best to say only that the various plant types, impacts, formats, and entry levels appear often enough in different combinations to be considered essentially independent of one another (beyond the exceptions already noted).

A PRELIMINARY EVALUATION OF OUTCOMES

The patterns identified above constitute a sort of rough map of where experience with public involvement in electric facility planning has accumulated during its first, largely experimental generation. Against this background, the discussion can now focus on the question of how satisfying the respective interaction experiences were in the eyes of their utility sponsors. The appropriate data from Table 4.1 can be analyzed in much the same quantitative fashion as the foregoing parameters -- i.e., by setting up a matrix in which the three outcome types are located on the horizontal axis and all other categories on the vertical. That is done in Table 4.5.

Looking first at the bottom line, one immediately encounters an impressive result: with positive outcomes indicated in 41 out of 58 cases, the cumulative satisfaction rate is slightly greater than 70 percent. Note also that: (1) nine of the 17 non-positive cases fall into the mixed/ indeterminate category, so that just one in seven experiences can be described as sorely disappointing; (2) only two sub-categories have a rate of positive results

---

* For example, at the two early levels of planning committees were used more often than dispersed forms of interaction (71% vs 57%), whereas at the two late levels the reverse was true (committees 50%; dispersed 67%).

Table 4.5
Overall outcomes of interaction experiences

NET ASSESSMENT:	MOSTLY POSITIVE	MOSTLY NEGATIVE	MIXED/ INDETERMINATE
YEAR:			
pre-1970	1	2	0
1970-1972	7	1	2
1973-1975	9	1	2
1976-1978	9	4	4
1979-1981	15	0	1
OWNERSHIP:			
investor	24	5	7
public	17	3	2
TYPE:			
nuclear	11	6	4
fossil	22	1	4
hydro	8	1	1
LEVEL:			
reconnaissance (1)	3	2	0
comparison (2)	7	2	0
assessment (3)	9	3	3
review (4)	22	1	6
IMPACT:			
environmental	14	5	2
socio-economic	15	2	4
mixed	12	1	3
FORMAT:			
committee	14	2	5
dispersed	19	3	3
hybrid	8	3	1
AGGREGATE NUMBER OF EXPERIENCES	41	8	9

less than 67 percent, and none falls below the 50 percent levels; and (3) the latest three year period (1979-1981) shows very little evidence of difficulty, i.e., only one mixed and no negative cases were encountered out of a total of sixteen.

These data certainly bely the pessimistic notion that getting together with the public will necessarily give rise to political havoc and operational chaos, as many in the electricity business still believe. To the contrary, the voice of utility experience seems to be saying that citizen participation in facility planning has proven to be a workable and productive alternative, on the whole, to the politics of confrontation. The analysis presented here may not be conclusive on this point, based as it is on a relatively small sample of sources within or close to the industry, but the evidence is nevertheless most encouraging. This is especially so in light of the fact that similarly positive indications have been forthcoming from other informal tallies of practitioner opinion. In a 1981 study, for example, the U.S. General Accounting office found that utility officials who had been involved in open site planning felt that it had saved time and money by making regulatory reviews more predictable and less contentious (though such benefits had not been documented in quantitative terms).* Several testimonials to this effect were also recorded, of course, in the proceedings of the EEI Workshop which comprise Part A of this chapter.

This is not to say, on the other hand, that the effectiveness of participatory planning can now be taken for granted. Too many episodes (30%) have failed to yield satisfactory results, and there is an obvious dim spot in the otherwise bright picture: the mixed experience reported for nuclear plants. Collaborative efforts associated with this facility type have encountered at least some difficulty in just under half of all cases, and certain program features are clearly more problemmatic than others. Inspection of the nuclear listings in Table 4.2 shows that experiences were non-positive in three out of four cases where committees were used as the only means of public representation; in five out of seven which focused exclusively on environmental impacts; and in eight out of ten in

---

* U.S. General Accounting Office, The Federal Government Should Encourage Early Public, Regulatory, and Industry Cooperation in Siting Energy Facilities, Report EMD-82-18, Washington, D.C. (1981).

which public access occurred earlier than level 4. One cannot read too much into these statements, of course, not knowing the circumstances of each case or the nature of the disappointment claimed. Still, they are hardly surprising given the rancorous history of the nuclear debate, and the extent to which generic issues in radiological health and safety have been intertwined with the siting and licensing process.

Another reason to interpret success/failure statistics conservatively is that, as discussions elsewhere in this volume clearly suggest,* interactive methods of conflict resolution are inherently delicate and face obstacles that are often severe, whatever the context. Apart from specific matters of program structure, most difficulty is rooted in the pervasive suspicion that has built up in the minds of interest groups and the public at large over the last two decades. The atmosphere has been so poisoned with polarization, non-communication, and adversarial behavior that it continually threatens to cast a pall over any participatory initiative, no matter how well-intentioned are the parties or how carefully structured is the process. There is also a good deal of inertia to be overcome, interestingly, in establishing a cooperative relationship with licensing agencies, whose personnel on a number of occasions have apparently proven reluctant to engage in informal dialogue with utilities and citizens outside the context of the usual procedural channels.**

Participation practitioners also face significant internal barriers, especially lack of commitment and understanding elsewhere in the organization. Tensions are known to arise, for example, when the perceived mandates of different departments are in conflict; when management and technical staff do not support participatory efforts or misunderstand the public's role; or when community affairs personnel feel torn between loyalty to the company and responsibility to groups with whom they maintain a collegial relationship. Such difficulties were highlighted at the Canadian utility workshop,*** where it was agreed that

---

* As indicated in Chapter Three, for example, the organization and operation of various kinds of citizen advisory committees can be highly problematic. See also Chapter Twelve.

** U.S. General Accounting Office (see footnote on preceding page).

*** See Chapter Five, <u>infra</u>.

above all else the climate within the company itself "offers the most immediate challenge" to future progress in the participation field.

These sobering remarks warn clearly against viewing the prognosis for citizen participation through rose-colored glasses, but do they refute the basically positive message forthcoming from this analysis? Hardly. In fact, in light of the impediments noted it seems even more significant to find that the collective attitude of those with first-hand experience is one of cautious optimism. Most seem to be of the opinion that benefits outweigh costs by a substantial margin, and that operational problems can eventually be overcome. Bear in mind that most program designs to date have been based more on common sense and guesswork than on well-defined guidelines derived from lessons of the past. In other words, practitioners realize that there is much room for improvement in the state-of-the-art, and for them, consequently, the relevant question is no longer if or why the public should be involved in facility planning. It is how the interaction process should be structured and managed so as to operate to the satisfaction of the company and, ideally, all affected constituencies.

If such a learning curve advances rapidly in the years ahead, so too -- in this author's opinion -- will the prospects for more widespread acceptance of the participatory concept within the electric utility industry.

5

Chairman's Report on the 1981 Canadian Utilities' Workshop on Public Participation

Robert C. Henderson

Canada's most obvious characteristic is its size, as one can discover by pondering any world map or the words of long-time Prime Minister Mackenzie King, who once said: "If some countries have too much history, we have too much geography." Too much geography has certainly affected the development of generation and transmission systems across the nation's ten provinces and two territories. Thousands of miles of transmission lines connect population centres -- generally located in clumps within a narrow strip along the United States border -- with generating stations, many of which are hydroelectric and located in the north.

Electric power for Canada's diverse industries and 25 million people is supplied primarily by thirteen utilities, of which ten are publicly-owned Crown corporations (or provincial commissions) and the remaining three are investor-owned. The bulk of production and consumption takes place in Ontario and Quebec -- the nation's industrial and population heartland -- where close to two-thirds of the total 375 gigawatt-hours of electricity flowed in 1981. As far as generation technologies are concerned, these are diverse and reflect the country's physical and economic realities. Quebec, British Columbia, Manitoba, and the Atlantic provinces rely mostly on hydroelectric resources. Saskatchewan and Alberta have built a predominantly thermal-based system by taking advantage of extensive coal deposits. Ontario Hydro has developed a mixed system

---

Mr. Henderson is Section Head and Acting Manager in the Community Relations Department, Route and Site Selection Division of Ontario Hydro, Toronto, Ontario, Canada.

with roughly equal contributions from coal-fired, hydro-electric and nuclear stations. Finally, the sparsely populated north depends mostly on small diesel generators.

The current economic downturn and increased conservation practices have reduced the load growth throughout Canada, resulting in the cancellation or postponement of many large generating projects. Nevertheless, billions of dollars are committed during the 1980s to extending the roughly 75,000 miles of transmission lines traversing the nation. It is safe to say that these new lines -- together with other proposals to construct generation and transmission facilities -- will command high public visibility. As they did in the 1960s and 1970s, affected individuals and groups will almost certainly challenge the need for system expansion and at the very least demand a say in the selection of sites and routes.

With public and governmental pressure having become a fact of life for utilities with development plans, most Canadian companies have elected to involve the public in some manner in the locational process. This was not always so; rather, it occurred as projects throughout the country were increasingly challenged and delayed, resulting in skyrocketing costs and threats to system reliability. Examples in this regard include, among others, the Bruce Generating Station to Milton line in Ontario; the 125-mile Calgary to Lethbridge line in Alberta; and the double 500-kV line feeding Vancouver Island in British Columbia.

As programs in public participation evolved independently across Canada, utility personnel increasingly felt a need to exchange information on experiences and methods. There was a growing recognition that public involvement in decision-making was a complex and important field and one with which the utilities had to come to terms. The people involved in this mushrooming field faced differing problems and needs, and realized that joint studies would result in a better understanding of the goals and techniques of public participation.

It was not until the late seventies, however, that some action was taken to create a vehicle for ongoing communication. This occurred at two national workshops for public involvement practitioners at which representatives of Canada's three largest utilities -- Ontario Hydro, Hydro-Quebec, and British Columbia Hydro and Power Authority (B.C. Hydro) -- agreed to establish a forum for dialogue among practitioners facing both linear and localized problems in siting. This group envisioned that such a forum would ultimately be "institutionalized" either through association with the Canadian Electrical Association (CEA) or as an independent group funded by member utilities. The CEA alternative was thought to have particular merit in that virtually all Canadian utilities are members and the organization is set up along functional lines. In any event, it was agreed that a logical first step would be to assemble a larger group of key people in the industry, with the objective of pinpointing "who we are, what we do, and what we could be doing."

A planning committee was thus assembled to prepare for this event, which would have a workshop format (i.e., without outside speakers) to maintain informality and encourage maximum utility attendance. The target group was working level, i.e., those responsible for design and implementation of public involvement programs, together with key specialists involved in related activities within the utilities. Senior management and non-electrical utility staff, it was felt, would be out of place in the initial "nuts and bolts" discussions.

Plans called for a three-day program that would facilitate active participation of all delegates in meeting four basic objectives. In simple terms, these were:

* to share experience on "state of the art" and corporate philosophies and programs;
* to identify common problems;
* to improve strategies and techniques;
* to promote ongoing communication.

Based on these objectives a theme was established for each day. The first would concentrate on identifying personnel, on what each company was doing in the field of participation and how it was organized for the purpose, and on comparing the Canadian situation with that of the United States. The second day would be devoted to examining the state of the art more closely, through case study presentations and subsequent discussion. And the closing session would explore broad issues and trends, common problems, and opportunities for continuing contact and dialogue (especially through establishment of some form of professional organization).

The workshop convened in Toronto on November 4-6, 1981, and was attended by representatives from nine of the thirteen major utilities across Canada (see Table 5.1). Originally it was hoped that all companies would send multiple delegates so the group would total 25-40, but this proved overly optimistic in view of the large travel dis-

Table 5.1
Canadian workshop participants

Ken Buller, Ontario Hydro
Community Relations Department

Bruce Bursey, Newfoundland and Labrador Hydro
Environmental Services Department

Jacques-andre Chartrand, Hydro Quebec
Communications/Programme d'equipment Information

John C. Dawson, British Columbia Hydro and Power Authority
Community Relations

Wayne Everett, Manitoba Hydro
Environmental Sciences Department

Don Falconer, Ontario Hydro
Community Relations department

Brenda Forbes, Ontario Hydro
Community Relations Department

Don Grant, Saskatchewan Power Corporation
Public Liaison

Bob Henderson, Ontario Hydro
Community Relations Department

Ron Kujala, British Columbia Hydro and Power Authority
Community Relations

Louise Malo, Hydro Quebec
Communications/Program d'equipment Information

Garry Prokop, TransAlta Utilities Corporation
(Calgary Power)
Generation Planning

John Robinson, Alberta Power Ltd.
Engineering and Construction

Gary Stairs, New Brunswick Electric Power Commission
Public Affairs

tances involved in several cases. The smaller number of attendees was not disadvantageous, however, insofar as there was correspondingly greater opportunity for in-depth consideration of problems encountered both inside the organization and in running programs in the community. Nor was any problem encountered in obtaining a cross-section of perspectives, stemming either from differing degrees of commitment to public involvement (a full spectrum was in evidence) or from utility size (which ranged from small investor-owned companies to public utilities supplying an entire province). Finally, a collegial atmosphere developed which in turn gave rise to a strong group commitment to further the interaction process begun by the work shop.

Against this background, a summary of the workshop proceedings can now be presented. This will be organized according to the three basic types of information that was exchanged at the workshop, relating to: (1) the broad features of industry and company programs in public participation; (2) the specific approaches to citizen interaction employed by selected companies in the context of actual siting/routing projects; and (3) the nature of problems, issues, and needs that arise in the operating milieu of the participation practitioner.

OVERVIEW OF CORPORATE APPROACHES TO PUBLIC PARTICIPATION

The opening day was devoted primarily to a series of nine presentations, eight of which described the status of participation programs in the organizations represented at the workshop. In a nutshell, these presentations revealed that:

* the scale of the participation function varies widely, as does the extent of concentration on generation vs. transmission projects;
* the responsibility for public involvement is located in a variety of departments (ranging from "Public Affairs" to "Engineering Operations" to "Environmental Services" to "Corporate Planning"), some of which carry out other tasks as well and some of which do not;
* the nature of environmental legislation and governmental policy (also variable across the provinces) seems to have a major influence on the level and extent of public participation in the decision-making process.

All this was obviously useful for background purposes and to help identify issues for the participants to deliberate in the subsequent days of the workshop. For example, interest was generated in delving further into organizational matters, and in exploring techniques (including funding of intervenors) for interaction with native and other special interest groups.

Also beneficial in this regard was input from the ninth and final speaker, who complemented the utility-specific information provided by the others with generic data on industry activity, as collected in a 1980 survey by Ontario Hydro. In this survey a cross-section of 61 North American utilities received a questionnaire which probed such topics as organizational structure and participation policies, role of the public in decision-making, techniques of interaction employed, evaluation methods used, and common problems and needs. Of the forty-two companies that responded, 33 said they were involved in locational studies and 22 indicated they involved the public to some extent in their projects. One-third of those in the latter category were Canadian utilities.

What follows are synopses of the opening day presentations, beginning, in the interest of proceeding logically from the general to the specific, with the results of the Ontario Hydro survey.

Review of North American Utility Experience With Public Participation*

The survey showed no pattern of differences between Canadian and U.S. utilities as a whole. Differences in the degree of public involvement were more between utilities themselves rather than on a regional basis. No formal lines of communication between utilities were indicated -- each utility appeared to be working in isolation. The average experience in public participation was 6.4 years indicating a still emerging field of study. The point at which most utilities involved the public was when lines or sites were identified.

Half of the 42 utilities that responded to the survey did ask the public to review their proposals, and did try to reflect public concerns. Of these, two utilities asked

---

* Presentation by Phil Wong, Senior Community Relations Officer, Community Relations Department, Ontario Hydro.

for input in identifying and selecting alternatives (consulting role). Three utilities engaged in a joint planning process, which gave the public responsibility for certain decisions while the final decision rested with the utility. Two utilities encouraged the public to provide input to corporate policy areas.

About three-fourths of the utilities that involve the public undertake some community profile work (i.e., attitudes, issues, services, demography). This is generally through personal contacts supplemented by literature research, census data and media clips. What's more, almost all utilities employed a wide range of techniques in their programs including public meetings, consultation with public interest groups, slide shows, media, community liaison and public hearings.

While the use of public meetings continues as a primary technique, many utilities are now finding they are not conducive to public feedback but rather result in polarization of opinions. As a result, more utilities are discovering that open houses and public information centres are more effective in encouraging two-way information. The most common public participation problems mentioned from the survey were polarization between the public and the utility, public apathy and lack of utility credibility. Despite this, a majority of the participants saw the public playing an increasingly important role in the planning process, and that expanding public participation into policy issues would assist in overcoming some of the existing problem areas.

Newfoundland and Labrador Hydro*

Newfoundland Hydro is a provincial Crown corporation with 950 employees and a total 6673 MW capacity in place and under construction. Its Environmental Services Department provides services to all aspects of project planning and implementation (planning, public participation, and monitoring during construction and operation). The department presently consists of ten people, most of whom are involved in monitoring the construction phases of projects. No one person has sole responsibility for public participation.

---

* Presentation by Bruce Bursey, Senior Ecologist/Assessments, Environmental Services Department.

Newfoundland Hydro was the first developer to go through the province's environmental impact assessment process. Four Environmental Impact Statements (EIS) have been completed since 1975, each showing an increased level of public involvement.

In 1980, legislation was introduced requiring public participation during the planning phase of projects. One public meeting is required during the preparation of the EIS, and if a public hearing is to be held, further public input is required. The method of involving the public, however, is determined by the developer and agreed on with the government before planning for the first project commences. The need for the project is a small part of the EIS. The level of detail required on the discussion of alternatives has not been settled.

Newfoundland has a core group of people who have been involved in a number of projects over the last five years but, by and large, the environmental movement is just starting. Justifying projects has not been an issue, but as environmental concerns increase, corporate policies are changing to meet them. At the present time, officials feel that the level of emphasis on public participation by Newfoundland Hydro is sufficient to match public sentiment.

New Brunswick Electric Power Commission*

New Brunswick Power is a provincial Crown corporation which reports through a chairman to the provincial cabinet. There are about 2,400 employees and assets of $2.3 billion and 2500 MW of installed capacity. There is no public utilities board, and therefore the onus is on the company to recognize public concerns and encourage its employees to get involved and consult with the public whenever possible. There is no one group charged exclusively with the responsibility of public involvement, but the main work in the area is done by the Public Affairs Division. In addition, one person in the Operations Division and two in Engineering and Construction are interested in public participation as it relates to their responsibilities.

There is no official environmental legislation in New Brunswick, other than an Order-in-Council which has been in effect since 1974-1975 requiring environmental considera-

* Presented by Gary Stairs, Communications Advisor, Public Affairs Department.

tions in planning. The provincial Department of the Environment has recently indicated public involvement should be included in planning studies.

The Point Lepreau Nuclear Generating Station, currently under construction in the province, requires an operating license from the Atomic Energy Control Board (AECB). During planning of the facility, a citizens' review committee was formed complementing a community affairs office established at the site with three full-time staff.

The public has not expressed many concerns over transmission lines because most have been in operation for years and are on government-owned land.

Hydro-Quebec*

Provincially-owned Hydro-Quebec employs 18,000 people. Reporting to the Vice-President of Information are four directorates, including the Communications/Equipment Programs, where the staff involved in public involvement are positioned. Currently there are 25 people, nine of whom are consultants for generation and transmission line projects, whose sole responsibility is public participation. Their mandate is to provide all the required services to ensure the integration of communications in all activities, to conceive and implement public participation programs, to identify socio-political aspects and to ensure these are taken into account by the concerned publics at appropriate times, depending on the nature of the project, and to integrate public input into decision-making.

Government legislation requires Hydro-Quebec to file an EIS on generation projects of 10 MW and over, and transmission lines of 315 kV and up. While the legislation requires the company to describe its communications program, there is no requirement concerning the actual structure of the program. Nonetheless, Hydro-Quebec feels it is to its advantage to integrate public views into the planning process and hence smooth the government approval process.

In general, the loss of agricultural land is a major concern to the people and government of Quebec. Concerns about the effect of northern developments on native people and the environment are also given high priority. Projects are questioned by the public, although at present the

---

* Presented by Louise Malo, Conseiller en communication, Communications/Programme d'equipment Information.

utility does little beyond explaining what it is doing. Exports of power to the U.S.A., New Brunswick and Ontario are also being questioned, especially by those whose land the transmission lines must cross. The public in Quebec wants to be involved in major projects.

Ontario Hydro*

Ontario Hydro employs 28,000 people, has an installed capacity of about 20,000 MW, and is a provincial Crown corporation. The Route and Site Selection Division within the corporation has responsibility for acquiring approvals for transmission line routes and generating station sites. Community Relations, a department within the division, is responsible for preparing and implementing public involvement programs on a specific-project basis. Public Relations deals with corporate issues -- such as rate increases or acid rain -- rather than specific project information.

Community Relations has two sections: public involvement and community relations; and the community studies section. The latter is responsible for social impact assessment and the former is divided into pre-and post-government approval. The major work of the department for the next three to five years will be on transmission projects.

Ontario Hydro's public involvement programs often include liaison committees with local representation. Committee involvement varies with the scale of the project but, on major projects, the committee can play an important role in identifying alternative routes.

Consideration has been given to a corporate policy for public participation. Ontario Hydro is working with provincially-based groups to review its public participation process.

The Environmental Assessment Act of 1975 does not stipulate, but rather intimates, that a public participation program must be carried out for major projects. The proponent is ultimately held accountable through the provincial government review process or at a public hearing.

---

* Presented by Ken Buller, Supervising Community Relations Officer, Community Relations Department. Editor's Note: For a more detailed exposition of the Ontario Hydro program, see Chapter Six, *infra*.

As a result, Ontario Hydro is committed to a program of public involvement throughout the planning process.

Major concerns about transmission lines are frequently expressed by agricultural communities over loss of prime farmland and by environmental groups about visual pollution.

Manitoba Hydro*

Manitoba Hydro is a provincial Crown corporation with approximately 4,000 employees and an installed capacity of 3900 MW (3500 MW of which is hydroelectric). The Public Affairs Department gets involved if the media is interested, otherwise responsibility on a project level for public input lies solely with Environmental Services.

The utility's Environmental Services Department, consisting of eight people, reports to a general manager of Corporate planning. The department works with the engineering line organization on specific projects, the bulk of which are transmission projects in southern Manitoba. Communication in the north is largely handled by northern field officers. Generation projects (almost all hydro and in northern Manitoba) are handled by consultants.

The Manitoba Planning Act (1976) guides land use in the province. An Interdepartmental Planning Board, with senior representatives from government departments and other Crown corporations, reviews district land use plans and other projects in order to determine the legitimacy of project tradeoffs.

Native groups are not eager to participate in project studies for fear they will affect claims the groups are negotiating with the provincial and federal governments over other recently constructed Manitoba Hydro projects.

Saskatchewan Power**

Saskatchewan Power is a provincial Crown corporation with approximately 3,400 employees. Public participation is positioned in the newly-created Public Liaison Depart-

---

* Presented by Dr. Wayne Everett, Project Manager, Environmental Services Department.

** Presented by Don Grant, Director, Public Liaison Department.

ment. This department is divided into project relations (involvement directly related to projects) and community relations (with a promotional function).

The objective is an informed public provided with opportunities to contribute to the planning process. Saskatchewan Power attempts to take into account public needs and concerns, to encourage more people to participate, and to reach more acceptable decisions. Early criticism provides the corporation the opportunity of reconsidering project plans and proposals. As such, the process of decision-making can be as important as the decision itself.

Draft public involvement policy guidelines state that Saskatchewan Power will involve the relevant public in the planning of major programs and projects which will integrate economic, technical, environmental and social considerations. The public has a right to be involved during all stages, and their contributions to corporate decision-making, while advisory, are highly persuasive. Public involvement programs will be structured according to the nature of the issues addressed and the publics affected.

Alberta Power Ltd.,* TransAlta Utilities Corporation**

There are three electric utilities in Alberta -- Edmonton Power, TransAlta and Alberta Power. Edmonton Power services and is owned by the City of Edmonton and generates very little power. TransAlta is an investor-owned utility which provides approximately 70 percent of Alberta's electrical energy requirements. Alberta Power, also investor-owned, operates primarily in northern Alberta but also services a couple of areas in the southeastern part of the province.

Alberta Power is a small company with 1,350 employees and an installed generating capacity of about 1000 MW. The utility's newly-formed Transmission Applications Group has seven people involved in right-of-way planning; site planning (substations) and writing applications; and designing and implementing public participation programs. The department's work has largely centred on 1434-kV single-pole lines located within existing road allowances. Alberta

---

* Presented by John Robinson, Right-of-Way Planner, Engineering and Construction Department.

** Presented by Gary Prokop, Projects Representative, Generation Planning Department.

Power expects that future 240-kV projects through more densely populated areas and more intensive agricultural areas will be much more complex.

TransAlta has an installed capacity of 4000 MW which is all thermal (mostly coal). TransAlta's Generation Planning group consists of mining (coal), technical and environmental personnel, and one social consultant. His responsibility is to act as a connecting link between the public and the engineers and miners within the company on generation-related projects. In the System Planning group, the right-of-way planner is responsible for public participation on a project-specific basis.

Alberta's social and environmental guidelines under the Land Surface Conservation & Reclamation Act (1973) are quite brief and not binding. The minister of the environment can ask for an Environmental Impact Assessment, although the utilities are not required to involve the public. However, they generally do anyway.

Alberta does not have a strong environmental movement. The province at present seems more geared to development, such that growth is not being questioned. Most public concerns relate to radio and TV interference of induced currents in fences and granaries located under the lines. The major social concern revolves around land acquisition and compensation.

British Columbia Hydro & Power Authority*

B.C. Hydro is a provincial Crown corporation with approximately 9,000 employees and an installed capacity of 7400 MW. Its Community Relations department develops and implements public participation programs. Department staff deal exclusively with future planning and work hand-in-hand with the planning engineers.

Some thought is being given to involving the public in corporate policy areas -- exports, debt management, rates -- by having three or four forums a year on various topics where planners and the public can interact.

The B.C. Utilities Commission Act is a relatively new piece of legislation regulating generation stations of 20 MW and over and 500-kV transmission lines. Two major hearings have been held under the Act to date. The public in

* Presented by John Dawson, Manager, Community Relations Department.

British Columbia frequently question the need for large development projects and want a greater say in corporate decision-making.

CASE STUDIES IN PUBLIC PARTICIPATION

The second day of the workshop began with the presentation of three case studies, which provided the attendees with a more detailed view of how selected programs had been developed and applied. This closer look at participation experience added considerably to the factual basis for discussion established the previous day, and helped bring into sharp relief -- through the typical question and answer process -- a number of key issues. Brief summaries of the descriptive content of each case are provided below.

Case Study No. 1 -- Supply to St. Catharines Area (Ontario Hydro)*

When it became apparent to Ontario Hydro that the St. Catharines area would need back-up capability to its power supply by 1984, a public involvement program was initiated in March of 1978. The proposal was to build 8 miles of 115 kV line with 2.6 miles on a new right-of-way, with the balance built on existing rights-of-way (partly through a park area). The study area was about 12 miles by 6 miles and covered parts of the cities of St. Catharines, Thorold and Niagara Falls, and the Town of Niagara-on-the-Lake -- an area with a population in excess of 250,000 people.

The goals of the public involvement program were: to explain the nature and purpose of the study; to have the public assist in the collection and verification of data; and to present alternatives for public comment prior to recommending a route for approval. Packages of information, including study area maps, were sent to interested individuals and groups at key points throughout the study. The mailing list included municipal government representatives, local Ontario ministry representatives, public utilities, umbrella groups, and area residents. When alterna-

---

* Presented by Craig Taylor, Supervising Community Relations Officer.

tive routes were identified, property owners were sent a letter and invited to attend a public information centre.

A number of techniques were used to facilitate public involvement. A municipal liaison committee of 20 people, with a local non-Hydro person as chairman, was established to represent a cross-section of local interests and to act as a local sounding board. Hydro's intention was to identify as many local concerns as possible and resolve them during the planning process. With the possibility of public hearings following submission of the Environmental Assessment to the provincial government, it was to Hydro's benefit to resolve as many potential problems as possible.

The committee was also taken on a tour of the System Control Centre, as well as of old and new transmission stations in the area. The information centres were held at local community halls over a period of a day or two at three points in the process. Photographs, topographical maps, and information panels were on display. Comment forms were available for those who wished to register concerns. The mailing list expanded to about 200 as the study progressed. Presentations were made to local interest groups and to the Niagra Escarpment Commission, which controls development on the escarpment. Regular status reports were distributed and newsletters and advertisements placed in local media.

The public involvement program required $50,000 of the total planning budget for the project of $250,000. The total project cost was set at $4.5 million. A two-year frame for public involvement on such a study is not normally the case. However, due to the social and land use complexities in this study, the time was necessary.

Case Study No. 2 — Cheekeye to Dunsmuir Study (B.C. Hydro)*

This project involved two parallel 500-kV single-circuit transmission lines from the Cheekeye substation on the mainland near Squamish, north of Vancouver, to Dunsmuir substation on Vancouver Island near Qualicum Bay, north of Nanimo. The first circuit is scheduled for completion in 1983. The project will be a combination of overhead transmission lines and submarine cables crossing Malaspina and Straits of Georgia.

---

* Presented by Ron Kujala, Senior Community Relations Coordinator.

The Cheekeye to Dunsmuir project was not typical of transmission projects undertaken by B.C. Hydro in the 1970s because the line was one of several methods assessed to supply Vancouver Island with power. Other options included fossil-fueled thermal generation on the island itself, and a policy to control the demand for future energy use. The decision to go with a transmission line was made before the public had an opportunity to become involved.

There was no established mechanism for B.C. Hydro to go out to the public to discuss its load forecast. The decision on method of supply and study area was made on technical grounds and a consultant was hired before the public became involved. This laid the groundwork for later public discontent with the project and subsequent calls for a hearing.

Phase One of the planning process consisted of an internal assessment of systems requirements, need for a new line, and preliminary investigation of the technical, economic, social and environmental feasibility of transmission alternatives. Terms of reference were established, a study area was determined, and in July, 1976, consultants were hired to prepare a report. This was followed by a press release and distribution of an information package to the public.

After being reviewed by B.C. Hydro in October, 1977 the Phase One report was sent to the provincial government Land Use Committee Secretariat (ELUCS) for review. The report was also distributed to provincial and municipal governments, regional interests and interest groups. Consultation meetings were held in local areas involving the consultants, B.C. Hydro and the concerned groups. Following nearly a year of meetings, a corridor route was recommended and a report submitted to the ELUCS and government resource agencies.

In November, 1978, the provincial Cabinet approved the corridor route in principle. Following this, B.C. Hydro thoroughly evaluated the selected route. In the sensitive area of the corridor route on the Sechelt Peninsula, a number of alternative routes were defined and B.C. Hydro and the consultant worked with the communities to find an acceptable route. In fact, 60 percent of the community relations effort on the project went to finding a route through that area. As a result of public concern further research into the submarine cable technology was undertaken, to enable the crossing from Texada Island to Vancouver Island directly.

Throughout 1979, many discussions were held with concerned groups, several of whom called for a public hearing. Despite these calls, however, the ELUCS gave final approval to the route in February, 1980.

Several major concerns were raised by the public throughout the planning process. One was that justification for the choice of the project to solve the energy supply of Vancouver Island was made without public input. Concern was also expressed about the effect of herbicides -- used on the rights-of-way to control vegetation -- on domestic water supplies. Next, because the route was through an environmentally and socially sensitive area, the size of the study area was thought to be too restricted. And in broader terms, the provincial government and B.C. Hydro were accused of withholding information and being manipulative.

In response, the government created a new department to deal with project justification under the B.C. Utilities Commission Act. At the same time, the public interest groups joined to form the B.C. Energy Coalition, a powerful group which may influence public attitudes toward energy policy in the future.

Case Study No. 3 -- Northern Grid Energy Development (Saskatchewan Power Corporation)*

In 1969, North Saskatchewan Electric, a wholly-owned subsidiary of Saskatchewan Power Corporation (SPC), was created to serve the electrical power needs of all areas of the province north of 55°N. Diesel generators and local distribution systems were set up with lines connecting to the south. During the 1970s, electrical consumption in the north increased and the loads are now near capacity. In addition, new mines are being established which are presently relying on diesel generation. In response, northern residents have expressed interest in an inexpensive alternative to costly diesel.

The social structure in the north is made up of three major groups: Metis and Non-status Indians; Status Indians; and the white population. In any area, therefore, there can be three authorities to deal with -- the Band (for the Status Indians), local advisory council (for the Metis and Non-status Indians), and the local community

* Presented by Don Grant, Director of Public Liaison.

authority (which represents all three). This variety of authority between communities and wide distribution of people makes the area unique to deal with. The public participation program needed to be developed with sensitivity to the special nature of these publics.

Three major components of the program were community research, meetings (workshops, public meetings, open houses and interviews) and media contacts. Prior to public announcement of the study, a socio-economic overview of northern communities was done to learn about issues, demography, leaders and the communications network. Later, community profiles assessed people's attitudes in more depth.

At workshops, people were oriented to the study and asked for input on program design. These workshops were effective in initiating public involvement and facilitating mutual education. Public meetings were held in 34 communities to convey general information and provide feedback. These were arranged through or co-sponsored by the local government body or band. Open houses -- the preferred type of meeting -- also provided opportunities for informal discussions. These were well advertised in local media and by personal contact, and comments were recorded.

Interviews by phone or in person assisted in creating public awareness and clarifying information and news releases, advertisements and media interviews played a similar role. They also helped establish the fact that northerners, while supporting the project, were generally concerned that large scale development would accelerate the rapid change facing northern communities and significantly affect these communities.

The next step in the study will take a careful look at the impact the lines will have on the environment and the communities. A number of community visits will be held to talk about local concerns. Later, an EIS will be composed and submitted to the Department of the Environment for review. This will detail environmental and social effects and narrow the corridors while providing opportunities for public comment. If necessary, a public inquiry may be held before the project is approved.

DISCUSSION OF PROBLEMS, ISSUES, TECHNIQUES AND NEEDS

When not listening and reacting to company presentations, the workshop participants spent their time exchanging views on a variety of common problems, emerging issues, applicable techniques, and possible areas for future research. Such discussion occurred within the assembly as a whole, at the end of both day one and day two, as well as in two sub-groups formed after the case studies had been delivered on the second day. And discussion was also evoked by a special presentation, on the third and final day of the workshop, by a consultant to Ontario Hydro who had recently completed a research paper reviewing current trends and future themes in the public participation field generally. What follows is a summary of the principal results to emerge at each of these focal points for deliberation.

Group Discussions of Company Presentations

The most extensive discussions took place in the small groups set up on the second day, each of which ultimately presented a report to the larger body. The major features of these reports are listed, respectively, in Tables 5.2 and 5.3. In comparing these reports in joint session, the group arrived at several points of consensus. To be effective in choosing techniques, for example, it was agreed that practitioners must try to better understand conditions in the study area as well as the nature of their publics (including, among other things, the extent to which different constituencies might wish to be involved). The group also converged on the basic idea that public participation should be aimed at bringing information in, not pumping it out.

What is most striking about these overlapping summations, however, is the extent to which they are dominated by the issue of coping with internal problems. Nearly everyone had encountered difficulties in dealing with both technical staff and management elsewhere in the company, and felt that negative attitudes (moreso than organizational structure) comprised the major obstacle to building interior recognition and credibility of the participation function. It was also observed that changes in this domain happen slowly, and that is normally takes significant controversy (or the rejection of a big project due to public concerns) to drive home the necessity for citizen involvement. Such findings were further emphasized by an informal survey in which each of the 17 participants were asked to note the two most important problems they faced as practitioners, together with the two most important topics on which they would like to have further dialogue. Tabulation of the 34 responses (17 x 2) in each area showed that:

Table 5.2
Contents of group no. 1 report

A. Major Problems Members Face In The Working Environment

1. Attitude of management and the utility as a whole toward public participation.
2. Need to involve senior management in field work as an educational process.
3. Initiating public participation -- should be early in the process.
4. State of the art of the discipline -- relationship to Social Impact Assessment (SIA).
5. Ability to influence.
6. Ability to release information to the public.
7. Conflicting public values, conflict resolution.
8. Refusal of certain key public groups to participate during the study, preferring to use political influence on the final decision.

B. Techniques

1. More so than specific techniques, the group felt there was need for a clear planning process to identify issues and publics and to facilitate understanding of the public participation process within the utility.
2. Need to evaluate public participation programs.
3. Need to define community profiles as to purpose, intent, use and value in designing the public involvement program.

C. Areas for Future Research

1. Evaluation of programs and monitoring of impacts.
2. Techniques for conflict resolution, i.e. mediation.
3. Public participation problems in large urban areas.
4. Participation by public in corporate policy areas.

Table 5.3
Contents of group no. 2 report

A. Major Problems Members Face In The Working Environment

1. Corporate commitment to do public participation from key internal staff -- a slow evolutionary process.
2. Understanding of the process of public participation and what it can do.
3. Adequate resources to do the job (human, technical, financial, information).
4. Appropriate timing of the start of public participation.
5. Be prepared to deal with corporate policy issues.
6. Lack of model for public participation programs.
7. Monitoring of construction activities.
8. Suspicion or mistrust of the public as to corporation's motive to do public participation (credibility).
9. Possible forerunner of unwanted economic development.
10. Growing awareness and concern for the total effect of the project on the environment.
11. Difficulty in justifying changes in terms of reference of a study which is underway.
12. Corporate policy issues dragged into project issues.
13. Changing political directions resulting from public pressure and disagreement between government agencies as to what should be done.

B. Techniques

1. Information centres.
2. Public meetings.
3. Liaison Committees (internal/external).
4. Documentation of activities.
5. Workshops/seminars.
6. Developing a policy within the corporation.
7. Training.

(Continued)

Table 5.3 (Cont.)

B. Techniques (Cont.)

8. Editing of documents re consistent policy and completeness.
9. Orientation of internal public/joint decision-making; political lobbying.

C. Areas for Future Research

1. Concept of public participation model development.
2. How much does the public want to be involved? when?
3. Public participation in corporate issues - which ones? when? how?
4. Electrical and biological effects research - brought forward in a form the public can relate to.
5. Public participation in Social Impact Assessment (SIA).
6. Media involvement and a better understanding of the process.
7. Information handling and documentation techniques -- how to evaluate information and disseminate information.
8. Corporate statement of public participation.
9. Systematic evaluation of public participation programs.
10. Public education in terms of advantages of public participation.
11. Types of information pieces which are most useful.
12. International or interprovincial study of public participation techniques.
13. Techniques for developing a community profile -- how to validate information in profile.

* 32 of the 34 most common problems identified were internal problems, including lack of commitment to and understanding of public participation (noted by 16 people) and attitude and involvement of senior management (noted by 9 people);
* 31 of the 34 responses on further research pinpointed internal areas, including evaluation and monitoring of participation programs (18 responses), corporate policy-making (5 responses), and models of participation (4 responses).

Those in attendance at the workshop were thus acutely aware, to say the least, of in-house factors affecting the operation of public involvement programs.

Another point readily apparent to the group as a whole was that structuring an organizational policy on public involvement is a lengthy and frustrating endeavor, and is often an educational process for staff in all departments (who require some background to understand the policy document). Also highlighted as crucial in the formulation of a basic policy is the method used to develop the document, and particularly the amount of public participation in this process itself. It is difficult, though, to achieve a balance that offends as few internal people as possible, while retaining enough clout that the public has some confidence in the policy. A comprehensive involvement process while developing the policy may be too complicated and costly to deal with. A general rule of thumb should be to involve in the review those people who will eventually be affected by the policy or have to work with it.

What this discussion revealed, perhaps most significantly, was that a participation practitioner often finds it difficult to achieve a balance in loyalties. There is a certain responsibility to the utility who is paying the person's salary but also a responsibility to the public being dealt with. The objectives of these two sectors are often in conflict. Under such circumstances, the group felt, the practitioner must ensure that ultimate loyalty is to the <u>process</u> to ensure that opportunities are provided for input and that public concerns are heard.

Review of Research on Current Trends and Future Themes for Public Participation

The closing day of the workshop began with a briefing on the work of consultant Jim Morganstern,* who used as his jumping off point a summary of the Delphi study presented by James Creighton at the 1980 Futures Conference in Toronto. Creighton had described how citizen participation was institutionalized in the 1970s through regulation and staffing, but in such a way (i.e., on an issue-by-issue basis) that its overall impact on decision-making was difficult to gauge, giving rise in turn to uneasiness about the cost of public programs. The group sensed, nevertheless, that in the near term (1980-1985) there would be increasing political pressure to provide formal opportunities for public involvement, resulting in continued expansion of programs and more institutionalization at all levels of government. However, the process was seen to be most effective for local rather than policy issues in this period. Later in the 1980s, participants felt participation would be used to a greater extent in policy, but still reluctantly by some agencies.

In a more specific vein, Jim Morganstern indicated that his research had identified several future themes for public participation. His initial theme, and that which received the most elaboration, was that emphasis will be placed on more sophisticated frameworks to evaluate public participation programs. He asserted that, in the last five years, evaluation has been oriented to case studies and the use of personal opinion as to whether the process worked. Evaluation checklists were established but these were hard to follow closely. The emphasis has been on goal specific assessments with rigorous methodologies to assess whether the program met certain pre-specified goals. Data was gathered mostly from interviews. Lately, evaluation is being placed into the context of the whole public participation program and its contribution to the agency's decision-making process. In this way, the agency can use the evaluation to begin to develop internal evaluation processes.

Mr. Morganstern welcomes efforts at goal specific and framework evaluations and feels public participation prac-

---

* See J. Morganstern, Current Issues and Future Directions in Public Participation, unpublished report to Community Relations Department, Ontario Hydro (October, 1981).

titioners have the knowledge and ability to integrate the involvement process with decision-making. They have a long way to go, however, and lots of problems to surmount. Among the questions he posed in this regard were the following:

* What is it we are evaluating? Is public participation a means or an end in itself? Did it make an impact on the planning effort? Did the techniques work and was it cost-effective?
* Why are we doing the evaluation? Do we need to prove to someone that we're accomplishing something, or are we trying to learn how to do public participation better? This is critical in determining the evaluation method chosen.
* When do you evaluate, i.e., simultaneous with program implementation or afterwards? This too will influence the results.
* Who should evaluate? An internal evaluation has greater access to data and is less threatening. An external evaluation, however, may result in the injection of new ideas and will be attractive if the purpose of the evaluation is to learn how to better involve the public.
* What are you allowing the evaluators to do? Can they tell you how to better meet your goals or can they also question the goals themselves?
* How comprehensive is the evaluation to be? Will the evaluation neglect some goals because they are too difficult to measure?

The group had no immediate answers to these queries, but did point out that a practitioner must be aware of his/her reasons for doing an evaluation. This will have a bearing on the type of evaluation methodology used. Some system for program evaluation must be set up that is ongoing and will grow with the agency instead of just responding to a certain set of goals.

A second broad topic on which Morganstern had several things to say was that of techniques for participation, which he saw as becoming subject to critical research and refinement. Rigorous testing of the assumptions that underlie the use of citizen advisory committees, for example, will be common. What's more, innovative techniques will be assessed in terms of compatibility with traditional programs and how they fit into the decision process. Techniques will also be viewed with an eye toward consistency with the principles on which public participation programs are based. Mediation, for example, could encounter difficulties in this regard because only bargainable issues can be dealt with, there is no control over the timing of its introduction in decision-making, and confidentiality is often required.

Community profile development was also identified as an area of future interest. Agencies will seek a more sophisticated understanding of their publics, especially characteristics that affect ability to participate and how different groups will participate. Here it was noted, for example, that citizens may pull out of the process for political reasons, fearing that by participating in advance they will be ineffective at later hearings if opposition becomes necessary. Morganstern thus felt proponents will pay more attention to community structures and look at why groups decide to opt out of participating. Programs, he argued, must be designed to fit the nature of the target public.

The third and final general subject of the Morganstern presentation was the application of public participation. Among the comments made in this respect were the following:

* We will see the "maturing" of public participation and its integration as an essential component into the mainstream of planning practice. Social Impact Assessment (SIA) and public participation will need to be integrated, with a weighting of qualitative and quantitative information.
* Public participation will increasingly be used in contentious decision-making process. This will necessitate the further development of specialized techniques and strategies. The credibility of the proponent will be an issue.
* More agencies will explore public participation in policy development. Implementation will be problematic requiring care, and training courses with public participation will increasingly be used by government agencies.

Morganstern also felt that current criticism of the mandatory requirements for participation in the regulatory process will result in changes to the process. Is a hearing an adequate means for public input? What about standing, range of issues considered, intervenor funding? Changes to the nature of the public hearing process will likely occur (e.g., less formal hearings and more attention to participation that precedes the hearing). Public participation will be structured to fit into the regulatory process. While public participation is not likely to eliminate the requirement for a hearing in major projects, it does allow the proponent to identify the unresolved issues. A sound public involvement program demonstrates to the regulatory body the proponent's willingness to respond to public concerns. Proponents will be more likely to accept the idea of public participation if they see it improving their chances of getting through the regulatory process.

To this the group added that agencies as well must face the political realities inherent in decision-making. Often agencies have a problem determining what the social impacts are and making value judgements on them. They must then open their planning process to the public, let the public speak for themselves and have the politicians make the final decision.

FUTURE OPPORTUNITIES FOR DIALOGUE AND INFORMATION EXCHANGE

Participants as a whole felt that the workshop had been productive and should be repeated regularly in the future. Many had attended seminars on public participation in general, but found a workshop with others in the same field more productive. It was agreed that while this first session had explored some of the general trends and problems being faced in electrical utilities across Canada, the next sessions should build from this and be more specific. Those utilities who were not able to attend the workshop were to be provided with the proceedings and other information to encourage them to participate in future sessions.

It was also agreed that a 5-person planning committee -- a representative sub-group of the workshop participants -- should be formed to "start the ball rolling". This committee would take the issues expressed at the workshop, begin developing a format for the next workshop, and develop a strategy for dealing with possible representation in the Canadian Electrical Association. Bob Henderson of Ontario Hydro was appointed first chairman of the subcommittee, and John Dawson of B.C. Hydro agreed to host the next workshop in Vancouver.

The group listed a number of topics as agenda items for that next workshop. These included generating stations; involvement of native people and special interest groups; process issues (rather than internal problems); more workshops/planning sessions; policy issues; and intervenor funding.

PART THREE

Selected Public Involvement Programs

6

The Evolution of Public Involvement in Project Planning at Ontario Hydro

Robert C. Henderson

In many ways, Ontario is the keystone province of Canada. It is centrally located along the northern rim of the Great Lakes, thus linking the historically affluent east with the resource rich west. It is the most populous of the nation's ten provinces, with 8.7 million people or over one-third of all Canadians. And with 413,000 square miles, stretching 1,000 miles from north to south and nearly the same from east to west, it is the second largest province -- roughly equal in size to Texas and California combined. Because of these factors, Ontario has always played a major role in the country's economy, the development of its institutions, and the formation of its public policies.

The majority of the Ontario population live in the southern cities which comprise Canada's industrial and financial heartland. This region is the home of major automotive, steel, chemical and, lately, high-technology industries. The largest city and provincial capital, Toronto, is the head office location of many of the country's largest financial institutions, including the large chartered banks, insurance and management companies and the Toronto Stock Exchange. Toronto is also the home of Ontario Hydro, which provides over 95 percent of the electricity consumed in the province and -- with almost 25,000 megawatts of installed capacity, assets approaching 20 billion dollars (Canadian), and 30,000 employees -- is one of the largest power companies in North America.

Mr. Henderson is Manager of the Community Relations Department, Design and Development Division (Transmission) of Ontario Hydro, Toronto, Ontario, Canada.

Ontario Hydro serves its customers through a transmission network spanning 40,000 kilometers and a generating system comprising 68 hydroelectric, five coal-fired, and four nuclear power stations (see Figure 6.1). Each generating source currently accounts for about one-third of the electricity produced, although it is anticipated that nuclear power capacity will increase to about 50 percent of the total within the decade. This will occur as nuclear units currently under construction come on line and allow the phasing out of other, less efficient coal-fired stations. As to the future, load growth has declined to the point that another major power plant will not be required until the late 1990s, and all siting studies are therefore on hold. Planning for the expansion of the transmission system remains in an active state, however, so work to identify additional routes and facility locations is continuing.

For more than ten years now, Ontario Hydro has involved the public in the process of planning for major extensions of the generating and transmission network. This paper describes how citizen participation evolved into an integral part of that process, in terms of philosophy, technique, and organization. The paper also recounts how public involvement was utilized in actual projects, particularly those associated with the location of transmission facilities (on both an individual and system-wide basis).

ORGANIZATIONAL BACKGROUND

It would be difficult to overestimate the importance of electricity in the development of Ontario throughout the 20th century. Lacking major coal, oil, or gas deposits, the province decided in the early 1900s to turn to its extensive supplies of water power, the cornerstone of which has always been the Niagra River and its famous falls (whose water is shared with the State of New York under treaty arrangements between Canada and the United States). To tap this abundant energy resource, a public utility was established in 1906 under the Hydro-Electric Power Commission Act. As of 1974, upon passage of the Power Commission Amendment Act, that utility received its present name of Ontario Hydro.

Sir Adam Beck, the turn-of-the-century manufacturer and politician who was the founding chairman of the organ-

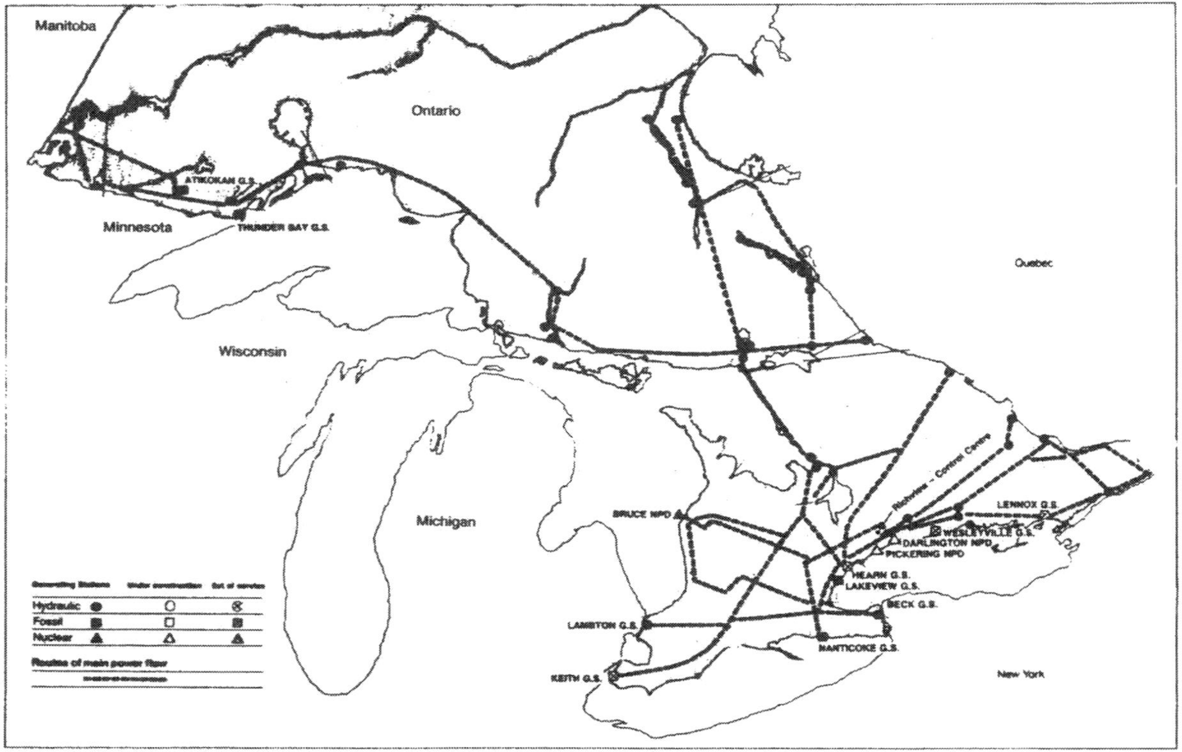

Figure 6.1 Ontario Hydro — Major powerline routes and generating stations

ization, would have difficulty comprehending the extent of subsequent growth if he was still alive. In 1908 the company entered into an agreement with fourteen municipalities under which it assumed responsibility for the generation, transmission, and sale of power at the lowest feasible cost. That same agreement exists today, but it has been extended to 321 cooperating municipal utilities, for whom Ontario Hydro acts as a central supervising body with power to approve and control certain features of their operations. Hydro also supplies power directly to about 100 large industrial customers and 770,000 rural households, again under the mandate to keep costs to a minimum.

Hydro is a Crown corporation which is self-sustaining, deriving no revenue from taxes, but whose capital borrowings are backed by government guarantees (and require Cabinet approval). The utility is held accountable to the Ontario legislature* through the Minister of Energy, with which Hydro files an annual report. The Ministry determines the broad policies within which Hydro operates, and regulates it according to the provisions of the Power Corporation Act of 1970 and the Power Commission Amendment Act of 1974 (which affected Hydro's form of administration but not its basic rights and obligations). Proposed applications for rate increases are reviewed by the Ontario Energy Board, another government agency for which the Energy Ministry is responsible. After intensive public hearings, the Board submits non-binding recommendations concerning the proposed increase to the Minister, who makes the final decision.

Major construction projects undertaken by Ontario Hydro, such as the building of a new generating station, are generally reviewed by the Ontario cabinet and are sub-

---

* In all Canadian provinces, it should be noted, government is based on the British parliamentary tradition. The Queen, represented in Ontario by the Lieutenant-Governor, is nominally the head of state, but power resides with the premier and his ministers of the Crown. These ministers are appointed from the ruling political party -- the party which elects the most representatives to sit in the provincial legislature -- by the Lieutenant-Governor on the recommendation of the premier.

ject to the Environmental Assessment Act of 1975.* This law mandates a detailed study process to gauge the environmental and social impact of large development proposals. The Ministry of the Environment reviews Environmental Assessment Reports filed by Hydro and decides whether the project's impact is such that public hearings are necessary before the Environmental Assessment Board. The Ontario Cabinet reserves the right to reject or modify board decisions. The government may also exempt proposals from all or parts of the Act.

Finally, government oversight of Ontario Hydro has been extended in recent years to issues in long-term power supply planning. This was precipitated in 1974 when the utility issued a report entitled "Long-Range Planning of the Electric Power System," which predicted a sharp increase in the demand for electriaity such that a doubling of system capacity (with an emphasis on nuclear generation) was projected for the subsequent decade. This projection engendered much opposition from groups at the regional and local levels, who were critical of the lack of public input and attacked the report on a number of fronts, including need, capital costs, and environmental impacts. In response to this controversy, the provincial government established a Royal Commission, chaired by Dr. Arthur Porter, to publicly review and appraise all aspects of Ontario Hydro's expansion plans, including the commitment to nuclear power. From 1975-1980, this Porter Commission examined Hydro in exhaustive detail, greatly increasing public knowledge of company operations and planning.

Internally, Hydro is divided into three main functional areas -- Operations, Engineering and Services, and Planning and Administration -- each with an executive vice-president. As shown in Figure 6.2, Engineering and Services includes a Design and Construction Branch, of which the Route and Site Selection Division is a part. The over-

---

* Hydro also must adhere to a multitude of provincial laws and regulations in areas not directly related to energy or the environment. This includes land expropriation and compensation legislation and Ontario Municipal Board regulations. Hydro actions are also governed in certain areas -- principally with respect to nuclear generation and power exports -- by Federal regulatory bodies.

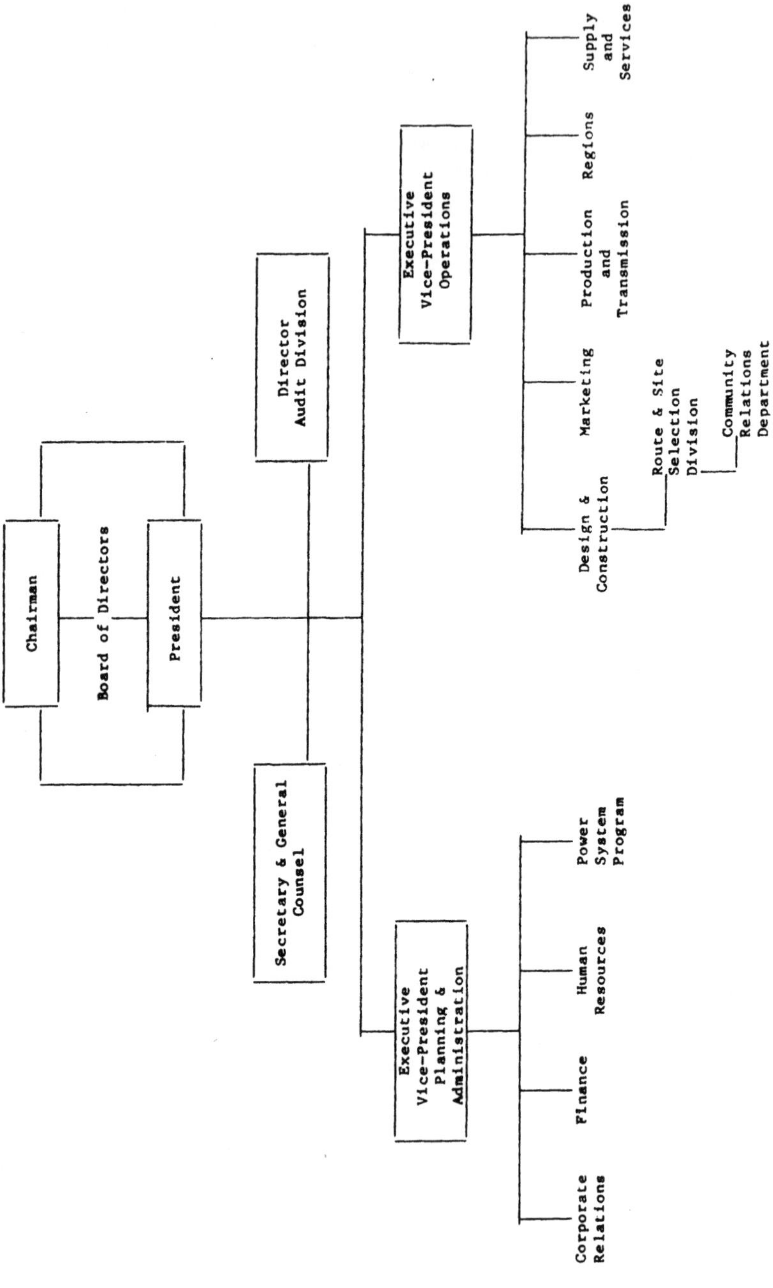

Figure 6.2 Ontario Hydro corporate structure

all responsibility of this Division is basically two fold: first, to identify transmission line routes and power plant sites; and then to obtain all local, regional, and provincial approvals so a project can proceed with property acquisition and construction. With such a broad mandate, the staff of the Route and Site Selection Division of necessity comprises a mix of generation and transmission engineers, land-use and environmental planners, and -- perhaps unlike similar departments in most other utilities -- community relations personnel responsible for public involvement. The latter are included because, since its formation in 1975, it has been Route and Site Selection policy to encourage public dialogue throughout a study, enabling interested individuals and groups to contribute their personal perspective, values and any specific details about their local geographical area. In addition, the public increasingly has been given the opportunity to consider and comment on the need for the project and the system plans which could meet the need.

Within the Division, the Community Relations Department is responsible for the development of public participation programs and techniques on a project-to-project basis. The basic objectives which guide their efforts are the following:

* provide the corporation with an understanding of the public's interest and ideas on project planning;
* through public information programs, promote the constructive contribution of citizens in the planning process;
* establish a cooperative, rather than adversary, relationship early in the planning process.

In pursuit of these goals the Department coordinates with regional staff who are under a separate Branch (other than Design and Construction), and with Public Relations, Government Relations and Public Hearings, which are in the Corporate Relations Branch under Planning and Administration. Public Relations in Ontario Hydro deals with corporate issues such as acid gas emissions and nuclear safety rather than project-specific information.

The Community Relations Department has two sections -- the Social and Community Studies Section (responsible for social impact assessment), and the Public Involvement and Community Relations Section. The latter is divided into two areas. One is responsible for the development and implementation of public involvement programs leading to

Environmental Assessment approval. The second area develops and carries out communication programs designed for individual landowners and municipalities where the facilities will be built. This program builds on the public involvement activities which occurred during the planning process.

EARLY DEVELOPMENT OF A PARTICIPATION STRATEGY

Ontario Hydro's public participation programs and techniques have evolved from the same reason as those of other utilities -- in response to dramatic changes in the needs and expectations of society. By the early 1970s, the environmental movement which blossomed in the late 1960s was having an impact on the company. There was not only an increasing public awareness of the environment but also a rural concern about land-use policy and the encroachment of industrial and commercial development into agricultural areas. Generally speaking there was a growing desire by the public, and particularly the special interest groups, to have a direct say in matters that would impact on their lives and on the communities in which they lived.

Strong opposition to two projects underway at the time brought the need for public participation home to Hydro planners. The first, a proposal to build a transformer station in downtown Toronto, met with extreme opposition from local residents. At one point a group even staged a sit-in on the executive floor of the Corporation's head office. Plans for the station were changed, and changed again, and have been mired in a seemingly endless quest for local and provincial government approval ever since.

The second project to face angry and widespread opposition was a proposed 500 kV transmission line to link the Nanticoke coal-fired generating station on Lake Erie with the Pickering nuclear station on Lake Ontario, 140 miles away. Responding to public pressure, the Ontario government appointed Dr. Omond Solandt to head a public inquiry reviewing the route chosen and to examine another Hydro 500 kV line proposal as well. This line would link the oil-fired generating station on Lake Ontario near Kingston to a transformer station just east of Toronto. Both lines, needed to strengthen Hydro's bulk power grid and move "locked in" power into the system, crossed areas of prime farmland.

The need to justify its route location before an independent body forced Hydro to revise its planning process, which the Solandt Commission criticized as being insensitive to public concerns. The corporation soon came to recognize and accept the fact that it had to provide the opportunity for the public to become involved in the process. Prior to this, virtually all facets of power system planning other than broad policies set by the government were determined in-house. What little external planning there was consisted almost solely of contact with various levels of government to avoid planning conflicts. Typically, Hydro reviewed project plans with a liaison committee consisting of representatives of provincial government ministries potentially affected by the plans. Other regulatory agencies, including federal authorities, regional and county planning officials and conservation authorities were also consulted regularly. Only after alternatives had been considered and a preferred route chosen were local municipalities contacted to determine if the project conflicted with any planned developments. The general public was not invited to contribute to or comment on the proposal.

Once the need to involve the public was recognized, Hydro's first step was to develop a set of program objectives. Far-reaching for their time, these were:

* to provide opportunities for the general public to learn about the proposed study, required facilities and associated Hydro practices, policies and procedures;
* to enable interested individuals and groups to contribute a local perspective and a set of values and specific details about their locale, which could be reflected in the environmental study methodology;
* to provide a forum for presenting and discussing alternative transmission corridors and routes considered by the study at key stages;
* to demonstrate to the public that the routes under consideration were the best available.

The method of achieving the objectives was thought at this time to lie through a series of public meetings throughout a study area at key stages of the project. Plans called for the meetings to be held at the beginning of a study, approximately halfway through, and at its completion.

Hydro's first set of public meetings were held in connection with a transmission study to find two separate routes for double-circuit, 500 kV lines connecting the giant Bruce Nuclear Power Development on Lake Huron to a transformer station just west of Toronto. These initial meetings, although advertised on the radio and in local newspapers, were not well attended: the twelve meetings involved drew a cumulative total of only 870 people. Nevertheless, a second series of public meetings was scheduled once alternative corridors were identified. Again radio and newspaper ads were used to encourage the public to attend. In addition, maps showing alternative routes were mailed to each rural resident in the study area. This meant producing upwards of 5,000 pieces of mail.

Once alternative routes were identified and people realized they may be affected, cumulative attendance at the 12 meetings jumped to over 1,900 people. Local residents were suddenly concerned, and some were very hostile to a new line crossing their area. But the meetings were dominated by a small minority opposed to transmission lines anywhere. This small vocal group appeared at each meeting and dominated the question period to the exclusion of the local residents. Attendance at some of the meetings numbered in the area of 350-400 people and the meetings often ran late into the night. The situation worsened further in the third set of meetings, which attracted over 3,300 people (almost four times the number at the original set). By this time press coverage was invariably emphasizing the confrontational aspects. Hydro and its officials were often on the front page and portrayed in an unfavourable light.

By 1974, with more groups and individuals demanding a role in the planning process, it had become obvious that the public meeting concept was not producing the sort of constructive interaction originally anticipated. Rather, the meetings most often resembled hostile debating societies which encouraged only emotional arguments, and where minority positions were given too much attention to the detriment of residents of the affected community. What's more, the general public couldn't understand the technical and financial data presented, and Ontario Hydro learned little from those attending because there was no opportunity to talk to people individually -- or calmly. The process simply didn't work to anyone's satisfaction, except perhaps that of a vocal minority who rejected the need for system expansion and demanded Hydro study "other" alternatives. In the end, the project was delayed over seven years as intervenors battled its construction before government boards, at expropriation proceedings, and in the courts.

As a result of this unhappy experience, Hydro began to move away from public meetings as its major vehicle for citizen involvement, and toward the information centre concept. This involved use of a local auditorium, church hall, municipal office, shopping mall or school for one to three days at various points in the study process or project life cycle, whenever a need arose to share or gather information. The centres were designed so that people could drop in and discuss different aspects of the study with the planning staff, working their way through displays covering the scope of the project, alternative locations, and so forth. At these centres, by the way, many people became involved not only in the specific problems of routing a transmission line, but also in issues concerning need, property policies, and construction and maintenance practices.

The principal advantage of the information centre was that it allowed the members of the public to meet technical professionals on an informal, individual basis, thereby reducing hostility. This allowed fully-rounded project information to be provided for the first time, both orally and in printed and map form. In turn, the public was asked to provide information, including both fact and opinion, about local conditions, the process for comparing alternatives, and the alternatives themselves. It was felt that if the public knew about the project and contributed data to aid in the selection of the final location, the recommended proposal would enjoy wider acceptance, even if it was impossible to satisfy everyone.

Also during 1974, Hydro commenced a major review of its public communications techniques more generally, which at that time embraced speaking engagements, field contact work, news releases, and unmanned displays as well as public meetings and information centres. With the help of outside groups and individuals involved in the planning process as observers or participants, the company sought to analyze the goals and techniques inherent in the early public involvement programs, and to assess both the benefits and problem areas associated with participation efforts. The review objective, basically, was to identify ways to make the planning and interaction processes more effective.

Overall, the review indicated that the Hydro initiative in public involvement had been reasonably successful in communicating information to a wide variety of people, the result being a greater degree of public understanding about individual projects. The number of people who became involved was significant and the process stimulated a regular flow of two-way information, comment and criticism. And as a general rule, Hydro was receiving support from various constituencies (including the media) for being more open in its planning. It was recognized that the utility was providing an opportunity for communities to become well-informed about a study and to contribute meaningful and useful information to the planning staff. Even government review boards commented favourably on the public participation program, although their praise was qualified by suggestions to the effect that the public should become more involved earlier in the process.

From Hydro's standpoint, too, it was apparent that the initial program had fallen short of objectives in some respects and therefore needed modification. For one thing, the managers realized that complete public acceptance of a particular project was not an achievable goal, despite public involvement in the route or site selection process. As some projects moved to the late stages, for example, political resistance was still being encountered from affected property owners. The company also learned that public distrust of Hydro engendered by previous practices was difficult to overcome. Frequently, as a result, the methods for conducting a study were not fully appreciated, and criticisms of the procedures and execution of the public involvement program were encountered. In addition, the need for the project often was not properly understood or accepted, especially by some people who opposed any further development of power facilities and used the forums for public involvement to attack Hydro expansion in general.

On the basis of this evaluation, the review team developed a set of recommended guidelines to be followed in future participation programs. The principal recommendations were as follows:

* Hydro should encourage dialogue with the public during the early, middle and late planning stages and continue until completion of the line/site location study.
* For transmission route studies, Hydro should substitute informal information centres for formal public meetings. A combination of meetings and information centres would be used for generation site selection

studies, and public meetings would continue to be used during projects involving specific generating stations.

* There should be more public discussion concerning Hydro's long-range planning needs in addition to discussion of specific project proposals.
* Hydro should actively promote a better community understanding of environmental study procedures.
* There is a need to develop a method by which Hydro and the communities potentially affected could engage in a more organized dialogue.
* The concept of citizens committees should be adopted on a trial basis. These committees would examine specific issues associated with the project and give advice to the Hydro project team.

Early in 1975, improved guidelines for project planners based on these recommendations were adopted by Hydro management.

Another key event of 1975 -- which was very much a watershed year for public participation in Hydro planning -- was the formation of the Route and Site Selection Division. The company's experience to date and its growing recognition of the importance of environmental matters indicated that a major reorganization was necessary of the various staff involved in siting and obtaining approvals for new facilities. To that point, for example, responsibility for planning new transmission lines had been allocated to varying degrees among five departments within different divisions: Transmission Route Planning, Forestry, Community Relations, System Planning, and Property Acquisition. As a result, staff involved in major expansion projects (and public participation, as of 1971) also had other department duties. When priorities came into conflict, predictably, department duties often took precedence over project-related tasks.

Upon the creation of the new division, however, no longer were these employees confronted with conflicting duties and priorities experienced when they were widely dispersed in various departments throughout the organization. The reorganization thus represented corporate recognition of environmental concerns and need for public involvement. Public participation was now organizational policy, and Hydro was ready to invest the time, money and other resources to ensure it would work.

EARLY EXAMPLES OF INVOLVEMENT PROGRAMS

The Route and Site Selection Division came into being as the Corporation was projecting -- as noted earlier -- a sharp increase in the demand for electricity. To meet this demand, studies were beginning in several parts of the province to find sites capable of supporting some 12,000 megawatts of new generation, to be developed (along with associated transmission lines) in three stages of 4,000 megawatts each. As part of the environmental assessment component of these studies, citizens committees were established, contacts with municipal officials were initiated, and meetings and information centres were held to communicate with the general public. Such activity soon came to a head, however, as all work related to additional site identification was put on hold in response to declining load growth and intense political pressure for a review of expansion policies (via the Porter Commission).

One generation study aborted during the mid-70s was known as the North Channel Study, which sought acceptable locations for nuclear units along the north shore of Lake Huron (see Figure 6.3). Public involvement began in 1974 with meetings of representatives from ten provincial ministries for the purpose of identifying governmental interests. Status reports were sent to and continuing contacts were made with the ministries to maintain input to the study. Meetings were then held with local municipal councils and with the councils of native Indians living in the area.

Public meetings and information centres, widely publicized throughout the study area, encouraged the public to become informed and to participate in the public involvement program. News releases, feature articles and advertisements were issued on a regular basis to the media. Information kits were also distributed and mailed to all those who expressed an interest. Personal contact was made with numerous members of the North Channel community to inform them of the project, and all verbal and written requests for information were answered wherever possible. Every effort was made to gather the names and addresses of those who had requested information and/or had attended public meetings and information centres. These mailing lists, comprising about 650 individuals and groups, were used in the distribution of project status reports, which were issued at key points in the study process. Finally, offers to address groups and associations on details of the study were made. As a result, 13 requests were received

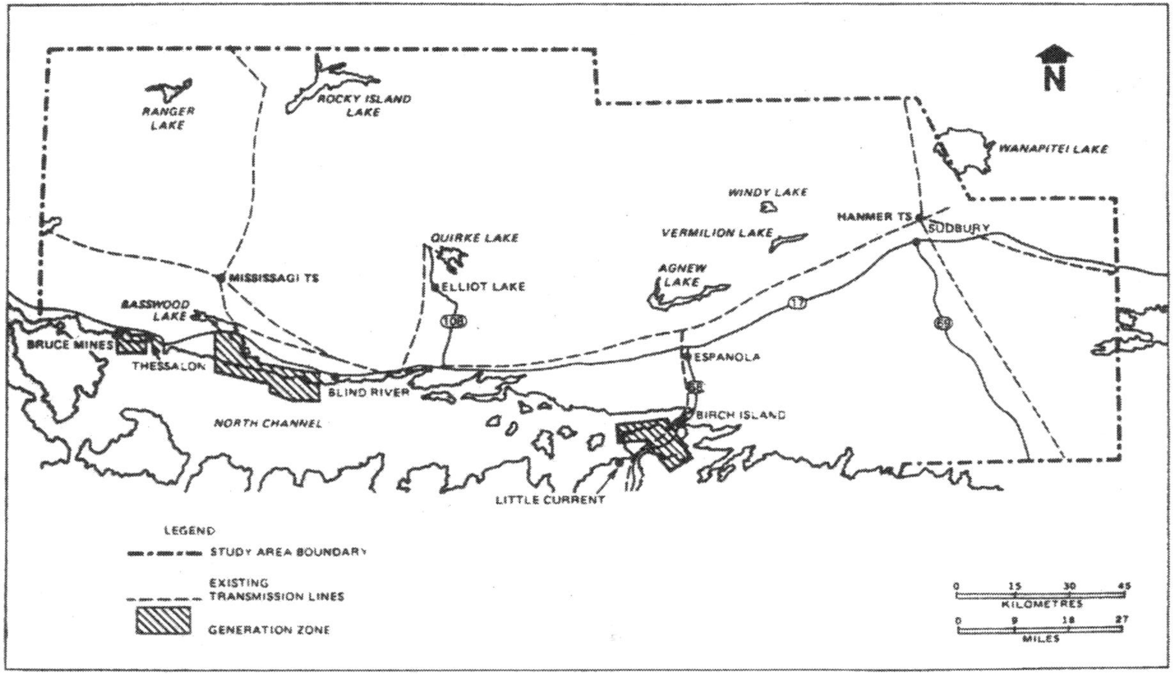

Figure 6.3 Study area boundary

and presentations were given by Hydro representatives to service clubs, school groups, historical societies and other groups in the North Channel community.

In order to increase involvement with the public and to provide maximum opportunity for interaction and dialogue in a more formalized setting, Hydro formed citizen working committees. Considering the size of the study area and the dynamics of establishing effective communications with large groups, it was considered advisable to form two working committees, one at Blind River (West Citizens' Committee), representing the east and central parts of the District of Algoma, and the other at Espanola (East Citizens' Committee), representing part of the districts of Sudbury and Manitoulin. An advertisement was placed in all local newspapers listing the names of the committee members, groups represented and the role and objectives of the committees.

A total of 40 committee meetings were held prior to the tabling of committee recommendations. To assist liaison and communications, four of the 40 meetings were held jointly. At two of these joint meetings, speakers representing opposing points of view were invited to ensure that committee members had a balanced outlook. Following each meeting a release of the proceedings was distributed to the local media, and many of the local newspapers published these releases or their own reports.

Because the proposal was abandoned before Hydro began to study the specifics of the project, it was difficult to evaluate the effectiveness of the public involvement component. The study process was acceptable to those groups and individuals who participated and the site selected aroused very little public opposition. But there was growing contention, both with environmental groups in the area and other groups throughout the province, on a set of larger questions. In the matter of need, for example, opponents argued that conservation could save the utility millions of dollars in capital construction costs (and preclude disruption to the environment) by reducing load growth. What's more, Hydro was coming under severe criticism for its choice of nuclear technology for baseload generation. To what extent these issues might have been hashed out constructively in the context of that particular study is, at this point, purely speculative.

Not similarly problematic, on the other hand, were Hydro's programs to involve the public in transmission planning efforts under the new guidelines developed in

1975. A good case in point was one of the first projects undertaken by the Route and Site Selection Division -- the location of a new 230 kV transformer station in downtown London, a city of 262,000 people located in southwestern Ontario. In view of strong prior opposition to a similar proposal in downtown Toronto, it was understood the project would get timely government approval only if it had a broad base of public support and understanding. In this case, the public participation program met or bettered the expectations of everyone involved.

In May, 1975, Hydro arranged with the London public utility to make a joint presentation to the City Council, in order to define the need for the new transformer station and a transmission line to feed power to and from it. After the council meeting, which was attended by community representatives invited by the mayor, Hydro staff met with interested citizens and formed an ad-hoc committee to come up with firm recommendations, hopefully in support of the proposal. This 14-member London Hydro Working Committee included three Hydro representatives, several community association leaders, representatives from the local utility, an environmental protection group, a downtown alderman, an interested citizen with no group association and a professional planning consultant hired by the City of London. The consultant was selected by the working committee with the understanding she was to act as watchdog for city planners and assist the committee in investigating and evaluating alternative plans. Members of the committee were expected to represent their group's position and concerns during the process and maintain regular communication with them, so the recommended route for the line and transformer site would be acceptable.

The committee met 27 times between January, 1976 and March, 1977. Working closely with Hydro planners and Community Relations personnel, the committee reviewed and agreed upon the need for the facility and the size of the study area. The meetings were open to the media and the minutes were made available at City Hall. Special effort was made to encourage public understanding by holding one or two day information centres in schools and community halls, at which people in the study area were invited to contribute suggestions and "speak their minds" on the proposal. These centres were advertised in the press and publicized through flyers delivered door-to-door in the affected area.

The principal recommendations from the committee to Hydro were:

* London needed a new heavy-load transformer station in the central area;
* the capacity of two other transformer stations (Nelson and Clarke) serving the area should be increased in the meantime to meet immediate needs.

The committee also set down guidelines for the choice of the station site, describing the size of the study area and the criteria to be used for identifying possible sites. For example, a site had to accommodate a station that would meet the power needs of central London to the year 1996, and would have to be on vacant land, land occupied by abandoned or unused buildings, or on land containing an industry slated for relocation. In addition, each candidate had to be at least 180 feet by 60 feet, exclusive of land for set-back or landscaping; have access for high and low voltage lines; have access to water and sewage; and be free of legal entanglements. The site would also be evaluated according to the station's potential effect on residential areas, businesses, future development, and, of course, people. The station itself would have to conform with city planning policies, and measures to minimize neighborhood disruption during and after construction would have to be considered.

City planning staff listed 15 potential sites and the public added seven others. The committee rejected 17 of the 22 sites for various technical, legal and social reasons, after public information centres had been held to obtain public comments. Attempts were made to secure options on the five remaining sites, but it turned out that only two options could be negotiated at a suitable price and within an acceptable length of time. The recommended site was chosen because it was the less expensive to buy, was more economical to supply with power, and was more acceptable to the local community. Even so, another round of information centres was held to obtain comment on the recommended site and the associated transmission line alternatives.

The work of the committee led to fairly quick approval of the project by the London City Council and by the various provincial government ministries. And from the experience Hydro learned some valuable lessons about citizen committee studies:

* The working committee members understood and accepted their role in the public participation process.

* The work of a committee is most efficiently done when the various interests on the committee state clearly what their goals are and when Hydro can state clearly how it will use their input.
* Technical information is not easily assimilated by a lay public and written information should be geared to the lay audience. The committee can develop the necessary technical understanding quickest and with most confidence if it has access to non-Hydro expertise.
* A working committee functions best when the people involved are able to understand and accept diverse views.
* A certain degree of frustration and an inefficient use of working committee time occur when the committee as a whole deals with issues of concern to only a few of its members.
* Information centres can be an efficient public participation technique if there is a clear idea of what is wanted from those who attend.

There was another sidelight to the London experience. During the study process, one other transformer station in the area was looked at by the committee since it was to be upgraded to fit into the new system. Complaints were raised during the investigation which in turn led to quick corrective action by Hydro -- providing yet another beneficial side effect of public cooperation.

CURRENT APPROACH TO PARTICIPATION

Building upon several project experiences such as the London study, Hydro's public involvement program has evolved in such a way that a variety of refined techniques are now available to meet the needs of specific projects. As can be expected, new methods are usually introduced on a trial basis and evaluated, and those which prove effective are incorporated into route studies as circumstances indicate. The techniques currently used are listed in Figure 6.4 according to their respective time periods of introduction, and those most commonly employed are described briefly below.

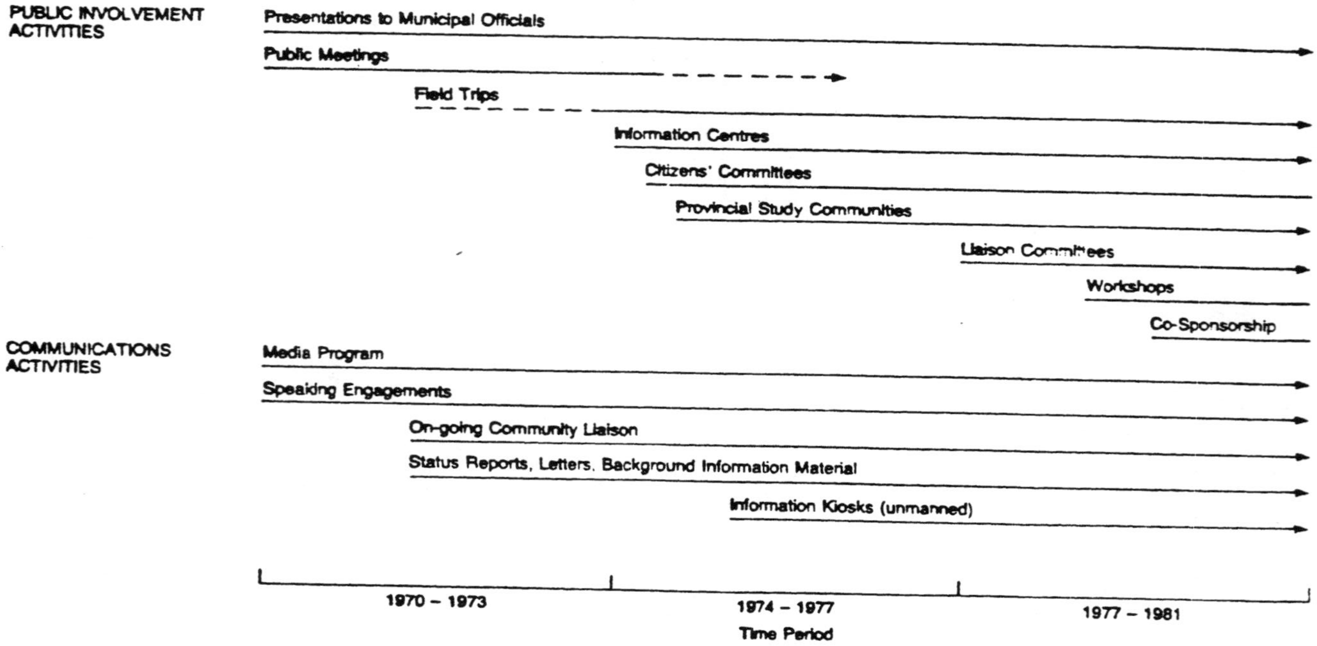

Figure 6.4 Evolution of public involvement techniques at Ontario Hydro

Citizens Committees

Of particular note is the role of the citizens committee, which has increased to the point where on several major transmission line studies citizens have been directly involved with the project team in plotting alternative routes. The contribution of the citizens committee significantly enhances the quality and public credibility of the study, since it incorporates diverse values into the process. This is recognized by regulating authorities.

A citizens committee is comprised of local citizens who reflect a cross-section of community interest. The committee, which is voluntary, works with Hydro's project team in the route selection process. Through a series of meetings and workshops, the committee and project team perform distinct functions contributing to a recommended proposal. Typically, the committee is involved in reviewing the study data and providing new information if necessary; identifying the types of areas to avoid with the new facilities; ranking these areas in order of priority; assisting in the identification and evaluation of alternatives; and selecting the preferred alternative. The citizens committee members need not have experience or training in planning matters, although they normally have a very strong interest in preserving one or more elements of their environment. Hydro approaches and invites individuals representing prominent groups and organizations to sit on the committee. They are then expected to keep their parent body informed on a continuing basis of the committee's work and bring back to the committee the advice of the parent body. These individuals are compensated only for travel and accommodation expenses associated with the study.

It has been found that the effectiveness of a citizens committee is increased if its size is held to about 20 members. For a large study area, more than one committee should be formed. When several committees are necessary, close coordination between the groups is essential. This is best accomplished through periodic joint meetings to review information and priority setting and through the sharing of meeting minutes.

Liaison Committees

Occasionally a liaison committee, also reflecting a broad range of community environmental and local government interests, is formed rather than a citizens committee. This occurs when:

* the potential participants desire only a monitoring role;
* there is reason to believe that the impact on the community may be relatively minimal;
* the options available for the proposal are limited.

The role of the liaison committee is usually to be a 'sounding board' for the project team and to act as the primary means by which information is exchanged with the parent organizations. The views and concerns of various interests represented on the committee are voiced by its members during each stage of the study. General agreement in the committee is sought by the project team before progressing to the next step. The committee, though it does not itself make decisions, influences key decisions made by the project team.

Workshops

A workshop is a seminar or series of meetings designed for intensive study, work or discussion among the participants. A workshop differs from a meeting in the following ways:

* the number of participants is smaller (no more than 10, including Hydro staff);
* the task to be accomplished is relatively specific;
* the format, although carefully structured, is less formal.

Workshops may be used within the citizens committee framework when small group discussions are more effective than a meeting of the entire committee (e.g., ranking priorities within specialty areas).

Information Centres

Periodic information centres, or open houses, can increase the general public's awareness of and involvement in the route selection studies. They are used at any or all of the following stages:

* at the commencement of a study to announce the program and to provide background information to the public;
* when alternative bands, routes, or sites have been identified and there is a need to gather public commentary as part of the comparative evaluation;
* when a preferred band, route or site has been identified and there is a need to obtain public commentary on the result of the study in support of the work done by committees and/or a need to inform the public;
* after government approval to convey to the public the details of the approved project; to identify conditions additional to those of the original recommendation; to describe project construction schedules and property acquisition policies; and to introduce staff who will handle the implementation program.

Information centres, staffed by the project team and local Hydro staff, normally run one to three days and are held in church halls, municipal offices, schools or local meeting halls. The location and hours of operation are well advertised. The centres usually contain graphic and audio-visual displays and printed background information. Public comments are invited and documented and personal contact is made in follow-up discussions.

Tours of Existing Ontario Hydro Facilities

Project committees, community groups, property owners, government officials and other interested parties are occasionally invited to tour Hydro facilities. This can prove beneficial in numerous ways. Key publics can gain a better understanding of the impacts, size and nature of the facilities. People from an area where a study is underway have the opportunity to discuss impacts of a proposed facility with other people from areas where similar facilities already exist. This has the potential to alleviate concerns and provide a perspective of both positive and negative impacts.

Field Contacts

Face-to-face and telephone communication with individual project publics plays a crucial part in any successful study. Personal contact is used to:

* obtain information for use in the project;
* introduce a new study to key community leaders;
* create a broader ongoing understanding of the project;
* clarify any areas of misunderstanding.

Hydro staff use personal contact to keep in touch with community concerns and to monitor the effectiveness of the public participation program.

Media Programs

Project media programs are extensive and consist of any or all of the following:

* news and feature story releases;
* advertisements;
* letters to the editor;
* press conferences, briefings and tours.

Media communication is geared toward milestone events within the project life cycle and the completion of the study. The media is used to publicize specific events and to respond to unscheduled or unforeseen project-related activities.

As might be expected, there is no standard way to carry out a program of citizen participation. The selection of appropriate methods for a specific project will depend, for example, on the objectives of the participation program as determined in light of an understanding of the communities to be involved. As shown in Figure 6.5, individual techniques offer different degrees of opportunity for interaction, involvement, issue resolution, and mass outreach. Other factors that weigh heavily in program design are the basic approach to the study, the target dates that must be met, and the resources available. Of particular concern in this respect is the length of time, and therefore the amount of money, required to complete a study with continuous public involvement. Major studies could last up to two years, a significant period considering the

TECHNIQUE	OPPORTUNITY FOR INTERACTION	OPPORTUNITY FOR INVOLVEMENT	OPPORTUNITY FOR ISSUE RESOLUTION	OPPORTUNITY TO REACH LARGE NUMBER OF PEOPLE
Citizens Committee	High	High	High	Low
Liaison Committee	High	High	High	Low
Workshop	High	High	High	Low
Information Centre	High	High	Low	High
Tours of Existing Facilities	Moderate	Low	Low	Low
Field Contact	High	High	Low	Low
Media Program	Low	Low	Low	High

Figure 6.5 Ranking of public involvement techniques

lead time in Canada to put a major nuclear station fully into service is about 17 years, and about eight years for a major transmission line.

In preparing to develop a participation program, staff of the Community Relations Department first collect certain baseline information through research, discussion with other company employees, and interviews with key individuals within the affected communities. The twofold purpose is to develop:

* a community social profile which identifies community issues, interest groups, local leaders, communication techniques, demography and the economic base of potentially affected communities;
* a sense for the needs of the public, that is, the degree to which different publics want to be involved in the study process.

This information is then analyzed in conjunction with the scope of the project and the environmental study process, and a draft program is prepared which would include the following elements:

* the appropriate participation methods and communication techniques;
* the schedule and sequence of events to coordinate with the overall project plans;
* the proposed activities of project staff at each stage of the study;
* the listing of the resources (i.e., capital manpower, materials) required to complete the program;
* an outline of indicators by which the success of the public participation program will be evaluated.

This program is then reviewed with project team members to ensure compatibility with the other components of the route selection process.

An integral component of the program, it should be noted, is an evaluation of its effectiveness. Attempts are made to determine the extent to which objectives were met, the degree to which participants were satisfied with the process and its results, the effectiveness of the public participation techniques, and what suggestions participants have for improvements. Evaluation occurs both during the course of the study and upon its completion. The most frequently-used methods of obtaining the views of the public are group discussions, personal interviews, questionnaires and comment forms. Often, a Hydro staff member not associated with the project team documents the group discussions in order to encourage participants to be frank.

Although the formal evaluation component is still in its infancy, some of the benefits are obvious. It provides valuable information on how participants in the study view the corporation's attempt to reflect their concerns. It provides management with another perspective on the work of staff other than what is reported in the media. But, perhaps even more significantly, it allows a self-critique for the staff responsible for the program. This, in turn, is the catalyst for improvements to the approach taken to involve the public.

NEW FRONTIERS IN PUBLIC INVOLVEMENT

Aside from the ongoing development of participation methods, Hydro's public involvement activities today also include a program to open up the route planning process in new ways. More specifically, steps are being taken to involve the public in the question of need for transmission facilities, a key public concern which throughout the 70s tended to cloud the issues surrounding route selection. In an attempt to separate the need and routing issues, Hydro decided in 1981 to initiate interaction with the public at the <u>system</u> stage of planning, at which point need determinations are typically made by a utility. This was done in the context of a recommendation by the Royal Commission on Electric Power Planning -- and an announcement by the Minister of Energy to the same effect -- that Hydro should proceed with the planning and approval process for a second 500 kV transmission line connecting the Bruce Generating complex to southwestern Ontario.

Following the announcement, Hydro began a public involvement program aimed at the following objectives:

* to ensure that the people of southwestern Ontario were aware of plans for expansion of the bulk power transmission system;
* to ensure that the people of southwestern Ontario were provided with opportunities to express their views and influence the selection of a system plan;
* to establish a foundation for effective two-way communication with the communities and individuals who would be involved in future hearings and/or route stage studies.

The first step in pursuit of these goals was the distribution of a company report entitled "Southwestern Ontario Evaluation of Alternative Transmission System Expansion Programs." The report identified six general plans, each representing a technically feasible scheme, and outlined the technical, economic, and environmental factors relevant to their evaluation. The report did not state a Hydro-preferred alternative. Rather, the intention in making the report available to the public was to seek the comments of interested individuals and groups before completing an Environmental Assessment which would recommend a specific plan.

Approximately 2,500 copies of this report were distributed, and summaries were mailed to the following groups:

* Cabinet ministers
* Conservation authorities
* County federations of agriculture
* Direct industrial customers
* Directors of planning
* Electrical utilities
* Federated nature clubs
* Government ministries
* Interested publics
* Leaders of opposition parties
* Libraries
* Members of Parliament (MPs)
* Members of Provincial Parliament (MPPs)
* Media
* Municipal clerks
* Provincial organizations
* Regional ministry officers
* Working Group members

By mid-June, presentations were underway with accompanying information display to electrical utilities and direct industrial customers, planners, resource managers and other interested groups. Further presentations were made, in seven locations during mid-July, to public officials which included MPs, MPPs, mayors, reeves, wardens, regional chairmen, elected representatives from electrical utilities, conservation authorities and the Ontario Municipal Electrical Association. The presentations were designed to provide these people with information about the study in order to enable them to comment or provide written feedback to Hydro. It was also an opportunity to establish with the officials a solid foundation for their future involvement in hearings and route stage studies.

Following the initial presentations, a series of sixteen information centres were held throughout southwestern Ontario, from July to October, 1981. These open houses were established primarily in towns located within the future study areas for the six alternative system plans. The centres, which were open from 1:00 pm to 5:00 pm and 7:00 pm to 10:00 pm, consisted of 28 display panels, related to the need for the study, the alternative system plans, the environmental, technical and economic comparisons, the public involvement opportunities, the future study areas, the approval process and the overall project schedule. Project literature and general information pieces were available, as well as a comment form with a return envelope to facilitate input from those attending. Several Hydro representatives were always on hand to provide a review of the information displayed on the panels and to answer questions.

People requested information and introduced their questions objectively, which resulted in a constructive exchange of views between Hydro representatives and the public. Many visitors showed a keen interest in the planning and approval process. Initially, the level of understanding about the difference between the "plan" and "route" stage varied widely. Individuals who expressed concern about how their property was being affected assumed Hydro was at the route stage. As they became increasingly aware of the plan stage, their inquiries became more general in nature.

During this same three-month period Hydro convened a series of meetings with working groups, which were to be a key component of the public participation program. With the study encompassing 21 counties, it was determined that there should be five groups and that for convenience their meetings should be held in five towns near the geographic centres of the regional boundaries. To assemble the groups, Hydro staff contacted approximately 200 organizations and sent copies of a background information package. Once delegates were appointed, all necessary information was forwarded to them. The first working group meetings began in early July and membership lists were finalized shortly thereafter.

The following terms of reference for operation of the working groups were proposed and generally accepted by the representatives:

* The basis for study shall be the evaluation of alternative transmission system expansion programs in southwestern Ontario.
* Ontario Hydro will present and discuss the rationale for need; however, the evaluation of alternative transmission expansion programs should proceed on the basis that need will be determined at hearings of the Environmental Assessment Board (EAB).
* The environmental study process and information base endorsed by previous working groups shall be considered as given by the 1981 working groups.
* The working groups shall review and comment on the six system alternatives and the relative importance of trade-offs involved in the selection of a preferred system plan. In addition, working group members may propose alternative transmission system plans for review and discussion.
* The time frame for the public involvement program will be an intensive period of approximately four months.
* The working groups will consider preparing written submissions which identify and represent their views. These submissions will be included in Ontario Hydro's EA documentation.
* Working group members are responsible for representing the views of their parent organizations; however, the personal views of individuals are welcome.
* Working group members are responsible for keeping their parent organizations informed on the study process.
* Participation in the 1981 working groups shall not preclude members from becoming involved in the EAB hearings.

With previous working groups having been chaired by Ontario Hydro, some concern was expressed about the chairman's objectivity. It was suggested that a Hydro employee, much the same as a naturalist, farmer or businessman, would show bias toward his particular area of concern. As a result, an independent neutral group -- the Huron Erie Land Use Studies Group (HELUS) -- was employed to chair and take notes at all working group meetings. HELUS is an organization composed of public-recognized professionals and academics with a concern for the broader aspects of the environment in southwestern Ontario. A major part of its role as chairman was to bring discussion to a resolution which was understood and endorsed as the will of the meeting.

The working group meetings were designed to be an opportunity for in-depth discussion in an interdisciplinary setting and to facilitate the recognition of trade-offs through discussions among various interests. The first part of the program was designed to familiarize group members with the content of the evaluation report. The purpose of the second part of the program was to have a well-informed group discuss the key trade-offs involved in comparing the plans. The following outline of meeting agendas was proposed by Ontario Hydro and followed by the working groups with minor modification:

* Meeting 1 : review and discuss background information, including terms of reference and individual responsibilities.
* Meeting 2 : trip to Richview Control Centre in Toronto to observe the operation of the bulk power transmission system, discuss purpose of study and review need for additional facilities.
* Meetings 3 & 4 : Nomination and selection of vice-chairman; review environmental, economic and technical evaluation of alternative plans.
* Meetings 5 & 6 : comparison of alternative plans and presentation of group views to Hydro.

A media program with several components also formed an integral part of the public involvement program. Each component was designed to meet a specific objective. For example, press briefings were held early in the study to give media representatives an opportunity to become familiar with details of technical and environmental aspects of the study and to have an opportunity to discuss the study with Hydro representatives. These press briefings involved a slide presentation and a typical information centre set-up of 28 information panels. Throughout the study, Hydro regional staffs maintained contact with the media, keeping them advised of various meetings and other events occurring in the program.

In addition, the electronic and print media were employed by Hydro to increase public awareness of the study and to outline the opportunities for public involvement. This program involved 60-second spots on three television stations, together with advertisements in weekly and daily newspapers and in farm publications. The advertising told the public of the study, of information centres in the circulation area of the paper, and of the availability of the "Let's Talk Transmission" information kit which provided details on the study. A box number was provided in the television spots advising the public where to write for the kit and coupons were provided in the advertisements in weekly, daily and farm papers.

The public involvement program concluded in November with the submission of the Environmental Assessment report. Hearings were held under the Consolidated Hearings Act in March of 1982 and approval to proceed to the route location stage was received in November, 1982.

The route stage of the study was a massive undertaking, requiring concurrent studies in a 5,800 square kilometre study area, to identify preferred routes for four transmission lines (over 400 km) and a site for a transformer station. The study was made all the more difficult by economic pressures that required it to be completed within a 14 month period.

For a study of such scope, Ontario Hydro implemented the most intensive public involvement program it had ever attempted, employing all the major program tools proven successful in earlier studies, including working groups, liaison committees, information centres, presentations, workshops/demonstrations, media campaigns and bulk mailings. (See Figure 6.6.) When the study concluded with the submission of an Environmental Assessment in December, 1983, we looked back over the year-long program and found that there had been over 200 study-related public events, 2,000 newspaper articles clipped, and over 100,000 pieces of project information distributed.

As a result of the public involvement program, Ontario Hydro went into public hearings in 1984 with full knowledge of the publics affected by the recommendations and the public concerns, and with an environmental assessment document that explicitly reflected the contributions of the affected public. In this respect the process has benefited Ontario Hydro, the public, and the government which has to approve the project, and hopefully will result in relatively speedy approvals. At the time of this writing, approval has been received to construct the lines in eastern Ontario, and the hearing process is still in progress in southwestern Ontario.

Component	Scope
Four citizen working groups, representing local governments, agriculture, recreation and environmental interests met with the study team throughout the study to offer information, rank environmental priorities and identify community concerns.	• four groups formed of 35-50 members each • met 12-15 times each
Two liaison committees addressed the more specific concerns of siting a transformer station and routing a line near a major highway.	• two groups • 16 and 30 members • met 10 times each
Media campaign in community press, TV and radio announced the four key stages of the study and showed more information could be obtained at information centres, project meetings or presentations.	• over 230 1/4 page newspaper ads placed • over 300 radio spot ads • over 2,000 newspaper articles monitored, most were balanced
A media contact specialist travelled throughout the study areas updating local media, being available to present Hydro's views.	
Information centres and presentations in community halls presented detailed study information and provided opportunity for the public and affected property owners to contribute to environmental data.	• 90 one-day information centres in community halls
Mailings of project newsletters, notices of public hearings, summary of environmental assessment.	• mailing list of over 35,000 people

Figure 6.6 Public Involvement in Ontario Hydro's Southwestern Ontario Route Stage Study

CONCLUSION

The foregoing discussions amply demonstrate that public involvement is now an integral part of the planning process for all new transmission projects at Ontario Hydro. Through experience the Corporation has learned that a successful participation program helps reduce the possibility of a lengthy approval process, thereby reducing lead time and cost. In fact, it is now generally accepted that with citizens working closely with Hydro planners to identify preferred transmission line routes and/or transformer station sites, the need for formal hearings for local projects has almost been eliminated. Another positive spinoff of public involvement is that the utility has been forced not only to justify its route/site selection process, but also to present its position on issues beyond the scope of individual projects. At the information centres, committee meetings, and other forums, people often question Hydro's policies on power exports, rates, marketing, conservation, and nuclear generation, and such matters must be discussed before the process can go forward. This has been beneficial to the public and to the Corporation, which has become more responsive to the needs of those involved in the process and has shown it is prepared to discuss their concerns.

Despite those advances, however, it must be stated that the public involvement program is still subject to some important constraints. For example, although project publics are engaged in a form of joint planning and play a key role in generating an ultimate proposal, they do not have any responsibility for making a final decision. Rather, the choice of which alternative is to be recommended for government approval is made exclusively by the Board of Directors. This practice has been criticized by the Royal Commission on Electric Power Planning, which recommended in 1980 that Hydro "should adopt joint planning processes whereby real decision-making authority is shared with, and in some cases left to the initiative of, citizen representatives." But this is a major step which the Corporation is not yet prepared to undertake. In this respect, the process of "opening up" has generally gone about as far as it can within the corporation.

Another and perhaps more significant limitation concerns the fact that openness to the extent it is practiced in transmission planning has not been applied to major areas of corporate policy-making, where actions continue to be developed and implemented in a traditional manner.

Participation programs associated with facility location are not designed to directly address issues and concerns about policy decisions, so the public can only raise them in the context of regulatory proceedings such as the annual rate review hearing. But these hearings are very legalistic and necessitate the assistance of attorneys to present evidence, a highly formalized approach which tends to discourage citizen participation and portray Hydro as a highly bureaucratic, unapproachable organization. Thus, a marked contrast exists between the modus operandi of the Corporation in the route selection process and in policy formulation, and this has created confusion and frustration in the minds of many Hydro publics.

This, then, is where Ontario Hydro stands today, approximately a decade after it created the Route and Site Selection Division and gave it a mandate to go beyond public meetings in defining an active role for concerned citizens in project studies. On the one hand, the Corporation clearly has come a long way in developing methods of interaction; on the other, it is apparent that pressure remains for further application of collaborative techniques. Most observers would probably agree in this respect with the assessment of the Porter Commission, which acknowledged that since 1972 Hydro "has utilized some imaginative approaches to involving the public in its planning process," yet recommended that the corporation be "encouraged to continue and, where necessary, to expand its participation program to ensure that the public is fully involved." This may well occur, because the foundations for constructive relations and future cooperation with the public are now firmly in place.

7

Citizen Participation in Decision-making at Portland General Electric

Ruthann Mogen

From the vantage point of an electric utility company, society has changed a great deal. The public used to be complacent about how their lights came on. But now they are not only interested, but also intelligent and concerned about how we, the utilities, affect their lives. As their interest was piqued, they knew they could affect us long before we gave them credit for it. We called them "small, special interest groups" and did not take them seriously. As a result, in the last decade utilities everywhere suffered large financial losses from public opposition to the construction of new generating facilities. In part because of public opposition, moreover, the nuclear power industry came to a virtual halt.

The effects of public intervention, legislation, and regulation in the 1970s had Portland General Electric, an investor-owned utility, in danger of a financial tailspin. Stock prices dropped 30 percent, Standard and Poor's downgraded the company's rating, and stock dividends decreased. In addition, millions of dollars were wasted fighting the public's will. There are some who would argue that the expenditure was worthwhile. Most of those people are still clinging to the concept that money spent on a good idea is not wasted. But when the money is gone and the good idea is only noted as a loss on the financial statement, it is

---

Ruthann Mogen is Manager of the Community Relations Department. Jerry L. Bingold, Branch Manager of the Community Relations Department, assisted in the preparation of this article.

time to reevaluate management's criteria for decision-making.

Long-time utility employees pride themselves in their engineering competence and their commitment to reliable service, but they have resented the intrusion of outsiders with little engineering expertise. While this resentment is understandable, and certainly at times justified, it has been detrimental to the interests of the industry, the ratepayer, and the general public. We have come to understand that real tradeoffs between the production and distribution of electric power and the environment do exist, and that the choice of which tradeoffs should be made is a political, not a technical, decision.

This is the story of how PGE changed its attitude and its method of interaction with the public. In particular, it describes a program for seeking public opinion in connection with a proposal to build a second hydroelectric station at Willamette Falls, near Oregon City. The original powerhouse, a wooden structure called the T.W. Sullivan Station, had been built across the river in 1899 for the purpose of sending electricity to light the streets of downtown Portland, 14 miles to the south. That made history as the first long-distance transmission of electricity in the United States, and it changed the way utility companies did business. Thus it is particularly apropos that another proposed project at Willamette Falls, ninety-one years later, was again to mark a new era in the conduct of PGE's business affairs.

A NEW MANAGEMENT BRINGS ATTITUDINAL CHANGE

Historically a leader in the development of hydroelectric generation and high-voltage transmission in the 1970s, PGE also built the state's first nuclear power plant. Today, the company serves half a million customers in an area of 3,350 square miles, providing 40 percent of Oregon's electrical needs. We operate eight hydro plants, ranging in capacity from 15,000 to 333,000 kilowatts, and several thermal plants including the 1,110-megawatt Trojan nuclear plant; the 530-megawatt Boardman coal plant; and the 534-kilowatt Beaver combustion turbine plant. PGE employs 3,200 people.

As president and chairman of the board, Frank Warren had led the company through many successful building years. In the early 1970s this expansionary mode was still acceptable within the company, but external to it the message was "no more building." This became painfully apparent in 1972 when the company, having begun construction of the Trojan station, sought government approval for its second and third nuclear plants. It was at this time that we began to experience planned, organized opposition. During a speech made to PGE employees in 1974, president Warren said that "the utility business is certainly an exciting place to be today; I wish it weren't so exciting." Thereafter the company struggled unsuccessfully with public intervention in the regulatory process for another eight years until, in 1982, the license application was withdrawn from the Nuclear Regulatory Commission.

During this troublesome period the company's attitude began to change, and eventually so did its leadership. In 1980, Robert H. Short, former newspaper editor and lobbyist, was elected chairman of the board, and he quickly moved to change the thinking of this engineering-oriented company. The effects of those changes were dramatic.

Short had already brought in some new officers to help him deal with the company's poor image and deteriorating financial conditions. Among these was James W. Durham, Oregon's Deputy Attorney General, a man who characterized himself as a "street smart" lawyer and who had a good sense about public policy. He knew that a credible image is earned, not bought. And he knew that an image can change only when the attitudes of top management change, and when management is committed to a new approach. Short created an Office of External Affairs and assigned Durham as senior vice-president. Doing so brought together the Public Relations, Public Affairs, and Legal Departments, as well as many community relations efforts, under one senior officer. Public Relations had the responsibilities of advertising, media relations, internal communications, and community and civic relations. Government relations was assigned to Public Affairs.

During this period of reorganization, Short commissioned a study of our corporate image. Many in the Company were truly baffled by the loss of public support in the 1970s. To find out just how great this shift was and whether the trend would continue, the study was broader and more inclusive than had been attempted in the past. It included two major elements. First, a random sampling of PGE customers and others was conducted in order to test our image and proposed methods to improve it. Second, internal and national interviews were conducted by a management consulting firm with experience in strategic planning. In addition, a management task force was established, with

Short as chairman, to develop an action plan based on the results of the study.

On the positive side, we learned the public perceived us as being competent, knowledgeable, and concerned with conservation. They also thought our employees were friendly and courteous. On the negative side, though, the public seriously questioned our credibility and honesty and our interest in them. They felt we did not provide enough information nor did we listen to their concerns. Related to this was a perception of a "management vacuum" at PGE. No one appeared to be listening, so no one must be there making decisions.

Our homework also showed the percentage of the public favoring public takeover of our company increased from less than 20 percent in 1968 to 50 percent in 1979. One apparent reason was that public acceptance of our Trojan nuclear power plant was diminishing. In relation to this, the public perceived PGE as an advocate of nuclear power over other energy resources, and as without "sufficient concern for the health and safety of the public or the environment." Another factor was that the public perceived us as making excessive profits because we charged "high rates" and were presumably "attempting to profit from the energy shortage." Finally, the survey revealed that morale among our employees was uncomfortably low. Although they liked and supported the company, nearly half of them were reluctant to admit they worked for PGE.

We had not noticed our white hats turning gray. It was not surprising the public was unhappy without rate increases, but the dramatic social change was unexpected.

Short reacted to the results of the study with a commitment to change. That commitment came in 1980 when our Board of Directors adopted a corporate resolution of openness, candor, and a greater receptiveness to the public. The resolution read as follows:

WHEREAS:

1. Good corporate citizenship and credibility require the trust and confidence of the general public; and
2. Energy development and utilization have a significant effect on our society; and
3. Our decisions on energy facilities and energy management should not be taken unilaterally without consideration of the interests and concerns of other parties.

IT IS RESOLVED THAT:

> Because it is consistent with the corporate objectives of profitability and a reasonable return for shareholders, the officers and employees of Portland General Electric Company shall establish and maintain effective two-way communication with out customers, investors, public officials, community leaders, and other members of the general public, informing them of developments relating to the Company and the utility industry and learning from them their opinions, views, desires, and concerns about the conduct of our business.

With such words, the company sent the outside world a clear signal that it was changing its attitude toward dealing with the public. Another meaning implicit in the resolution, as it was in the reorganization and hiring of new officers, was that major decisions at PGE would be made according to political criteria rather than strictly engineering criteria. Nor would decision-making be free from external intervention. Such changes did not occur overnight, but the company had eventually come to realize that the extent of its influence depended upon its ability to develop a favorable corporate image and a greater sensitivity to evolving public interests.

SEEKING PUBLIC OPINION: THE CASE OF WILLAMETTE FALLS

Almost immediately, PGE had an opportunity to demonstrate its commitment to the open style of work announced in the foregoing resolution. On the drawing boards and ready for the permit process was a proposal to build a hydro plant on the east shore at Willamette Falls, ten miles south of Portland. Technical studies showed that additional power could be generated and that the project was economically competitive with alternative supply technologies. The proposed project would have a peak output of 60 megawatts and an average output of 35 megawatts. If approved, construction could begin within a year at a cost of $70-80 million.

Previously, we would have tried to sell the public on the project through education and (if necessary) confrontation, but in this case we chose a new method. Before deciding whether to proceed with the project, we wanted the public's opinion on all the issues involved. It was decided that the three nearby communities of West Linn, Oregon City and Gladstone be established as the target area because of the project's effect on local taxes, businesses and traffic, as well as on historic and environmental values. The assumption was that broader regional questions, such as the effect of the project on the overall northwest fisheries, would be addressed in the normal course of the federal licensing procedures. Discussing the project with tri-cities area residents before beginning federal licensing would give PGE more opportunity to identify and, perhaps, resolve some local problems before seeking a license.

On the positive side, our preliminary internal assessment of community issues indicated that the plant would produce power for nearby industries, would increase tax revenues in the local jurisdictions, and would provide employment opportunities. On the negative side the plant would dry up the Willamette Falls for many months of the year. The Falls are attractive and lie in the midst of three communities with great historical value. In addition, the Willamette River has the second largest run of anadromous fish in the state, making the issue of fish bypass and protection a critical one.

In May of 1980, Don Barney & Associates, a local consulting firm, was retained to assist us in developing and implementing a public involvement program for the proposed plant. The PGE team assigned to the program consisted of the consultant manager and coordinator from the Public Affairs Department; the project manager; an engineer with expertise in planning, design, and power supply from the Corporate Planning Department; a fish biologist from the Environmental Sciences Department; and a media and publicity person from the Public Relations Department. Working together, the consultant and the internal task force first developed a list of objectives to guide the company in its dealings with the public, and then spelled out a plan for seeking citizen opinion.

The public involvement program for Willamette Falls was developed with five basic goals in mind. These were as follows:

* <u>To provide information to directly affected communities:</u> this would include data on local and regional energy requirements as well as on the proposed facility and its effect on environmental (especially

fishing), economic, and social concerns in the community;

* To create a public role in PGE decision-making: the process would solicit and consider the community's response to key elements of the project, including energy need, land use and environmental issues, mitigation of adverse effects, and relationships to community planning and development (jobs, tax base, growth, etc.);
* To develop a prototype public participation process: the broad outline of this prototype would be used by PGE as a framework for citizen involvement in future construction projects;
* To improve PGE's performance as a corporate citizen: by involving the public in major decisions and by responding to community needs and interests, the company would reinforce its commitment to citizen involvement in decision-making;
* To improve PGE's corporate image and build credibility: with a better understanding of energy and utility problems the public would have a greater motivation to support energy development projects.

To achieve these goals, the PGE task force and its consultant carefully defined a multi-faceted strategy to reach the community and seek reaction to the proposed plant. The strategy comprised two basic parts: 1) formation of a Citizen Advisory Committee (CAC); and 2) implementation of six specific outreach measures (including a School Program, a Listeners Bureau, Community Workshops, Public Forums, a Project Information Office, and a Media/Publicity Program).

The program called for numerous persons within the company to be used in support of the task force or as staffing for the Listener's Bureau. However, although the company was always to be represented by employees at community events and meetings, it was determined inadvisable for PGE to directly sponsor citizen committees and public forums (involving outside experts on pertinent issues). Consequently, responsibility for the external process of community interaction was given to the consultant Don Barney & Associates.

A final noteworthy point is that the consultant's proposed methods to obtain broad-based public reaction to the Willamette Falls project met at first with some internal skepticism. The Engineering Department initially claimed, for example, that the public was not capable of making technical decisions and should not be asked to do so. Company acceptance of the proposal, however, was eventually realized through the active support of Robert Short and the Office of External Affairs.

THE CITIZEN ADVISORY COMMITTEE

The Citizen Advisory Committee (CAC) was the key technique employed by the consultant to represent various interests and community points of view on the project. As the first step in the selection process, the consultant (after discussion with PGE) undertook an effort to interview local opinion leaders in the tri-cities area, including elected officials, school administrators, business people, and citizen activists. The objectives of these interviews were threefold: to describe the process to community leaders; to develop a feeling for community concerns about the project; and to identify potential members for the Citizen Advisory Committee.

When interviewed, those on the original list suggested additional contacts so that, eventually, 80 individuals were interviewed. This in turn gave rise to a list of nearly 50 candidates for CAC membership, which served as the basis for the consultant's nomination process. Local citizens repeatedly mentioned were contacted by phone or in person by the consulting firm to determine their interest in serving on the committee. Interest in serving was enhanced by an Oregon City press conference held by board chairman Short, who stated his firm commitment to give the committee's findings high priority in the utility's decision on the project.

In determining the CAC's composition, care was taken to ensure representation for the full range of interests and values of the community, and to ensure the public believed the committee represented all groups that saw themselves as potentially affected by the proposed project. In the end, a list of 19 names was recommended to PGE. All were approved by the company, which had membership selection veto power but did not exercise it.

We announced the committee membership in September of 1980. It included representatives of the Oregon City/Gladstone/West Linn business, education, and trade union community. Besides two fishing group representatives, at least eight other members claimed long personal histories of recreational fishing near Willamette Falls. Historical interests were also represented, as were those interested in community expansion as well as those concerned with limiting growth.

The CAC members received a letter from the chairman of the board asking them to serve. The letter also outlined the tasks the committee would perform, which were to:

* review the technical elements of the proposal;
* examine the economic and social benefits and costs, and evaluate the tradeoffs involved when such interests as energy, environmental concerns, recreation, local history, and future community development are considered;
* provide a forum for the public to comment directly on the proposal;
* monitor other public discussions of the proposal held locally, including public workshops and meetings planned between November and January of the following year; and
* provide the chairman's office with the findings by January 30, 1981.

In cooperation with PGE, moreover, the consultant developed a handbook for the committee to use as a base of information on energy needs, alternative energy resources, conservation, and the Willamette Falls project. The handbook was a starting place and attempted to lay out all the issues that had been identified during the preliminary interview stage. This information was greatly expanded during the course of the committee's meetings.

After an initial briefing from PGE and the Oregon Department of Fish & Wildlife, the group broke into subcommittees. The three areas to be covered were: energy needs, community effects, and fish and recreation. Each subcommittee scheduled a panel discussion with resource people from various agencies. After all three subcommittees had hosted such a meeting, reports were given to the full committee. Following these preliminary reports, the subcommittees met and approved final draft reports with recommendations to the full committee. At the last meeting of the CAC, the final drafts were reviewed and modified. The committee then voted to submit the modified versions to PGE, along with a 30-page main report of conclusions and recommendations.

The committee functioned independently to the extent that it set its own pace and schedule and decided what information it should collect. Staff services for the CAC were provided by the consulting firm PGE had retained, which handled speaker scheduling, preparation of minutes, distribution of materials, and other logistical tasks requiring contact with the committee. The consultant also helped the co-chairmen develop the process of decision-making. As originally planned, though, the CAC maintained an arms-length relationship with PGE itself, whose staff people attended the meetings and worked with the committee but only as a technical resource. Company representatives answered the committee's questions on all aspects of the project promptly and completely and provided translations of complex technical material into lay terms. Members of the committee were also led on a tour of the site as well as the hydro plant and associated fish ladder in operation across the river.

Throughout its four month process of deliberation, the committee received reports on other activities in the community and had the opportunity to hear from concerned citizens. At several of the panel discussions, for example, there were questions or comments from the audience. In addition, all of the ten Advisory Committee meetings were public meetings, with advance notice sent to interested people and the press. The committee held a widely advertised public meeting on January 7, 1981 at the Oregon City High School. At that meeting, attended by approximately 80 people, reports from the Community College Forum, the School Program, and the Project Information Questionnaire return were given. The initial subcommittee reports were reviewed and the audience was asked to comment. Most of the discussion centered on the fish issue.

One of the Citizen Advisory Committee's most important functions was to serve as a clearinghouse for information. For one thing, group members not only listened to public comment but also researched, distilled, and absorbed a large amount of material on their own. In addition, the committee was continually provided with notice of all speaking engagements in the community about the project and invited to participate in other public information activities. Several of the committee members took part in those activities, either as panel members or audience. Press coverage of CAC activities also served as a means of getting information to the public.

OTHER OUTREACH TECHNIQUES EMPLOYED

Don Barney & Associates conducted six additional outreach methods, as described below. The first four methods provided two-way communication links between PGE and the communities. The remaining two methods were designed only to provide information to the public.

Listeners Bureau

The function of the Listeners Bureau, staffed by PGE employees from several departments, was to obtain information on the process of public involvement and on the project itself from interested community groups. The group was not referred to as a speakers bureau to emphasize that the intent was to listen to what community members felt about the project rather than to deliver the traditional "tell and sell" speech. We did, however, develop a slide show about the project to be used in conjunction with an opening presentation. A brochure was also produced, which was left with audience members after each speaking engagement. A mail-back questionnaire in the brochure covered some of the issues surrounding the plant.

PGE staff handled the assignment of speakers, but the consultant had responsibility for training and for scheduling "listening" engagements (which eventually numbered 21 in the tri-cities communities). Participant selection and training for the Listeners Bureau, it should be noted, were not traditional. Members were selected for their ability to listen and were required to attend a series of training sessions where they received background information on the project and instruction on how to deliver a talk. They also discussed techniques that could usefully be employed in the various types of engagements that were scheduled. An especially important phase of such training was conducted by a psychologist on the topic "How to Diffuse Hostility."

Community Workshops

Intended as an opportunity for local people to get more detailed information and to have an opportunity for in-depth discussion of the project, three workshops were planned, one each in West Linn, Gladstone, and Oregon City. The community education programs in the three school districts were identified as possible sponsors. Superintendents and community school coordinators in each of the school districts were contacted. Although West Linn and Gladstone showed limited interest, in Oregon City the superintendent and the community school coordinator agreed to sponsor a workshop. There was an announcement of the workshop in the community school mailer, which went to every household in the district. There were newspaper ads, flyers sent to 500 people on the mailing list developed by the consultant for the project, and posters displayed in Oregon City businesses and in public buildings. People were asked to preregister so that materials could be sent to them in advance. PGE and Fish & Wildlife personnel were asked to attend to act as technical resource people.

Even with a small turnout, the workshop provided some good discussion of the project. Because the group was small and the setting informal, everyone had the opportunity to discuss points of interest. The group came up with a list of concerns that was forwarded to the Citizen Advisory Committee.

Public Forums

One of the traditional ways of getting information into a community is a large public meeting. In the tricities area, Clackamas Community College had the best facilities and was known for sponsoring that kind of meeting. College officials were approached during the early stages of the process to gain their cooperation in sponsoring two forums, one at the beginning and one toward the end of the process. They agreed to sponsor one forum in December.

Under the guidance of the College's Community Services Department, a panel discussion of the project was planned. Representatives of PGE, the Citizen Advisory Committee, the Oregon Department of Fish & Wildlife, and the Oregon Department of Energy presented information on the area's energy needs, the alternatives available to meet them, the effects of the proposed power plant on the community, the fisheries problems, and other environmental questions.

The community college felt strongly that it needed to have not only the appearance of neutrality, but as little actual contact as possible with PGE and Don Barney & Associates. A College Advisory Committee was set up to organize the event, and the college used its own publicity department to publicize the meeting. To help the publicity effort, the consultant put posters in downtown Oregon City businesses and sent a notice to the project mailing list. The meeting was included in newspaper ads and radio announcements placed by PGE in November and December to publicize public forums and CAC meetings.

There were about 50 people in attendance. After the presentations, there was a question and answer session which centered on the fish issue. At the end of the forum a completed questionnaire, prepared by the college and expanding on the one used by PGE, was collected from participants. The results went to each Citizen Advisory Committee member and were part of the community college's report at the January public meeting.

School Program

In attempting to enlist the support of the school districts for a school program, interviews were held with the three school district superintendents. The Oregon City school superintendent and district vice-principal in charge of curriculum expressed interest. Because of full schedules for many grades, only one group was selected to participate: the PEAK program for the 7th, 8th, and 9th grades (PEAK -- Providing Education Alternatives for Kids -- is the Oregon City School District Talented and Gifted Program). This age group was considered to be a good target population because they were old enough to review and absorb the material on the project, but were still young enough to take material home. In this way, not only was the information getting to the students, but also to their parents.

Using a decision-making model developed for the students by a PGE staff educator, a program was planned covering two full days of intensive work with a week between for outside research. The three teachers who were coordinators for the program in the junior highs were given two evenings of orientation. One evening covered the technical aspects of the project and its effects on the community, and one evening was a review of the problem-solving process.

The students spent the first day of the process hearing presentations from many of the same resource people who had made presentations to the Citizen Advisory Committee. During the next week, the students researched energy questions, toured the existing T.W. Sullivan Plant at the falls, and arrived at their conclusions. After that research, students who had worked together in small groups made presentations to their peers. Each presentation was followed by a question and answer session that was often spirited.

To conclude this program, a representative group of students from each junior high school gave a five-minute presentation to the Citizen Advisory Committee at the January 7 public meeting, recommending that the plant be constructed with adequate fish protection. Judging from the turnout of parents at the meeting, as well as from the comments of the students and their teachers, the experience was worthwhile for all concerned.

Project Information Office

Beginning November 10, 1980 an information center sponsored by PGE was opened in downtown Oregon City on weekdays from 10:00 a.m. to 4:00 p.m. In addition to a full-time staff person who could answer questions (a local woman), the office had a scale model of the proposed project, a slide show, sketches of possible fish protection devices, and drawings of potential recreation lands associated with the project development. The office was intended to be a "drop-in" center to provide information, and to heighten awareness of the project among people who worked or shopped in Oregon City. It also was a contact point for the press or for people who knew about the project from other sources but wanted more specific information on meetings of the Citizen Advisory Committee, the location of forums, or other aspects of the project.

Foot traffic was slow at first and mainly confined to press people and those who heard about the exhibits at a Citizen Advisory Committee meeting. However, as the process went on and press coverage increased, more people came into the office. By the end of the three-month period, up to eight people a day stopped in to ask about a meeting or to look at the displays. Telephone inquiries also increased.

Office staff also kept in close contact with local press and public officials to make sure they knew meeting schedules, and organized a shopping center "blitz." Four high school students worked from 11:00 a.m. to 4:30 p.m. handing out PGE information brochures at two area shopping centers. That effort, combined with handouts at the Listeners Bureau engagements and elsewhere, resulted in the distribution of 2,000 brochures.

Finally, staff at the Project Information Office assisted in the tabulation of mail-back questionnaires in the brochure as well as those PGE sent separately to some 12,000 customers in its service area. The return of completed questionnaires was good -- about 10 percent.

Media Publicity

One of the most important elements of the program was getting information to the three communities via press coverage and other publicity. The company worked hard to provide the media with timely notice of PGE activities and events, particularly the tours of the T.W. Sullivan plant, Listeners Bureau presentations, workshops and forums, and the opening of the Information Office. Don Barney & Associates, acting as staff for the CAC, was the committee's primary contact with local and metropolitan press people. The consultant provided copies of announcements and minutes, and encouraged in-person coverage of Advisory Committee meetings.

Additional publicity came from some of the institutions involved in the events, which provided press releases, ads, or flyers advertising their event. Workshop announcements at the community school and Clackamas Community College were examples of this. A few committee members also used their own press contacts to get coverage.

The media program generated a large number of newspaper articles, including an extensive series in a newspaper outside the Portland metro area. Most of the coverage was of the Advisory Committee activities, although there were articles on other program elements as well.

PGE ARRIVES AT A DECISION

By the first week of February, 1981, PGE and received the report of its Citizen Advisory Committee and the opinions of all interested members of the community. The internal task force, which had continued preparations for licensing the Willamette Falls project during the public reaction period, spent the next month considering whether the proposal was still viable and if the CAC recommendations were attainable. The internal team then submitted its assessment report to management and on March 10, 1981, board chairman Short issued the company's decision in a public statement read before the CAC. What follows are excerpts from the task force report and the text of Short's remarks, which together summarize the outcome of PGE's initial experiment with public participation.

Excerpt From Internal Task Force Report (verbatim)

The Willamette Falls Citizens' Advisory Committee has recommended that PGE not build the project unless their 20 recommendations are accepted. Their recommendations reflect a need for this project, that all possible methods should be implemented to protect existing fish runs and the replacement of fish killed, and that PGE should continue to demonstrate social concerns in constructing the project. The major CAC recommendations were:

* Fisheries: PGE should design and construct fish protection facilities, begin hatchery construction prior to plant operation so that adequate fish are available to mitigate losses when the plant first operates, and guarantee to stop the turbines if excessive smolt depredation occurs based on a figure negotiated between PGE and ODFW and monitored jointly by the two organizations.
* Construction: If PGE constructs the plant, do so in a socially and environmentally acceptable manner and keep the citizens in the affected communities fully apprised of anticipated construction effects.
* Aesthetics: PGE should investigate the possibility for variation of the release of impounded waters upstream from the plant to provide increased downstream flows over the falls during dry periods.
* Historical: PGE should verify that no petroglyphs in the area will be disturbed by the plant's construction and agree to erect an educational display at a viewpoint over the falls that references the history of the falls.

The task force recommends that PGE management approve the following limited steps. (First), we should work with the PGE Fisheries Committee and the Oregon Fish and Wildlife Commission to establish: a) an acceptable smolt depredation percentage and monitoring methods; b) the need to shut down the turbines if this percentage figure is exceeded during plant operations; and c) the involvement of an ad hoc member on the PGE Fisheries Committee appointed by the Northwest Steelheaders. (Second), if we successfully resolve the technical fisheries issue, then we should commit to fund a hatchery siting study under the guidance of the PGE Fisheries Committee to determine if either existing Willamette River Basin hatcheries can be expanded, a new hatchery sited, or some combination of these to provide the necessary level of smolt production required to mitigate for operation of the Willamette Falls project. The hatchery facilities will be used to produce spring and fall chinook, summer and winter steelhead, and Coho, the species which now migrate past the Falls. It is estimated that this study will cost between $150,000 and $250,000 and take three months to complete. (Third), the Management Committee should reconvene, if these two issues are resolved, to further consider the project.

If PGE management elects to proceed with project design and construction, PGE should agree to the CAC recommendations, even though they will not take effect until after we have secured a FERC license.

On March 3, 1981, PGE agreed to provide the CAC with our response to their recommendations. The task force recommends that our response be a two-part process. Our initial response to the CAC should include:

1. thanks to the committee for their diligent work on the project and for leaving us an "open door" to make our decision on the project;
2. a commitment that if the project is developed, all of their recommendations will be met;
3. recognition that within their recommendations there are still several important issues that PGE feels must be resolved before we make a commitment to construct the project. These issues are: (a) agreement with the State of Oregon on an acceptable smolt depredation percentage and monitoring methods, and the need to provide plant shutdown as a fish protection measure; and (b) the ability to expand an existing hatchery or site a new hatchery to provide the required hatchery fish to mitigate for fish losses caused by the project;
4. a request for CAC concurrence that, if Items 2 and 3 are resolved, we have responded adequately to their recommendations;
5. a note that, based on compliance with all of their recommendations and current economics, the project appears to be economically justified.

Our longer term response to the CAC will be to periodically update them on the developmental status of the project.

Text of Final Decision Announcement by Mr. Short (verbatim)

Ladies and gentlemen of the Willamette Falls Citizens Advisory Committee . . . I can't thank you enough for the long hours, the many meetings, the weeks of research, your dedicated service to your community and Portland General Electric Company! We needed your help and guidance, and you gave it to us in an honest, responsive and constructive manner. You are a very special group of people.

The Willamette Falls proposal looks good on paper. But, we pledged to you and to this community that we wanted YOU involved in the decision of whether to build it or not. You told us the problems we faced and what to look for if we built the project. You told us we had fish problems we have not solved. And you told us you don't want us to proceed with the project if we don't solve these problems before construction begins.

When I appeared before the Tri-Cities Chamber of Commerce in December, I said there are two criteria for a "go" decision on the Willamette Falls project:

* that it be cost-effective;
* that the Citizen Advisory Committee recommends that we have community support for the project.

Even with the added costs to accommodate all of your recommendations, the project is still cost-effective. Economically, Willamette Falls is our most viable generating project.

That brings us to the second item: Did you say we have substantial community support? We believe we listened and heard you say "from an energy standpoint, yes . . . but from an environmental standpoint, NO." You pointed out very clearly that the project has little or no approval from either local or streamwide fishermen. You told us the Willamette is considered one of the mainstream rivers in the state, producing one of the largest and last runs of native anadromous fish, and the public just doesn't want any further development in this location which might disturb its present or its future. You asked us to reconsider the project -- not to build it -- unless we resolved the fish problems.

We are not going to build the project. As I have indicated, economics, is not the basis of this decision, and this was a difficult decision to make. We determined our social and environmental responsibility must be given more weight with this particular project.

We believe we could build the Willamette Falls project -- and protect the fish. But you told us you wanted a guarantee that it would work. Those guarantees are not possible based on plans alone -- the project would have to be built and tested in operation before we could make those guarantees. And even then, Mother Nature is mighty unpredictable at times.

Although we won't build the proposed Willamette Falls facility, we must develop renewable resources. The recently passed Regional Act and the Energy Facility Siting Council need for power standard require that PGE develop renewable resources before thermal generation. And, as you recognized in your report to us, there is a need for additional power generation. To help meet those needs, we will now be looking to an accelerated time schedule to study and evaluate the possible development of a second coal plant and other hydro projects. Remember, also, that PGE is not the only company operating under these mandates, and there is no law that says out-of-state or private interests cannot develop hydro plants on Oregon's waterways.

As we continue our search for new resources to meet your electrical energy needs and those of our customers throughout our service area, we will need citizen support. Most any development will have some adverse effect on the environment, but some are more minimal than others. We would hope that, through communication, you, the fishermen, and the general public will help us develop needed plants in those areas which are acceptable.

You have reported that your committee has learned a lot in the last few months. PGE did too. This has been a rewarding citizen involvement process. We have listened and improved our understanding. We believe we know each other better. Giving up the project has not been an easy decision for us. But we share your environmental concern. I said in December that our past energy decisions were not bad ones, but that we've practiced selective perception. We perceive you do not want us to build this plant. We hope we have demonstrated that we listened. Again, thanks to you who worked so hard to share in this community decision. (End of verbatim text.)

THE RESULT -- A NEW IMAGE

Was the project a success? Is PGE's new direction paying off? The answer to these questions is yes. We commissioned a tracking poll to measure our customer attitudes toward the company against our previous benchmark surveys. There was a dramatic improvement in PGE's customer relations climate during the two years following the Willamette Falls Public Involvement Project. Nearly two-thirds of our customers hold positive attitudes toward PGE. That's an increase of 20 percentage points in a three-year period.

Success was not determined by the decision to build the project or not. The success of the process was determined by the active support of the chairman of the board (Robert Short), the president (William Lindblad) and the senior vice president of external affairs (James Durham). These executives successfully established the tone of the program -- that PGE will listen and be responsive to what it hears.

Short made three public appearances in the affected community. He announced the citizen involvement process for the project at a press conference. During the middle of the process, he addressed the Tri-Cities Chamber of Commerce, where he presented a speech entitled "Public Participation in the Willamette Falls Hydro Project" (or) "How PGE Finally Got a Hearing Aid." After the Citizen Advisory Committee made its recommendations, he returned to their final meeting to announce management's decision not to build the project. Durham attended every meeting of the Citizen Advisory Committee, participated as a panel member for the public forum sponsored by the Clackamas Community College, and presented speeches in the community. Lindblad also attended Citizen Advisory and community meetings and hosted an appreciation dinner for Advisory Committee members at the end of the process. <u>It cannot be too strongly stated that senior management's visible and active participation gave credence to the Company's intention to seek and listen to public reaction to the project.</u>

The Willamette Falls Citizen Participation Program was not a "one-time effort" to change the public's perception of PGE. It is a way of doing business. The Willamette Falls experience taught us the methods to effectively involve the public in our decision-making. We have favorably influenced the public's image of the Company and we are continuing to use what we have learned.

Since that project, we have asked the public how they feel about a small hydro project on a river system dam we own on Mt. Hood. The public overwhelmingly favored that project. When we asked area residents to come tour the proposed site, we extended the invitation also to those fishermen who actively opposed the Willamette Falls Project. They favored this project. As an additional example, we recently asked the social service agencies to help us implement a new deposit policy. We had determined we were not good at screening those cases of exceptional need. We now allow extensions for deposit payments based on the recommendation of qualified social service agencies.

We have seen public opinion changing. CEO Robert Short was the first utility executive asked to address the Portland City Club, a local issue-oriented study club with several thousand members. A recent editorial further points out that we are doing things better by understanding the community's concerns. Because of public opposition a local regional government was unsuccessful in locating their garbage-burning "waste recovery plant" in Oregon City. The editorial suggested the governmental body should learn from PGE and talk to the public first.

Our new attitude does not connote weakness on our part, nor a lack of conviction in our decisions, present or past. The attitude is good business practice, because we cannot afford confrontation. We believe our image is changing, but it is a slow process. We spent too many years looking into a mirror rather than out the window. We cannot afford to slip back into our reflective mode. The image we want will be reflected in the public's knowledge and acceptance of PGE's sound business decisions.

8

Public Participation in the Tennessee Valley Authority's Energy Planning Process, 1983

Dawn S. Ford

The Tennessee Valley Authority (TVA), established in 1933, is a corporate agency of the United States Government. It is responsible for power production, environmental quality, and economic and resource development in the seven-state Tennessee Valley region. The agency serves approximately 7.8 million citizens who are either users of TVA electric power or recipients of programs funded through appropriated monies. It is administered by a three-person Board of Directors appointed to nine-year terms by the President of the United States. The General Manager is the principal administrative officer.

TVA is the nation's largest electric utility, serving a 78,000 square-mile area, and has undertaken a substantial nuclear power construction program. TVA supplies power at wholesale to 160 municipal and cooperative distributors and one privately-owned electric system which, in turn, distributes power to over 2.8 million customers in parts or all of Tennessee, Alabama, North Carolina, Mississippi, Virginia, Kentucky, and Georgia. TVA also supplies power to 50 industrial customers with large or unusual power requirements and several Federal nuclear, aerospace, and military installations. Financially, the power program is separate from other TVA programs. It is required by congressional act to be self-supporting and self-liquidating -- financed solely by the revenue from monthly electric bills. The TVA Act also requires the agency to ensure an

---

Dawn Ford is the former Chief of the Citizen Action Office at TVA, and is currently the President of Consumer Awareness Management in Knoxville, Tennessee.

adequate and reliable supply of electricity at the lowest possible cost. TVA is a grass roots agency with a long history of working in partnership with the people it serves. The Board of Directors has a history of commitment to public involvement, best expressed in 1953 by TVA's second Chairman of the Board, David E. Lilenthal:*

> We who believe devoutly in the democratic process should be the first to urge the use of methods that will keep the administration of national functions from becoming so concentrated at the national capital, so distant from the everyday life of ordinary people, as to wither and deaden the average citizen's sense of participation and partnership in Government affairs. For in this citizen participation lies the vitality of a democracy.

During the past few years, the current Board of Directors has intensified TVA's efforts to involve Tennessee Valley citizens in the decision-making process and increase the accessibility of top management. A formal agency public participation policy was developed and it has been TVA's goal to be a leader among public institutions in encouraging citizens to offer ideas and comments on decisions that affect their lives. The agency actively seeks public participation in the planning and design of its programs and in decisions involving important policy consideration.

To encourage effective public participation, TVA involves the public early in the decision-making process and provides accurate, understandable, and pertinent information on issues involving agency programs. In order to respond effectively when particular issues arise, the public must be made aware of TVA's plans and programs and accustomed to expressing its views. Effective public participation in planning also serves an education and awareness function by conveying clear information about agency actions to the public at large and to leaders of all groups affected by these plans. This in turn has encouraged a reciprocal information flow and has strengthened relationships between the agency and its constituencies. In October of 1978, the Board established the Citizen Action Office (CAO) to work across corporate division lines, designing a variety of public participation programs. An

---

* David E. Lilenthal, TVA-Democracy on the March, p. 144, Greenwood Press, Westpoint, Connecticut (1953).

opportunity was thus created for critical publics to have immediate and direct access to the corporation. In turn, through monitoring public sentiment, the CAO analyzes for TVA managers the potential impact of citizen concerns on the agency and, likewise, agency decisions on the public.

After a brief description of the principles, policies, and organizational arrangements which together define the participation function at TVA, this paper reports on some of the key projects undertaken by the Citizen Action Office since its inception six years ago. These include general constituency building activities as well as efforts to include the public in agency deliberations on three specific issues: (1) where spent fuel from nuclear reactors should be stored; (2) how to implement certain ratemaking standards under the Public Utilities Regulatory Policy Act; and (3) whether to defer several nuclear units already under construction.

THE PARTICIPATION FUNCTION AT TVA

The Citizen Action Office is responsible for planning and directing programs to ensure that public concern and complaints receive an objective and responsible review within the agency. The office investigates and responds to expressions of public concern about any TVA program and may recommend corrective action to the responsible office manager. It serves as the primary agency expert on public participation techniques, public attitude assessments, and public meeting organization. More importantly, through bimonthly and annual public opinion reports to program managers, the analytical and other services of this office are used as a tool in long-range corporate planning.

A major function of the CAO is to coordinate all of TVA's public participation programs and recommend new processes by which citizens can become involved in TVA's decisionmaking process. TVA maintains a formal system of codes which convey major administrative policy determinations in specific areas of continuing TVA-wide significance. There is a code concerning agency policy on public participation which charges program divisions with the responsibility for consulting with the CAO early in the planning of their programs to discuss appropriate methods for planning public input. Public views have been sought on issues as far ranging as transmission line construction, land use, utility power rate structure, electric service practice standards, and capacity planning.

Obviously, it is neither desirable nor practical for the public to be involved in every decision made by management. However, at TVA it is standard practice to seek the public's views in making all decisions which are expected to affect important policy concerns. In developing any public participation program, both citizens and corporate managers must understand that public involvement is not synonymous with usurpation of the authority and responsibility entrusted to administrative officials. Public opinion becomes a factor in the decision-making process and not a "grass-roots" voting approach where those entrusted with decision authority abdicate their responsibility. Just as management would not make decisions without first obtaining critical technical studies, an effective public participation program should be premised on the principle that it would be equally imprudent to make important decisions without first obtaining reactions from those publics that have a stake in the issues to be resolved.*

In determining those issues on which to seek public input, several factors are taken into consideration:

- Is the issue significant?
- Is there adequate time for meaningful public involvement?
- Does the new policy suggest a departure from previously stated direction?
- Is there a real choice to be made?

When an issue is identified staff from the CAO meet with staff from the responsible program division, such as the Transmission Planning staff or the Environmental Quality staff, to determine which of the many publics will be affected by TVA's decision, whether the issue is a local one or one with Valley-wide impact, how controversial it will be, and, finally, which process would be the most effective in obtaining public input.

It should be emphasized here that public participation is hardly a science, so no pat formulas can be used in determining the type of participation process to be used.

---

* James L. Creighton, "The Limitations and Constraints on Effective Citizen Participation," <u>At Square One: Proceedings of the Conference on Citizen Participation in Government Decisionmaking</u>, Federal Interagency Council on Citizen Participation, Washington, D.C. (March, 1977).

Still, there are certain standard objectives a good program seeks to accomplish. These are:

* to ensure that public concerns are considered before final policy decisions are made;
* to increase public understanding of corporate policies;
* to identify a greater range of policy options or identify unforseen consequences of proposed plans;
* to ensure that the public understands the choices involved and the issues that must be considered;
* to convey to various publics information they will find useful in helping resolve critical issues;
* to ensure the public knows why it is being asked to participate and the exact process to be used;
* to provide feedback on the reasoning behind a final decision and on how public comments were analyzed and used in reaching that decision.

With these goals in mind, participation programs can be tailored on a case-by-case basis, selecting techniques of interaction most appropriate to the issues at hand. In some cases, for example, policy papers and public meetings can be used effectively in obtaining informed public comment, while in other instances an ad hoc advisory committee, roundtable, or nominal group approach may be more effective.

As a final comment on general approach, the CAO believes that for public involvement to be successful it must be viewed by utility staff as an exciting new challenge, not a threatening obligation. Particularly inappropriate is the traditional way of holding a public session, where defensiveness and the development of an adversarial relationship with anyone who is critical suggests to the public that the organization is willing to listen to outside ideas only as long as they are in support of the company proposal.* Every effort has been made to avoid such problems in developing TVA's public participation program. For example, conflict management training -- including instruction in non-traditional processes of communication -- was given to over 200 managers who normally engage in public outreach.

---

* James L. Creighton, Public Involvement Manual: Involving the Public in Water and Power Resources Decisions, p. 27, U.S. Government Printing Office, Washington, D.C. (1980).

THE CONSTITUENCY BUILDING PROGRAM

At the root of TVA's public participation program are ongoing educational and constituency building activities with a variety of citizen groups affected by and interested in the agency. This type of public participation is a positive approach rather than a reactionary method of holding public meetings after problems arise. If a dialogue with the public is conducted throughout the year and not just in connection with controversial issues, a framework is already established thus making it easier to work with the public during a time of crisis.

In order to obtain opinion from a cross-section of publics, major efforts have been made by the CAO to identify persons interested in TVA programs and policies. Over 7,000 names of elected officials, community and consumer group leaders and others have been computerized according to program interest, and have regularly received invitations to various public meetings as well as information packets on major issues, such as electric rate increases or land use management. These citizens are encouraged, in turn, to share their views and concerns with TVA by using a toll-free telephone system known as the Citizen Action Line (CAL), which has proven to be one of the most effective mechanisms for public communication employed by the organization. Since the lines went into service in 1978, CAO has heard from over 225,000 people, and the number of calls received daily has increased steadily. During the first year about 81 calls a day were received; since then, the daily average has more than tripled to approximately 300.

The lines are usually staffed by six full-time, specially trained personnel. In handling calls, the emphasis is on providing quick, reliable, and understandable responses to questions and suggestions cutting red tape and correcting situations as circumstances allow. Although the staff strives for one-stop shopping, the more technical calls are sometimes transferred directly to a program manager. Thus a citizen could actually call in and talk directly with the person responsible for the program about which they have a complaint, which also gives managers a first-hand opportunity to hear citizen concerns. Such contact and accessibility is thought to be important enough, in fact, that Board members answer the phone at some point each month.

About 80 percent of all calls to the CAL are for information, and they pertain to a wide variety of subjects. For example, to break down the caseload by topic for fiscal year 1982:

* about 58 percent of the calls pertained to information about TVA's lakes and streams;
* about 10 percent pertained to the large number of energy conservation programs available to Tennessee Valley citizens and industries;
* about 9 percent regarded power-related topics;
* the remaining 23 percent of the calls covered topics as varied as land-related problems and the Knoxville World's Fair.

Open six days a week, citizens are frequently urged to use the CAL and express their opinion during a particular public participation process. Often it is easier to motivate citizens to use the lines than to attend a public meeting.

For those desiring face-to-face contact, however, the Citizen Action Office over the past five years has been involved in conducting over 400 public forums, of which approximately 125 have been open meetings of the Board of Directors. TVA invited the public to attend these meetings before the Government in the Sunshine Act was enacted, and, according to a 1981 study by Common Cause, thereafter compiled the best record among 26 federal agencies for compliance with the open meetings provision of that law.*

Most of the Board meetings are conducted in Knoxville, Tennessee, the headquarters for TVA. However, on a quarterly basis the Board meets in communities outside of Knoxville, to provide local Tennessee Valley citizens a convenient opportunity to see how the agency conducts business. The sessions also provide an opportunity for citizens to talk directly with the Board of Directors about any TVA-related subject. Aside from these regularly scheduled meetings, the Board also schedules public "listening sessions," patterned after an informal townhall format, at which citizens are encouraged to share their views and opinions with visiting Board members. Designed to increase the opportunity to hear public concerns in a more informal atmosphere than a Board meeting, these forums are held in various locations throughout the Valley area.

The CAO also plans regular "constituency meetings" between the Board and various publics. A full day is de-

---

* Douglas B. Feaver, "Sunshine Act: Not an Open and Shut Case," The Washington Post, p. A17 (September 29, 1981).

voted to learning about TVA programs in attendees' area of interest, and to discussing upcoming issues among themselves and with the Board and key staff. Among the groups that have been involved in these meetings are leaders of environmental organizations, representatives of the elderly, the League of Women Voters, representatives of agricultural, economic and industrial development interests, elected officials, leaders of the recreation and tourism industry, chambers of commerce, and educators. One of the basic tenants of constituency meetings is that the leaders of the groups invited to meet with the Board of Directors will in turn share the information they have received with their memberships. Thus, the agency is able to communicate indirectly with literally thousands of citizens.

The knowledge that direct access to top management is available results in citizens feeling confident they will be listened to and fairly treated at all levels of the company. This is a somewhat different approach than the use of advisory committees. In effect, each constituency group comprises a type of advisory committee without a formal structure. So instead of relying on the opinions of 15-25 people, the agency is able to establish a dialogue with hundreds of citizens representing a wide cross section of interests. Further, this dialogue takes place in a setting which is not confrontational.

Contact with a group does not end with one meeting. In fact, many groups have requested annual constituency meetings, and one ad hoc advisory committee has developed. The Board frequently solicits new ideas from participants and some good programs have developed from these suggestions. Sometimes regional meetings are held with leaders of one particular group prior to their annual meeting with the Board of Directors. Others have been invited on tours of various TVA facilities, and all groups are kept informed throughout the year of new program initiatives. Program divisions are also responsible for working with leaders of various public interest groups on an informal basis throughout the year. By providing an ongoing flow of information about issues that are often very complex, citizens are better able to provide input when asked to take part in a formal public participation process in connection with a particular decision.

A good illustration of how this approach works involved a year-long effort, beginning in 1978, to determine how spent fuel produced from TVA's three operating nuclear power reactors should be stored. Facilities at the plant sites for limited, interim storage of the fuel until it could be processed were already built or under construction. In the mid-1970s TVA began to design and implement a comprehensive spent fuel management strategy aimed at increasing the efficiency of its spent fuel storage facilities. However, President Carter's 1977 decision to indefinitely defer reprocessing of spent nuclear fuel meant that it was necessary for TVA to prepare to store even greater quantities of used radioactive fuel.

A TVA proposal to build a central storage facility that could be made available to other utilities or Federal installations generated much controversy. Although some citizens agreed with the concept, many citizens opposed the transportation of spent fuel over highways to a central facility, and thus favored onsite storage. TVA staff met informally with a variety of consumer groups to discuss the proposal and methods for public participation. In June 1979 TVA released to the public a preliminary option paper that discussed two alternatives for storing nuclear fuel: facilities at each nuclear plant site, or one central facility to serve all of TVA's reactors.

A copy of the study was mailed to about 600 citizens with a cover letter from the TVA Board of Directors which asked for citizen comments. One week after release of the study, a TVA nuclear engineer answered questions from the public for two days on TVA's toll-free Citizen Action Line. Several environmental groups were contacted by the Citizen Action Office regarding their interest in meeting with TVA staff to discuss the study. None accepted the offer, but some did comment they appreciated the gesture. At the same time, the option paper was given to six nationally known experts with various views on nuclear power to review TVA's options and ensure that all issues were adequately addressed. Their views were included verbatim in the final option paper that was released in October 1979.

Public meetings, designed to assess public opinion of away-from-reactor storage or onsite storage of nuclear fuel, were held in three Tennessee Valley cities in November 1979. The meetings were moderated by community leaders with TVA staff giving opening presentations. Two of the independent experts then summarized their review of the study. The audience was then asked to break into smaller groups, to enable everyone to have an opportunity to speak and ask questions of the experts or TVA staff. In at least one instance, the group decided they would remain together in order to hear each other's comments. All comments received at the meetings were forwarded to the TVA Board.

Public opinion was overwhelmingly in favor of constructing storage facilities at each plant site. The Board's final decision, which took into consideration public input, was to construct storage facilities at each agency plant site. That program is now underway.

PUBLIC INPUT ON RATE CHANGE

A second major example of how TVA has used citizen participation in power planning is a multi-year program used in connection with implementation of the Public Utility Regulatory Policies Act (PURPA). Part of the National Energy Act of 1978, PURPA required utilities to consider nine service practice standards on the basis of their effect on conservation of energy, efficient use of facilities and resources, and equity among electric consumers. In addition, the following six ratemaking standards were to be considered: (1) cost of service, (2) declining block rates, (3) time-of-day rates, (4) seasonal rates, (5) interruptible rates, and (6) load mangement techniques. TVA also added a seventh standard for consideration: a special additional charge for new, electrically heated or cooled, energy inefficient homes. PURPA did not mandate the implementation of the standards. The TVA Board of Directors would make the determination as to their appropriateness.

Any changes that might have resulted after consideration of the standards would have been a major policy change for TVA and its power distributors, affecting all ratepayers. Consequently, an extensive public participation program was developed by the CAO in cooperation with TVA's Office of Power. This process began in January 1979 and ended in May of 1982. Actually, two individual processes were conducted at different times -- one in connection with the service practice standards and the other in connection with the ratemaking standards. However, they were very similar in style and content.

Because the public had little information on the issues at hand, which were extremely complex, it was important to undertake a consumer education program if TVA hoped to obtain informed public comment. Toward this end, the following steps were taken before hearings were scheduled:

* A copy of the proposed standards with staff recommendations was mailed to leaders of constituency

groups that would be affected by the actions as well as individual citizens who had shown past interest in ratemaking. A cover letter was included which explained the complete public participation process.

* TVA experts worked on the toll-free Citizen Action Line on specified days to answer questions consumers might have about the material they received.
* In addition to notices in the Federal Register, extensive media coverage described the standards and provided information about the public participation process.
* Twenty-five libraries were selected throughout the Tennessee Valley to receive the initial documents as well as subsequent material.
* Workshops designed to assist citizens in understanding the issues were conducted by TVA staff in five major Valley cities. At each workshop, a nationally recognized economist addressed the issue of rate reform from a national perspective. This input provided an opinion other than TVA's, lending more credibility to the sessions.

A public comment period was then established, and citizens were urged to submit written comments as well as oral testimony at hearings to be scheduled. The official record, including verbatim transcripts of the hearings and copies of all written comments, was made available to the public at eight TVA offices throughout the Valley.

Hearings were conducted in the same five cities in which workshops were held. Two hearings took place in each city, one during the day and one in the evening so that all citizens could conveniently testify. Citizens were asked to contact the CAO in advance if they wished to make an oral statement at the hearings.

In connection with the process conducted on the electric service practice standards, arrangements were made, at TVA expense, with seven law firms for service as public counsel to represent the interests of consumers who otherwise could not afford to participate effectively in the hearing. In connection with the ratemaking standards, arrangements were made to engage independent specialists in economic statistics and other disciplines relevant to ratemaking, to assist consumers in effectively participating in the proceedings. Suggestions were solicited from the public for names of experts with whom they would want to talk. A committee of TVA staff and power distributors selected the economists based on their resumes. Once the experts were hired, TVA made arrangements for consumers to meet with them but otherwise had no other contact. Their purpose was to provide objective assistance to the consumer.

The public participation process conducted in connection with PURPA was the longest and most thorough ever conducted by the agency, and it had a significant influence on TVA decisions regarding the proposed standards. The staff incorporated public views and comments in recommending a determination on each standard, and the Board of Directors changed several determinations based on additional input from the public. Particularly beneficial, it should be noted, was the commentary received in connection with standard no. 3 (time-of-day rates) and standard no. 7 (special additional charge for new, electrically heated or cooled, energy inefficient homes). In both cases, before issuing final determinations, the Board personally conducted additional public meetings to deal further with questions about implementation that had been raised at the hearings.

As originally proposed, for example, standard no. 7 would have applied a special additional charge to the electric bill of new detached, single-family, electrically heated or cooled dwellings first connected to the TVA system after the effective date of the standard, unless the residences had been surveyed and found to comply with TVA weatherization standards. After the public hearing, the proposal was revised into the form of a special hookup charge, but further comment persuaded TVA that it was not appropriate to implement the standard in either form.

Interestingly, dialogue with the public continued to some extent even after TVA issued its "Final Determinations" (copies of which were mailed to all individuals who had testified or expressed an interest in the issue, as well as to selected public libraries). In particular, there were further discussions on the implementation of time-of-day rates, which TVA had decided would be mandatory for customers with demands in excess of 500 kW and voluntary for all others. When it became apparent that there was not a common understanding about how such rates would affect individual consumers, the Board solicited further public comment and subsequently concluded that the new rate structure should be optional for industries with demands over 5000 kW; that each distributor would have the option of offering time-of-day rates to its other customers on a voluntary basis; and that a timetable for mandatory application to all residential, commercial, and industrial facilities would be established after further discussions.

DEFERRING NUCLEAR PROJECTS

A more recent and easily the most ambitious and extensive TVA program in citizen participation centered on the question of whether to defer several nuclear units already under construction. In 1974 the TVA power system had seven nuclear projects underway or planned, involving a total of 17 reactors. By 1979, however, construction schedules were being adjusted to reflect a forecasted slowdown in load growth, and work was suspended on part of the capacity in the last three of the projects. Such revision in construction schedules was common in the electric utility industry at that time, as it became apparent that new power plants --planned in some instances as long as 10-15 years in advance -- would not be needed as quickly as had originally been thought to meet projected growth. This general slowdown was especially consequential for nuclear projects, with their high capital costs and long planning and construction lead times.

Having made some initial adjustments to this situation, TVA as of late 1981 faced one of the toughest policy decisions in its recent history: to accept the prospect of surplus power supplies, or to defer all further construction of two units at the Hartsville site (near Nashville, Tennessee) and one at Yellow Creek (near Iuka, Mississippi). With high interest rates and tremendous increases in construction costs, TVA's nuclear construction program had become the largest single driving force in rate increases to consumers. At the same time, in a period of economic recession, a shutdown would impact thousands of construction workers and the economy of those communities in which the plants were located.

To assist the Board in considering the difficult tradeoffs involved, a group was organized from various staffs to compile the best economic information available both within and outside TVA. As a guide for the Board and the public, that information was analyzed thoroughly to calculate the relative benefits and costs to TVA and the community of continuing or deferring construction work. The results were presented in a report, "Review of the TVA Load Growth/Plant Construction Situation," which analyzed three basic alternatives:

1. continue construction on the three nuclear units and reduce coal-fired generation, or retire existing coal-fired plants as surplus capacity develops;

2. defer construction of the three units and maintain the flexibility to restart construction or replace with coal-fired capacity (if high load growth occurs) or defer construction indefinitely (if low load growth occurs);
3. continue construction of the units and, if possible, develop arrangements to exchange power with other utilities for the period when TVA would have excess capacity.

Since the outcome of this issue would affect millions of TVA ratepayers, the Board of Directors wanted to get a broad sample of public comment before making a decision. The process used to obtain this input was critical since the issue was highly controversial.

To invite comments from the public, an extensive outreach program was developed by the Citizen Action Office.* Among the major elements of the program were as follows:

* The initial background report was mailed to more than 2,000 public officials, community leaders, consumer groups, and representatives of industry throughout the Tennessee Valley. A cover letter requested that oral comments be submitted via the toll-free Citizen Action Line, and all written comments be sent to the Citizen Action Office. A 30-day public comment period was established and immediate acknowledgement was made to each submitted letter.
* Document rooms in four Valley cities -- Knoxville, Chattanooga, and Memphis, Tennessee, and Muscle Shoals, Alabama -- were established so citizens could easily obtain copies of the staff report, read internal staff studies and memorandums, or any document concerning TVA's nuclear construction program. Copies of the documents were mailed to the public upon request.
* Extensive media releases urged citizens to call the toll-free Citizen Action Line and express their views concerning the construction program. a total of 2,735 calls was received on this issue.

---

* Tennessee Valley Authority, <u>Citizen Participation: A Case in Point</u>, Knoxville, Tennessee (1982).

As was its custom, moreover, the TVA Board held a public meeting on February 3, 1982, to receive additional verbal or written comments on the nuclear construction schedule. More than 150 people attended and 28 citizens submitted statements.

Still another big feature of the participation process on the construction issue was the scheduling of five public round-table discussions, under the sponsorship of the Tennessee Valley Inter-League of Women Voters. The League had recently completed an extensive study of TVA and expressed a particular interest in public education on nuclear issues. As an impartial and informed group the League was in a good position to lead meetings that would minimize potential conflict and result in more effective sessions for the participants. Had TVA not previously established a continuing dialogue with the local leagues, however, it would have been difficult to benefit from the organization's assistance.

With TVA assistance, meetings were held in Memphis, Oak Ridge, and Nashville, Tennessee; Decatur, Alabama; and Tupelo, Mississippi. The League handled all invitations and press releases. To assure a cross section of opinions, invitations were extended to 40-50 leaders of interested groups ranging from industrialists to environmentalists to participate on a discussion panel open to the public. TVA was represented at each meeting by a member of the Board of Directors. More than 600 citizens attended and participated in these meetings, which were a critical factor in the success of the entire process.

To assimilate all comments received via the toll-free telephone lines, correspondence, and public forums, the Citizen Action Office staff developed a series of summary reports. These summaries were submitted to the Board of Directors and the General Manager, and were made available to the press along with an appendix containing copies of the approximately 900 letters TVA received on the issue together with formal statements presented at public meetings. Subsequently, a concise report was developed summarizing all substantive comments received from the public and providing TVA's response to these comments. The Citizen Action Office staff distributed the report, "Review of and Response to Public Comment," to power distributors, local elected officials, those citizens who sent technical comments or suggestions, and Federal officials.

Citizens commenting on the nuclear deferral issue could be generally divided into the categories of ratepayers, public officials, industrial leaders, and power distributors. Of the citizens supporting the decision to defer, the majority were either ratepayers or representatives of industry. Industry views were expressed primarily by directly-served TVA customers and the Tennessee Manufacturer's Association, composed of more than 750 members. Of those who supported continued construction, most were employees at the plants and their relatives or elected officials in those counties which might suffer some adverse economic impact if the plants were deferred. The most frequently heard comments among these opponents to deferrment were:

* concern about the money already spent by TVA at the nuclear construction sites;
* fear of job loss, either at a TVA plant or in surrounding areas because people anticipated local economic depression if the plants were deferred;
* concern over lack of power, particularly during extremely cold weather.

Those citizens who supported deferment generally cited high electric rates and nuclear safety as reasons. Storage of spent nuclear fuel was also a concern to a large number of callers.

All comments were carefully considered by the Board prior to making their decision. This finally came on March 2, 1982, when the members voted 2-1 to defer construction of Hartsville A Units 1 and 2, and Yellow Creek Unit 1. Although there were many citizens who did not agree with this decision, no one voiced any complaints about the process used to obtain public input. In fact, at the Board of Directors meeting held on the subject on February 3, 1982, a woman in the audience stated that regardless of the outcome of the decision, she was pleased that TVA had sought public involvement and that no group could be critical of that process. Such sentiments were echoed in several published reactions to the participation process including the following:

* ". . . hearings in various parts of the region and a February 3 public meeting of the TVA Board will give the facts a thorough airing . . . Such an open debate is a refreshing contrast to the way questions about nuclear reactors and new generating capacity ordinarily are decided in other parts of the Nation." ("TVA Gives Its Ratepayers A Look At

Nuclear Option," January 19, 1982, editorial from the Louisville Courier-Journal).

* "Your commitment to solicitating public comment prior to decision-making . . . is most encouraging and, we hope, will be continued in future considerations of such far-reaching magnitude." (Memphis Area Chamber of Commerce).
* "The very process of involving the people more in TVA decision-making -- whatever decision might be ultimately reached -- should be a healthy one, exposing the Federal power and public works agency . . . to more democratic influences . . ." ("A Hopeful Sign at TVA," January 20, 1982 editorial from the Appalachian Observer).
* "TVA's willingness to explain the situation to the public and then listen to its response is much appreciated." (Constantine W. Curris, President, Murray State University, Murray, Kentucky).
* ". . . we have no complaints about the way the Board of Directors studied the situation, recognized its responsibility, and acted in the best interest of the public it serves." ("TVA's Decision," March 5, 1982, editorial from the Clarksville (TN) Leaf-Chronicle).

CONCLUDING REMARKS

TVA's public participation program is best described as a pervasive, broad-based outreach effort which allows management to receive input from a broad cross-section of citizens impacted by TVA. The program operates not only on an issue-by-issue basis, but also through constituency building activities which take place throughout the year. The latter provide citizens with enough baseline information about TVA programs and policies that when they are asked to participate in a process on any given issue they are better equipped to do so. In turn, the continued input from these groups serves as a barometer of public opinion for the agency.

Many utility companies believe public participation must be linked to a particular project or issue, on grounds that citizens will not participate unless the subject is controversial. It is true that it is often difficult to motivate people to attend public meetings. However, the programs developed at TVA have proven that citizens do have an ongoing interest in the agency. The key is to determine which publics are interested in and impacted by a corporation, identify the leaders of those publics, and then develop dialogue. In addition, accessibility and interest by top management are critical to the success of this type of public participation program. Leaders of public interest groups are interested in talking directly to leaders of corporations, for these are the policymakers.

There are, of course, some problems to overcome in establishing a constituency building program. During the first meetings a major portion of the session is often a learning process, since many of the participants initially do not have much information about agency programs and policies. In the beginning TVA set the agenda for the meetings, but it was quickly determined that the sessions were more effective if the groups were involved in this process. Likewise, there has been a change in how the meetings are conducted, with fewer presentations by TVA and more opportunity for discussion and input by the participants. Finally, consideration is being given to integrating some of the constituency groups via joint meetings between, for example, environmentalists and industrialists. In the past the meetings have been made up of a homogeneous group and problem-solving and input were somewhat one-sided, since the group shared similar views. It will be valuable to bring various groups together so they can discuss the trade-offs involved in various decisions.

Clearly, public participation is not an exact science, and there is a need to experiment with innovative methods of obtaining public input. Thus in the future, while TVA's program will continue in much the same way as it does now, it will also evolve as the issues of concern and affected publics change with time.

9

Public Participation in Routing Transmission Lines: A Program Born of Adversity

Dan McConnon

If Earth Day, 1970, symbolized the emergence of an environmental revolution in this country, it also marked the beginning of a most difficult period for power companies planning major expansion projects. Public opposition to these projects has become a fact of life, as illustrated by such cases as the Seabrook Nuclear Plant in New Hampshire, the Tellico Dam Project in Tennessee, the Diablo Canyon Nuclear Plant in California, and the Kaiparowits Project in Utah. Nor have high voltage transmission lines escaped political resistance. Consider, for example, the controversies that not long ago surrounded 765 kilovolt (kV) transmission lines in New York and Ohio, or the 500 kV lines proposed by the Bonneville Power Administration in the Pacific Northwest. But perhaps the most dramatic illustration is to be found in the Coal Creek Project in the upper midwest. Almost from the time the project was first announced, a section of associated transmission line became the subject of bitter dispute in Minnesota, ultimately resulting in violence and sabotage.

What is it, fundamentally, that sparks such rancorous contention? Perhaps it is concern for the environment, or a distrust of political and legal institutions, or even disregard for established norms of law and order. But what may be a common element in all power facility controversies is a renewed desire for self-determination that is part of a larger trend in the United States -- evident in many

---

Dan McConnon is the Manager of the Environmental and Lands Division of United Power Association, Elk River, Minnesota.

national polls during the 70s -- toward increased preoccupation with self. During that so-called "me" decade, this self-awareness on the part of individuals was manifest in demands to have a say about their lifestyle and future, to control their destiny to any extent possible, and to determine which events will personally affect them either directly or indirectly.

Whatever the origins of adverse public reaction to electrical facilities, in the case of the Coal Creek Project it is clear that a lack of opportunity for meaningful public input was an important source of problems. This recognition has prompted United Power Association (UPA), one of the Project owners, to change its relationship with the public at large and, more specifically, with those property owners whose land may be crossed by power lines. The lessons learned from the Coal Creek Project have prompted UPA to make allowances in the planning and siting of transmission facilities for full and complete public and landowner input, at a point in the process where input is meaningful.

The purpose of this paper is to examine the major issues raised during the Coal Creek Project's history, the causes underlying the controversy, and the lessons learned which may have application for projects planned in the future. A public participation program adopted by UPA as a result of its experiences with the Coal Creek Project is described in detail, particularly as it interacts with the state siting process. Finally, a successful high voltage transmission line project in which UPA utilized its public participation program is described and evaluated.

HISTORICAL OVERVIEW OF THE COAL CREEK PROJECT

As shown in Figure 9.1, the Coal Creek Project as finally designed and constructed consists of several major components. These include a 1,015 megawatt (net) two-unit power plant near Underwood, North Dakota (the Coal Creek Station); a lignite coal mine adjacent to the plant; a 436-mile long 400 kilovolt (kV) direct current (dc) power line extending from the generating station to just west of Minneapolis, Minnesota; two 345 kV and one 115 kV alternating current (ac) lines which deliver the power from the dc line into the existing transmission system; and miscellaneous converter and substation facilities. At the present time, the power plant, the mine, the dc transmission line, the converter stations, and the two 345 kV ac lines are fully

Figure 9.1 Major features of the Coal Creek Project

operational. The only segment of the Project left to complete is the 115 kV ac line from the Minnesota terminus of the dc line to a substation near Delano, Minnesota. Although the project consists of several distinct elements, it is only a small portion of the dc line in Minnesota that has created controversy and is thus of interest here.

The project is owned and operated by two Minnesota-based generation and transmission cooperatives -- Cooperative Power Association (CPA) of Eden Prairie, Minnesota and UPA of Elk River, Minnesota. CPA serves as the wholesale power supplier for 18 distribution cooperatives which provide electrical energy to about 138,000 consumers in southwestern and west central Minnesota. Until the Coal Creek Project, CPA had little generating capacity of its own. However, CPA owns and operates the transmission lines and all substations through which it receives power for its member cooperatives. CPA performs all of its maintenance and construction functions under contract with member systems, other utilities or independent contractors. CPA owns 56% of the Coal Creek Project.

UPA serves as the wholesale power supplier for 15 member distribution cooperatives. These member cooperatives provide power to 32 counties and over 183,000 consumers in central and northeastern Minnesota and a small area of northwestern Wisconsin. In addition to the Coal Creek Project, UPA owns and operates 226 megawatts (MW) of base load generation to meet normal demands and 87 MW of combustion turbine generation to meet peak demands. The base load generation consists of 177 MW lignite-fired plant near Stanton, North Dakota, and a 49 MW coal-fired plant at Elk River, Minnesota. UPA owns and operates about 2,700 miles of transmission lines, 182 substations, various maintenance buildings, and dispatching facilities at Elk River. UPA performs all of its own maintenance activities.

The genesis of the Coal Creek Project dates back to mid-1972, when CPA and UPA began discussing the possibility of jointly constructing a major generation complex to supply the power requirements of their rural electric distribution cooperatives, and to satisfy an expected 665 MW power deficit projected for the winter of 1978. Both Cooperatives recognized the need for a new source, since other suppliers who were providing them with power in 1972 could not give assurances power would be available in the late 70s and beyond. Also, the cost of purchasing power, even if it were available, was considerably higher than what self-generating costs were expected to be.

The Cooperatives authorized the necessary feasibility and engineering studies and, following their completion, submitted a loan application to the Rural Electrification Administration (REA) for the necessary funds. Following the preparation of a federal Environmental Impact Statement in February, 1974 in accordance with the National Environmental Policy Act, the REA granted initial loan approval and guarantees in the amount of $537 million. By July, 1974, land had been acquired in Minnesota for the converter station, grading had begun, and survey work and planning had progressed substantially. CPA and UPA also began to hold public meetings in each of the counties in Minnesota and North Dakota which would be affected by the dc line, and to acquire zoning permits where they were required.

During the summer and fall of 1974, opposition to the construction of the dc line developed in Minnesota. Some planning commissions and boards opposed a line routed through their counties and recommended the imposition of impossible or illegal conditions to permits. One county attempted to adopt new zoning regulations to exclude transmission lines altogether. Because of the impossible situation the Cooperatives found themselves in, they elected to have the State of Minnesota determine the route of the transmission line according to the 1973 Power Plant Siting Act (as amended), despite the fact that the line was officially exempted from the provisions of the statute.*

This law vested the state Environmental Quality Board with responsibility for locating all large generating plants and high voltage transmission lines (in excess of 200 kV), and also mandated "broad spectrum public participation" in that process. At the time of the Coal Creek Project, the EQB was implementing the Siting Act via a two-stage procedure, which comprised a corridor selection phase followed by route selection within an approved corridor.** At each stage, a citizens evaluation committee was formed as an integral part of the review activities. Such committees met frequently in the area of the proposed facility

---

* See L.B. Hartman and T. Simmons, "Public Issues and Regulatory Change: A Minnesota Experience," in Proceedings of the Second Symposium on Environmental Concerns in Rights of Way Management -- 1979, Electric Power Research Institute Report WS-78-141, Palo Alto, California (March, 1981).

** Corridor refers to a general location up to 30 kilometers in width, while a route is a specific location up to 1 kilometer wide.

in order to study the utility's application and alternative routes. Area residents were encouraged to attend these meetings to learn about the proposed power line and to share their views on it with committee members. Route proposals other than those submitted by the utility may be made by individuals, the committee or state government agencies.

Within this general framework, the licensing process evolved in the following manner:

(1) In April, 1975, the Cooperatives filed an application for a corridor certificate. Following a six-month period -- during which 82 witnesses submitted almost 1870 pages of testimony and 180 exhibits at public hearings, and during which a citizens corridor selection committee met numerous times -- the state issued the Corridor Certificate in October of 1975.
(2) On October 6, 1975, the Cooperatives filed an application for a Certificate of Need for the power line and associated facilities, in accordance with Minnesota statutes. Public hearings were held during December at which oral testimony from 22 witnesses was heard, including 15 persons from the general public. Sixty-five exhibits were received in evidence and the typewritten testimony covered 1151 pages. The Certificate of Need was issued by the state on April 2, 1976.
(3) Following the selection of the corridor by the state siting agency, CPA and UPA applied for the designation of a route for the line. A new series of public hearings were held by the state in March and April of 1976. In addition, a new citizen route evaluation committee was established to study alternative locations for the line, and both draft and final Environmental Impact Statements were prepared by the state on the final preferred route.
(4) After review of all reports and public input, the siting agency issued the route construction permit for the line on June 3, 1976.

In total, the Project was scrutinized in four separate administrative proceedings in Minnesota. Over 30 public information meetings and 50 public hearings were held; over 340 witnesses introduced 500 exhibits and testified for almost 6000 pages of transcript. Over 90 percent of the witnesses were members of the public.

Despite such thoroughgoing review, unfortunately, many people in west central Minnesota felt that state regulatory proceedings were "fixed", too complex for the layman and entirely for the benefit of attorneys, and provided little meaningful public input. Because of the perceived lack of public input, many strongly felt a total loss of control over the Project. An incident involving an incomplete transcript and additional hearings to correct the record added to the Project's problems during the siting process. Further complicating the situation was the fact that many people perceived that the Cooperatives were "taking" farmland without adequate compensation. This feeling persisted despite the results of the condemnation proceedings wherein the average award in Minnesota totaled $52,400 per mile for a strip of land 160 feet wide.

THE OPPOSITION AND THE ISSUES

One can infer from the above chronology that the Coal Creek transmission plan met with acute public resistance while in the regulatory arenas of Minnesota. What's more, following the issuance of the certificates and permits for the dc line, there were nine major lawsuits filed challenging state agency decisions. On March 1, 1977, the Minnesota Supreme Court consolidated the lawsuits and appointed a special panel of three District Court judges to hear them. On July 15, 1977, the special three-judge panel unanimously upheld the decisions of the state agencies. On September 30, 1977, the Minnesota Supreme Court upheld the decision of the special three-judge panel and, thus, the actions of the state agencies.

Utility projects elsewhere have had like experiences, of course, but what was unusual about this case was the extent to which opposition continued as the project moved into the construction phase. Beginning in October of 1977 and for a full year thereafter, CPA/UPA had to deal with difficulties imposed not only by Minnesota's harsh winter weather, but also by the crowds who engaged in physical abuse of workers and tried to physically interfere with construction activities. Local law enforcement officials were unable to maintain order and turned to the Governor's office for help. The Governor ordered the Minnesota State Patrol to assist local law enforcement officials where necessary. At times, over 200 members of the State Patrol were on duty in construction areas.

On several occasions violence broke out, resulting in several arrests. On one occasion, the State Patrol had to use mace to control the protestors. The next day pro-
essary. At times, over 200 members of the State Patrol
testers sprayed liquid ammonia fertilizer on the patrolmen in retaliation. Violence culminated one night when a private security guard and a sheriff's deputy were shot at by several men at a storage yard. The men were apparently intent on destroying equipment at the yard when they were surprised by the guard and deputy. Shots fired by the men missed the guard and deputy; however, one was wounded by flying debris and glass from their vehicle. Several arrests were made and one conviction resulted from the incident.

Despite the protestors' efforts, the Minnesota portion of the line was completed in October, 1978, almost one year to the day from the start of construction. The North Dakota segment of the line had been completed in 1977 with no significant problems on a normal construction schedule. The overall line was initially energized on October 17, 1978 followed by intermittent operation for testing of the converter facilities over a period of approximately six months. Continuous operation then began in May of 1979 and commercial operation began on August 1, 1979 following completion of Unit 1 of the Coal Creek Station.

Vandalism and sabotage during construction had been extensive. Heavy equipment belonging to the contractor was wrecked and line materials such as insulators and conductors were destroyed in storage yards. Before the line was completed, a lattice-steel tower was pulled down and destroyed by protestors. Even during operation, vandalism and sabotage continued to the point where a total of 16 lattice-steel towers were destroyed at a replacement cost of approximately $1.5 million. In addition, about 10,000 insulators have been destroyed for a total cost of $650,000. Security costs both during construction and operation have amounted to about $5 million, exclusive of those costs incurred by local, state and federal law enforcement authorities.

Against this background of severe public discontent, it is instructive at this point to turn attention to the substantive concerns raised by the citizenry during the project history. These fall into four basic categories:

* need for the project
* location
* land use
* health effects

Each is described briefly below.

As to the need for the project, those who opposed the Coal Creek dc line alleged the electricity to be delivered was for use in the twin cities of Minneapolis and St. Paul, rather than to satisfy rural demand. Opponents wanted the Cooperatives to aggressively pursue conservation to control electrical growth rather than allow it to continue unchecked. Opponents questioned whether alternatives to the proposed Project had been adequately considered during the planning phases. Such an alternative as building a generating plant within the Cooperatives' service area was more appealing to them than a mine-mouth plant with a long transmission line. Finally, opponents alleged the cost of power from the Project would be excessive and needlessly high.

Several questions were also raised with respect to the location of the dc transmission line. Extensive debate was held during the hearings and at public meetings concerning the advantages and disadvantages of property line routing versus diagonal routing. Opponents questioned why "prime" farmland seemed to be preferred to the exclusion of what they termed "wildlife land." Opponents questioned why more extensive use of existing rights of way could not be made. Further, a dispute arose about the location where the line crossed the border between the states of Minnesota and North Dakota. Opponents felt there was insufficient coordination between the states regarding the best border crossing point. Finally, opponents were concerned about the impact the presence of the transmission line would have on the value of the land it crossed.

Many farmer-landowners potentially impacted by the line raised further questions about the inconvenience the line's lattice-steel towers might create for their farming operations. Also questioned was the potential interference by the line with existing and potential irrigation in west central Minnesota. Landowners were also concerned because they felt the line would preclude the use of fixed wing aircraft for various agricultural operations. Further, landowners were concerned because they thought the electric field created by the transmission line might interfere with electronic monitoring equipment such as seeders used near the line.* Concern was expressed over the aesthetics of lattice-steel structures not being consistent with the landscape. Finally, it was suggested by several landowners that the transmission line be constructed underground to reduce environmental and aesthetic degradation.

Most problems in the above areas were ultimately resolved by regulatory or legislative changes or by acceptance of the line by residents who came to recognize that concerns originally expressed were not so severe as first thought. Not all of the issues raised by protestors have faded since operation began, however. One in particular remains to plague the Project and keep the controversy alive -- that of health effects. Concern in this area was initially raised in connection with ozone that might be generated by the transmission line. The Cooperatives sponsored an extensive study by the University of Minnesota which showed the transmission line contributes insignificant amounts of ozone to ambient levels.** Concerns were also raised over the biological effects of the electric fields created by the line, in light of various studies and reports dealing with ac transmission lines. As the Project evolved, attention was given to dc electric fields and, more recently, health effects alleged to derive from air ions in the vicinity of the transmission line. Landowner concerns have been trying to convince the state to revoke the construction permit because of such perceived adverse health effects. The state has taken several steps to investigate the health effects claim, including a study by a panel of independent scientific experts. Based on the recommendations of the science panel and other available information, the state concluded in December, 1982 that there is no significant risk of short or long term adverse health effects from the line to people or animals, and that the line's permits should not be modified or revoked.***

---

* R.J. Gustafson, V.D. Albertson, L.L. Kinney, Study of Operation of Electronic Monitoring Equipment Used on Mobile Agricultural Equipment in an Electrical Powerline Environment, University of Minnesota (September, 1979).

** S. Mains, "The Minnesota Power-Line Wars," IEEE Spectrum, vol. 20, no. 7 (July, 1983).

*** S.V. Krupa, G.C. Pratt, UPA/CPA High Voltage Transmission Line Potential Generation of Air Pollutants and Their Impact on Vegetation, University of Minnesota (January, 1982).

THE CONTROVERSY IN RETROSPECT

Anyone connected with the Coal Creek Project has undoubtedly come away with impressions as to what caused problems for CPA/UPA and what lessons should be learned from the experience. From this author's standpoint, there are several factors to bear in mind in interpreting the case and arriving at conclusions.

To begin with, it is important to realize that the Project was proposed at a time when Watergate and other external events had generally caused a public mistrust of government and big business. This country was in the midst of a social revolution in which many established practices and ways of doing business were seriously being questioned. In addition the Project coincided with a period in our history in which consumerism and public demands for a risk-free environment were very evident. The time was right for major energy projects to be challenged.

This was especially true in Minnesota, since another utility had proposed large transmission facilities and a major pipeline and interstate freeway had been completed in the general area just prior to the time the Coal Creek Project was first announced. These projects sensitized people and led to the rapid formation of opposition groups. Those who opposed the Project, moreover, were able to find strong leaders to take control. Although the leadership of the opposition groups changed often during the course of the Project, a new leader always stepped forward to fill any void. Leaders from outside the project area who had been engaged in various protest movements in the 60s were attracted to the controversy, often playing key leadership roles.

Another aspect of "the times" that helped make the Project so long and controversial was that people were demanding more of a say in the conduct of public affairs and in the direction taken by their own lives. This was particularly evident among the strong-willed people of several ethnic backgrounds who live in west central Minnesota. They generally felt a mistrust of "city people," especially those from the Minneapolis and St. Paul area where both Cooperatives are located. Farmers felt that this Project was one of several outside influences which were slowly deteriorating their way of life. Landowners who were closely tied to the land felt the power line an unnecessary intrusion on the land.

The last external major cause for the Project's problems was associated with the legal and regulatory framework in which the Project was conceived and planned. The Power Plant Siting Law was passed in 1973 and the Project was the first to be routed under it, and deficiencies in the law and regulations soon surfaced. Among the law's significant problems was its provision for "broad spectrum citizen participation." Many who opposed the Project misinterpreted this provision to mean "Do as I Say" and could not differentiate between participating in a routing decision and making the decision, which was the sole responsibility of the state. A similar problem had to do with the fact that the siting process assumed that no project could be stopped or significantly modified. The only matter to be decided was where the line should be located, not "whether" it should be built. This situation was misunderstood by the public and created further animosity toward the Project.

There is little doubt that both CPA and UPA were responsible for several of the problems experienced by the Project, as well. Initially, the Cooperatives exhibited a very inflexible position with respect to the location of the transmission line in farmland. In addition, they did not seek to communicate in any meaningful way and solicit input from the local units of government and the general public. Poor public relations stemmed from the fact that both Cooperatives were inadequately staffed to handle a project of this size. Representatives went in the field with inadequate training and background. Because of lack of appropriate staff and preparation, the Cooperatives made what some might consider to be an inappropriate use of consultants to speak on behalf of the Cooperatives at meetings and forums when Cooperative representatives should have been speaking.

The Cooperatives also misjudged the public's mood, taking the position that the opposition movement was a transient phenomenon which would eventually fade. Thus CPA and UPA did very little to challenge the perception by those directly affected that the line constituted an unacceptable intrusion on an existing way of life, which turned out to be the underlying basis for initial opposition to the Project.* While the effects of the line were specific to landowners in west central Minnesota, its benefits were general and not perceived to apply directly to these individuals. Consequently, the Cooperatives contrib-

---

* B.M. Casper and P.D. Wellstone, Powerline: The First Battle of America's Energy War, University of Massachusetts Press, Amherst, Massachusetts (1981).

uted to the Project's problems by rejecting contact with landowners in favor of complete reliance on the state's regulatory processes for routing the lines. As the Project progressed, outside dissidents took advantage of this situation and kept the controversy going, not because of any impact the Project had on rural Minnesota and Minnesotans but because, as social revolutionaries, they felt the Project represented the worst aspects of central versus decentralized electric power generation and big business in a capitalistic society, and therefore could be used to further other purposes.

At this point it is interesting, for additional perspective on the case, to contrast the Minnesota experience with that in North Dakota, where roughly two-thirds of the line is located. As previously noted, no major political problems materialized there; indeed, better than ninety-eight percent of the necessary right-of-way was acquired through voluntary negotiation with landowners. No significant acts of vandalism occurred in North Dakota, and no protests occurred during construction of the transmission line. There are several explanations for this lack of controversy, but the three main areas of difference relate to the people dealt with, law enforcement attitudes, the regulatory climate in North Dakota, and changes in the way the Cooperatives handled matters.

With respect to North Dakota landowners, they are more "cooperative-minded" simply because they are tied to a cooperative in one form or another. Further, North Dakota is an energy exporting state where residents are more accustomed to large energy facilities such as generating plants near the lignite fields, large hydroelectric dams, and long high-voltage transmission lines. They have come to accept such facilities as necessary for the economy of their state and the good of the region. One act of vandalism occurred in North Dakota which was promptly dealt with by local law enforcement officials and the judicial system. In Minnesota, on the other hand, law officers were hesitant, at times observing acts of vandalism, but taking no firm action, thereby almost inviting further destructive behavior.

North Dakota also has a power plant siting law under which the dc transmission line was routed. However, the laws and regulations are significantly different from those in Minnesota. The North Dakota process is concerned only with the acceptability of a route or routes suggested by the utilities or others, i.e., whether or not a proposed route meets pre-established criteria. In Minnesota, the process is more concerned with finding the "best" route, a concept which is desirable in theory but difficult to achieve in practice. As opposed to the Minnesota situation, the North Dakota siting agency is generally pro-industry and reasonably cost-conscious. This is because the same agency has the responsibility not only for locating large energy facilities but also for determining an appropriate rate of return to utilities for these facilities. The agency must be concerned about costs both to the utilities and to consumers.

Finally, because the North Dakota process occurred after the Minnesota process, both CPA and UPA had the opportunity of applying several lessons learned in Minnesota to the North Dakota process. For example, the Cooperatives were much more flexible in their dealings with the regulatory agency and landowners, and tried to work more closely with all parties.

Now, what conclusion can be derived from the overall experience with the Coal Creek Project? To this author, the following series of statements provides a good summary of the lessons learned:

* With respect to project need:

  - There must be a demonstrated need for the facility and the public and affected landowners must understand how it benefits them.
  - The question of central versus decentralized generation and alternative technologies must be addressed.
  - Also to be addressed are the use of conservation and load management.

* With respect to project design:

  - Expert analysis alone is insufficient since the problem is not solely a technical one.
  - Energy projects should have human and environmental design as well as technical and economic design.
  - Technical and economic factors are not absolutes but should be modified to accommodate human and environmental concerns. Project flexibility should be maintained as long as possible.
  - Agricultural and rural concerns need more recognition. Circumstances unique to a project area

require careful evaluation in the planning phases of a project.

* With respect to the regulatory process:

  - The regulatory process under which a project is routed needs to be streamlined and predictable.
  - A regulatory review process is more productive if it is accommodating rather than adversarial in nature.

* Finally, with respect to landowners affected and the general public:

  - Most members of the interested and affected public will cooperate if they believe they are being treated fairly. This requires earlier and more meaningful input to energy decisions and an open and straightforward information program. Substantial local public acceptance is essential for a successful project.
  - Trust and credibility are essential and must be built by the utility with the public and affected landowners.
  - Public input must be able to influence important choices.
  - Utility representatives should leave no questions unanswered.
  - Local authority over a project should be maintained to the maximum extent possible.

UPA's CURRENT PROGRAM OF PUBLIC PARTICIPATION

Because it was obvious that several problems experienced with the Coal Creek Project could have been significantly reduced or eliminated by establishing closer working relationships with involved landowners and the public, UPA adopted a formal program of public participation in early 1978. The program is designed not only to involve the public early in the planning stages of a new project, but also to interface with the state's siting process when necessary.

As outlined in Figure 9.2, UPA's program is divided into three basic stages. Each stage is further divided into a number of segments or phases. The pre-application

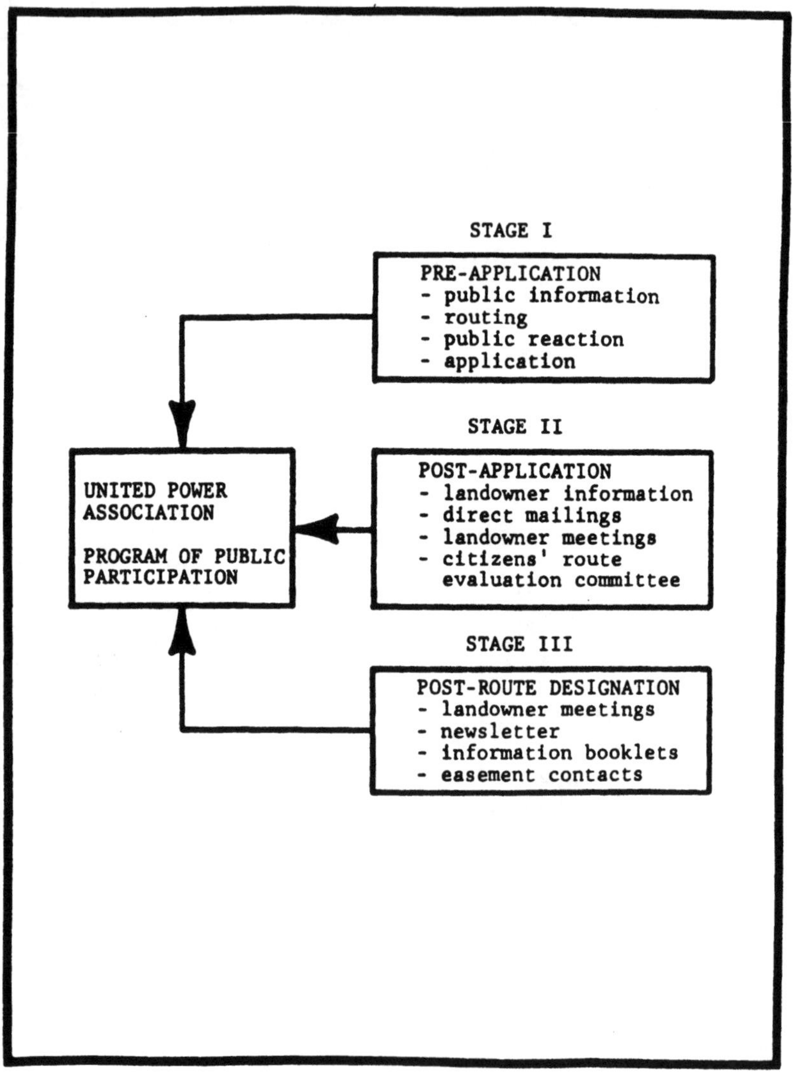

Figure 9.2 UPA program of public participation

stage consists of the public information, routing, public reaction, and application phases. The post-application stage consists of landowner information, and citizen committee phases. And the final or post-route designation stage consists of landowner contact and easement acquisition phases.

The public information phase of stage I has a number of purposes. It is intended to make the project known, to gather the concerns and preferences of those members of the public who will be directly affected, and to disseminate information on transmission lines in general. In this phase, no predetermined routes have been selected. A study area is defined and inventoried for natural, social, cultural, and agricultural features. Meetings are held with local units of government to inform them of the project. An attempt is made to meet with all organized units of government in the project area such as planning and zoning commissions, county boards, city councils, township supervisors, regional development commissions, and so on. Direct contact is established with members of the public by such means as a mobile van, school board meetings, farmers' union meetings, and direct mailings to consumers via the distribution cooperatives.

Based on the information gathered in the first phase, and other information such as state routing criteria and UPA internal guidelines, the second phase is completed in which planners put together all possible route and route segment combinations. Input from all departments of UPA is used to arrive at the alternative routes selected for further study. In the third phase, all routes are made public and information is solicited as to which routes or route segments are good and which are bad. Meetings are again held with every group that was met with in phase one. These meetings serve to keep the attention of the public and to provide them accurate information as well as providing a means of meaningful public participation and input. In the last phase of stage I, routes are selected, final information is gathered, and the application is prepared and submitted to the state siting agency.

Once the application is submitted and the formal state process is underway, the post-application stage begins. Here, the steps taken are dictated by the state siting agency, whose review process under the Power Plant Siting Act of 1973 is currently structured in the manner shown in

Figure 9.3.* First, the public is notified of the application through advertising in local papers and notices mailed to local government officials. The agency staff then holds information meetings in the affected area to explain the proposal and the routing process. In addition, a citizen route evaluation committee is appointed by the siting agency.

Formal public hearings are conducted by an independent hearing examiner. Prior to these hearings, the siting agency holds a second set of informational meetings in each county in which a route is proposed. At the meetings, the staff explains all proposed routes which will be discussed at hearings, gives background information, answers questions, and explains the hearing process. At the conclusion of the hearings, a transcript is produced and the examiner forwards his advice to the siting agency, in the form of findings of fact, conclusions, and recommendations. These documents, along with all exhibits from the hearings, an environmental impact document, and the citizen route evaluation committee's report, are reviewed by the agency before it makes its final decision with respect to the location of the proposed power line.

As all this is taking place, UPA's outreach efforts consist for the most part of keeping all those who may be directly affected informed on the progress of the state process. A mailing list is established of all those who own property along any route under consideration. Direct mailings of information about transmission lines and upcoming events associated with the state process are made. Affected landowners are invited to meetings about the project. In addition, Cooperative representatives work very closely with the state's citizen route evaluation committee to provide them appropriate information and guidance.

Stage III begins when the state siting agency designates the route for the transmission line and issues the construction permit. During this stage, a series of group meetings with landowners who will be directly affected by the transmission line are scheduled. These meetings are intended for discussion of structure locations and the exact centerline of the transmission line. Further ques-

---

* Siting procedures under the Act have been modified several times. In 1979, for example, the two-stage process was eliminated in favor of a single, combined proceeding covering the span of one full year.

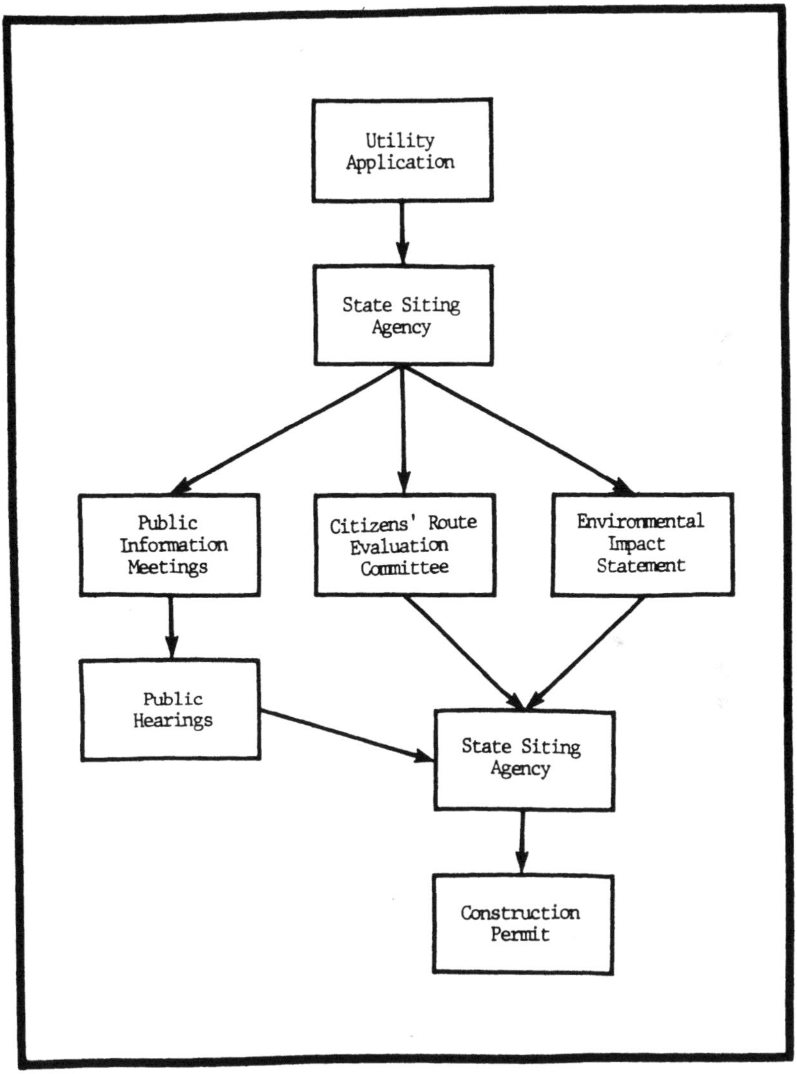

Figure 9.3 Minnesota's route selection process

tions almost always arise which may be answered at this time. Booklets containing information about such topics as easement acquisition, payment schedules, and construction practices are distributed to affected landowners. Stage III closes with landowner contacts for the purpose of negotiating required easements.

For lines under 200 kV, where no state siting process applies, UPA has established a careful planning process to chose line routes. UPA has formed a permanent committee known as the Transmission Facility Siting Committee. the Committee is made up of engineering, environmental, planning, construction and public relations personnel from the Cooperative. In addition, the Committee always includes representatives of the member distribution cooperative from the area of the proposed project. At a series of meetings, the Committee examines the area of the proposed project and identifies several route possibilities for the transmission line. One or more public meetings are then scheduled by the committee in the project area to inform the public about the project, answer landowner questions, encourage public discussion of the project, obtain information about the project area, and request public preferences for the routes identified by the committee or suggested by involved landowners. Sometimes two or three meetings are necessary to arrive at a consensus. However, once a consensus is reached, it is fairly certain that a route has been chosen which is acceptable to UPA from a reliability, cost, and environmental standpoint. The route chosen also has public acceptance because it has been selected by a process which has allowed for true and meaningful public input.

What have been the results of UPA's public participation program? To date, the program has been used on one high-voltage transmission project known as the Benton-Milaca Project, involving the routing of a 230 kV transmission line between two substations about 25 miles apart in central Minnesota. The first stage of the program began in January, 1978, and was completed with submission of an application to the state in October of that year. Hearings were held by the state in the summer of 1979 and the construction permit was issued in February of 1980.

The line is located in the service area of East Central Electric Association (ECEA), one of UPA's member distribution cooperatives. ECEA staff and resources were used extensively for Stage I activities. Major articles about the line were prepared and printed in several editions of the consumer newsletter published monthly by ECEA. Local newspapers along the proposed line route were given information about project meetings and reporters were kept up to date on the proceedings. Local radio stations in the area were kept informed about the progress of the project. UPA and ECEA staff members appeared on both news broadcasts and call-in programs to answer questions about the project. In addition, local civic and fraternal organizations within the transmission line study area were contacted and arrangements were made for speakers to discuss the line and the routing process.

A special committee made up of local individuals representing political, governmental and service organizations as well as farm groups, conservation and environmental groups was selected by UPA and ECEA to discuss the project and provide input. The groups were asked to identify concerns if the line was located on any one of several possible routes. As a result, the list of possible routes was narrowed considerably. UPA also used a recreational van, parked conveniently in major towns in the study area where people could drop in and ask questions during shopping trips. The van schedule was announced in local newspapers and in the ECEA monthly newsletter.

UPA and ECEA employees were also kept informed of the progress of the project. Information was provided for employee meetings explaining the routing process, need for the line, and other details. This was done so that UPA and ECEA employees could answer questions if they were asked. The ECEA Board of Directors also had considerable input into the preliminary planning and information process. Board members attended numerous official meetings during the routing process, and they individually made numerous personal contacts with area landowners. Meetings were held with each city council or town board in the study area to provide background information to local elected officials. As a result of all these activities, UPA had an exceptionally clear picture of where the line would be accepted and where it would not. Concerns brought out at meetings were considered and mitigating action taken to eliminate potential problems.

With this preparation, the application was sent to the state in October of 1978, beginning Stage II of UPA's program and the year-long state routing process in which UPA was heavily involved. The state selected a committee of local citizens to advise on routing issues and to suggest a recommended route. Many of these same people had already been contacted by UPA. It was no surprise, then, that the citizen advisory committee recommended the same route as did UPA in its application. The state held several other public meetings in the area where residents were able to come and express their concerns. At one meeting, a group of individuals from west central Minnesota who had opposed the dc line arrived to explain to local landowners how UPA and the state had treated them during the routing of the dc line. They were soundly rebuked by the local landowners and were told that no outside help was needed. Public meetings and hearings were almost uniformly quiet and orderly. Little came out that had not already been considered by UPA and ECEA.

To alert landowners to these meetings and to keep them updated as the process continued, UPA was able to identify affected landowners by their zip codes through ECEA's computerized billing system. Several letters were sent out during the course of the state routing process. Some of the letters just announced and reminded landowners of meetings that were upcoming. Others, from the general managers of both UPA and ECEA, explained or outlined details of the planned line. By the time the hearings were over and a decision reached by the state, UPA and ECEA had achieved total acceptance of the requested line route. Minor centerline corrections were made as a result of negotiations with individual landowners, but these were inconsequential.

The construction permit was issued by the state on February 28, 1980 and UPA easement agents began Stage III, the process of contacting landowners and negotiating easements. Of the 136 parcels needed for the line, 93.4 percent were acquired voluntarily. In the remaining parcels, the only issue unresolved was the amount of payment for the easement. All proceedings were amicable. The line was constructed during the summer and fall of 1981, with no unusual occurrences, and energized on December 23, 1981. Affected landowners and others in the area were kept informed of the progress of easement acquisition and construction of the line by way of individual letters and articles in ECEA's newsletter.

Our experiences with transmission lines lower than 200 kV have been equally successful. Since 1978, we have constructed about 188 miles of 69 and 115 kV transmission lines, for which a total of 641 separate easements were necessary. Of the total number required, UPA was successful in obtaining 623 easements or 97.2 percent by direct negotiation. Only 18 easements had to be acquired via condemnation, and the only issue in the majority of those cases was the amount of the easement payment.

CONCLUDING REMARKS

It is obvious that the public has become disenchanted with institutions over the past several years. Since utilities are major institutions, they have not been able to escape public criticism. Part of the solution to this problem involves providing the public an opportunity to exercise an influence over routing decisions. Concessions must be made at the front end of the routing process to avoid problems or conflicts at later stages of a project. UPA therefore makes an effort to approach the public early in the routing process.

Such a program of public participation as UPA's is not without potential risks or problems. More time is necessary at the front end of a project to develop the program and allow for public input, and this may present difficulties for the utility. Because more front-end time and information about a project is available, the opportunity is greater for opposition forces to develop and organize. Our experience has indicated, however, that if a project is sound, if it is adequately communicated to the public, and if the public is allowed to have meaningful input about project details, opposition will be inconsequential.

The benefits of such early and continuous contact with the public are many. Projects are more likely to be accepted by the public and local landowners, and right-of-way should be acquired with a fewer number of condemnations. Since landowners are aware of the project, right-of-way agents should be able to acquire more easements on the first contact and spend less time explaining a project, more than offsetting the time and expense of the front-end public participation program. Public perception of the utility should be heightened, thereby improving public and consumer relations. And finally, the utility can take opportunities presented by direct public contact to present accurate facts to the public, thereby reducing rumors, false information, and speculation. The program is working well for UPA and may have application in the operations of other utilities.

10

The Value-Sensitive Approach to Facility Siting: Two Examples

B. Fritts Golden and Patrick Kennedy

The siting of any major facility may become the subject of strident public debate, reflecting conflict among the differing values of society that would be affected by development. Unfortunately, the role of government in dealing with such intergroup conflict is not always clear. Ideally, public officials should have a beneficent intent and seek equitable solutions to problems, including suitable compromises to avoid intolerable loss. But in fact, government is far from being monolithic or neutral. Operating at its various levels and through a host of agencies and bureaus, it is often divided on issues and thus joins interest groups and the facility applicant in debating a proposed siting.

The number of actors involved and the variety of perspectives they bring to a siting decision create an uneven, fitful process, but this is not worrisome in its own right. To the contrary, open debate is a sign of a vigorous society, one that is able to shape and reshape itself through a clear articulation of values. An awareness of issues, usually based on values that strongly motivate people, is what gives direction and substance to the debate. Fears are voiced and responded to, aspirations are put forth, and needs are expressed. Controversy and conflict, in other words, are axiomatic in a pluralistic, democratic society.

B. Fritts Golden is a principal and Patrick Kennedy is a project manager in the consulting firm of Rogers, Golden & Halpern, Philadelphia, PA.

This is not to say, of course, that conflict is inherently constructive. Rather, steps must be taken to ensure that divergent views are expressed in a useful way and that their role in decision-making is clearly defined. This task logically falls to public and private planners who, in order to be effective in helping decision-makers arrive at reasoned conclusions in an expeditious manner, must be able to anticipate conflict and accommodate value differences throughout all phases of planning, including the earliest ones. This can be done both explicitly, through formal mechanisms, and implicitly, by the modes of planning and the siting strategies adopted.

Since 1975, the Philadelphia-based firm of Rogers, Golden & Halpern (RGH) has worked for four major clients to develop and refine an approach to facility siting that deals explicitly with value issues and is specifically geared toward conflict resolution. The goal has been to make available a methodology that simultaneously accounts for technical requirements of a proposed facility (engineering, economic, environmental, etc.) as well as for the concerns of people whose interests are potentially affected by development of the project. These concerns typically center on both socioeconomic and environmental effects. Like other siting techniques, the RGH approach involves a sequential narrowing of alternatives, beginning with the elimination of obviously unsatisfactory areas and then moving through a systematic evaluation of candidate regions, leading ultimately to the identification of candidate sites. What distinguishes the RGH method, however, is the manner in which preferences among both technical and non-technical criteria are established and applied, which is really the cornerstone of the siting process.

More specifically, RGH believes that in many cases the public can be expeditiously and usefully involved in helping to determine preferences among siting criteria. Despite what conventional wisdom may hold, citizen participation in site selection is not an invitation to disaster. Instead, a thoughtful interaction program serves to inform and educate the public, make use of its knowledge, and reveal concerns and issues that must be dealt with by the developer. Both the technical and sociopolitical aspects of siting can benefit from public involvement, although the more specialized tasks (e.g., data development) are conducted by a professional staff. Involvement often takes the form of an ad hoc committee established to review and comment on the technical work, help modify non-technical criteria, and in general work to determine what preferences and values should be incorporated into the decision.

It should be noted that genuine open planning, in which outside individuals participate directly in the search for a new plant location or transmission route, is practical only where the facility developer does not have to enter the marketplace to buy the site or right-of-way. In many states, electric utilities are fortunate that they or a public agency can exercise eminent domain to secure sites, but where a company has no alternative but to acquire property in a private transaction it may be foolhardy not to maintain business secrecy. Under these circumstances, the relevant siting criteria for use in the RGH methodology must be deduced from public policy statements, regulations, staff knowledge of the region, anonymous surveys, and confidential discussions with informed individuals and officials. Such has been the case in two of the four instances in which the methodology has been applied to date -- for the American Petroleum Institute (1978-1979), and for Detroit Edison Company (1981-1982).

Privacy requirements did not preclude early revelation of plans for the other two clients to date, however, both of whom were government agencies. In the more recent (1982) work RGH has developed a generic methodology for siting hydrocarbon facilities on the Canadian Arctic coast, under the sponsorship of Environment Canada. This methodology envisions full use of open planning, since the Federal and territorial governments are in a position to grant sites to industry in most of the Arctic. The same holds true, in essence, for the initial (1977-1978) RGH client in this series -- the State of Maryland. There, the Power Plant Siting Program has been authorized by the legislature to acquire potential sites for power plants under a unique land-banking statute. There is no obligation for a utility to use state-owned sites, although there is a presumption that agency approvals for designated projects at such sites would be readily forthcoming. At any rate, the situation clearly provides an opportunity for direct citizen participation in the site selection process. The purpose of this chapter is to describe the distinguishing characteristics of the siting methods developed for use in Maryland and Canada, including the role of citizen participation. But first it is useful to explain the conceptual basis of the RGH approach more generally, and how this approach deals with some of the practical questions that arise in siting.

PLANNING IN A VALUE-LADEN CONTEXT

For present purposes, values can be viewed as principles that are intrinsically desirable or of worth to an individual or group. Values are characteristics or conditions that we would not want to see diminished or altered. They influence how people deal with each other and the degree to which they are willing to enter into agreement on a subject or issue. In the case of facility siting, values are most usefully employed when grouped into logical sets that are internally supportive of each other. A set of ecological values, for example, might call for pristine waters, lack of noise, clean air, preservation of wilderness, and recreational access, among others. Similarly, social and economic values can be grouped according to sets. Despite such categorization, however, there are no sharp lines between sets of values. In practice, an individual is likely to subscribe to values found in each of these sets, if not all of them.

A most important distinction to be made at this point is between values and _preferences_, which are simply expressions of a personal or group priority among values. While preferences are constantly operating within each of us and in social and political groups to affect behavior and define goals, they usually do not require a highly developed presentation or vociferous defense. This comes only when they are challenged. Despite what appears to be a high degree of commonality in values in the population as a whole, the ranking of these values, in the face of a decision that will (or is perceived to) jeopardize a group's preferred values, leads to conflicts. Rarely will someone reject another's values, but often one will disagree with that person's choice as to which are the more important values. In some instances, we may even find ourselves discounting some values to bolster those we wish to have prevail. Thus, people who, in a non-threatening situation, would espouse strikingly similar sets of values may, as a result of preferences, find themselves diametrically opposed in a situation where two or more values cannot simultaneously prevail.

Conflict in facility siting, then, must be understood as originating not with divergent values but with differences in the order of preferences among values. Because of this, it is important to help people involved in siting recognize their own values and preferences and those of their opposition, in order to work toward the most satisfactory resolution of the issues. In making a decision to oppose, alter, or encourage a siting decision, the values potentially affected must somehow be ordered and preferences established. This is at the heart of any siting decision. Disregarding political power (which is obviously a significant factor), advocates of a particular decision are effective in the decision process only to the extent they can sharply define their values and preferences and clearly state why they should be given weight.

Given that the basic issue in siting controversies is conflict in value preferences, it is incumbent upon planners to devise or adapt processes that are sensitive to the question of what values are affected and what choices among values must be made. In making such choices, unfortunately, there is rarely a universal truth or an obvious commonweal. Lacking an objective standard, planners must be as systematic, thorough, and dispassionate as possible in developing options and in clarifying tradeoffs among alternative courses of action. More particularly, a decision on where to locate a facility must be predicated upon: 1) a clear and early definition of policies, issues, goals, and approaches; 2) reasoning that is evenhanded and intellectually rigorous; and 3) good documentation. Without these necessary characteristics, any siting methodology will fail to yield results that have credence with the public, industry, or government bodies.

Needless to say, the integrity of any site selection process also depends heavily on the existence of a technically useful and descriptive data base. Values can be employed effectively in siting only where there is information as to existing and expected future conditions. Typically, data may be drawn from existing sources as well as be developed specifically for the project at hand. These data will include a mixture of numerical and non-numerical information that will vary as to units of measure, age, and reliability, and they will be a mix of empirical measurements, estimates, projections, extrapolations from known points, and professional judgements. Amidst all this, a planner must be concerned with questions of both quantity and quality of the data set to be developed in the siting. One characteristic of great importance, for example, is the form data comes in and any limitations on its utility. In facility siting, which is primarily a locational matter, data is usually most valuable in a spatial (i.e., map) format, supported by tables and narrative.

Beyond this, it must be determined if an adequate amount of useful information is available for the tasks at hand. It is likely that existing information, with a modest collection effort, is sufficient for identifying potential sites. However, distinguishing among site options and executing detailed evaluations normally require more data than is to be found in prior studies and files. To keep the task of additional data acquisition within manageable proportion, therefore, planners find they must acknowledge the variation in importance or relevance of various factors. By establishing a critical "decision set" of key or vital factors, which is fairly small, or by ordering all potential factors in terms of significance, planning resources can be allocated efficiently in relation to the issues most in need of careful analysis.

This effort, of course, must rely on some means of establishing which factors are more or less important and what weight they will have in helping reach a decision. This points to a key question: who decides what is or is not significant and which factors will have more weight? Planners could do this in isolation, but it is certainly not advisable without informal contact with the body politic. A government agency, for example, may set forth policies and goals, but our system of regulation still relies on the private sector to interpret and make use of this information. Failure to clearly set forth such policies and goals, or to understand them, can easily result in siting proposals that are unsatisfactory to a given authority. Recognizing this, public agencies and developers increasingly are seeing the value of informal meetings to ensure that the private planning team is aware of governmental concerns. At the same time, the developer can begin to determine what physical, environmental, social, and economic factors will affect or be affected by the planned facility.

To apply this principle of consultation more broadly, many siting methods use what are known as 'nominal groups' or a 'Delphi' approach in choosing which factors are important and in determining their relative rank. Since the composition of the group used will obviously affect the results, one can have representatives of different value preferences act as a structured committee to arrive at a consensus. In doing this, it is important to focus attention on general concerns and specific tradeoffs, and to try to keep people from anticipating the final consequences of their intermediate choices. One approach is to have the group members abide by a rigid format for debating and voting on choices and to deal only with abstract values or factors, independent of a particular geographic site. Another party (the planner) takes the conclusions of the group and uses them to actually identify sites. This, of course, is only possible when seeking to identify potential or candidate sites, not when a specific site is presented for approval or disapproval. But if there is no guarantee that the final site selection will be free of controversy, it is clear that the process of elimination that precedes it will be responsive to the full range of concerns.

MINIMIZING CONFLICT IN ENERGY FACILITY SITING: THE MARYLAND CASE

In 1977-1978, RGH performed a multifaceted study for the Energy and Coastal Zone Administration within the Maryland Department of Natural Resources. At the core of this study was development of a method for screening regional data in order to identify coastal zone areas that were most suitable for major energy development, chiefly from a physical and environmental standpoint but with some attention to social and fiscal parameters as well. The screening was done for each of fifteen facility types designated by the client agency, ranging from pipeline landfalls to nuclear power plants. In the singular case of power plants, moreover, the process was taken to the point of identifying and evaluating actual sites. Eventually, four specific locations suitable for either coal- or nuclear-fueled units were identified.

From the outset, this methodology and its subsequent use in identifying candidate areas was geared to minimize conflicts. The viewpoints of the industrial and public sectors were translated into criteria (i.e., levels of acceptability applied to factors) which were then used, in combination, to determine the potential development areas.* Essentially, the screening process first set forth a par-

---

* The public concern criteria, which in nearly all cases applies to geographic data, fell into two groups. The first was labelled Exclusion Criteria because specific areas -- military installations, various other federal lands, and land already developed -- were deemed a priori not to warrant consideration. Potential Restriction Criteria, on the other hand, were associated with conditions that might restrict development but which could not be used to exclude a facility at an early phase. Wetlands, prime agricultural land, natural areas, flood hazard areas and the like were areas subject to this type of criteria.

ticular industry's requirements, then the study area (Maryland's sixteen coastal counties) was examined to find locations where these requirements could be met without violating a separate set of criteria based on public concerns. The study was fully documented in that all data sources, all criteria, and all intermediate decisions were included in the reporting process to the agency.

The Maryland planning effort was also an open one in that RGH met often with a broad-based steering committee, to obtain guidance in the year-long process of developing the methodology, determining which factors to collect data on, and establishing the criteria to be applied to each factor. This committee included representatives of the state (14 people), of local governing and planning bodies (5 elected officials, 8 administrative officials), of citizen groups (5 people), and of industry (6 people). The meetings of this committee gave people with a common interest in the region's future an opportunity to exchange ideas and to understand the basis of other opinions and viewpoints. An unanticipated benefit of this vehicle was that it led to a greater sensitivity on the part of steering committee members to the difficulties of arriving at equitable solutions, given legitimate competing values.

Beyond this, two aspects of committee operation are particularly worthy of note. First, in determining what were acceptable and unacceptable conditions relative to any one siting factor, a consensus approach was used rather than a formally structured process. This was practical because with no imminent siting decision to be made, it was possible to divorce the consequences of applying various criteria from the abstract discussions of the factors and criteria themselves. Second, the dissimilarities of experience and knowledge among those involved led us to abandon soliciting well-defined criteria. Rather, the usual procedure was for RGH to elicit general concerns and them frame them into specific criteria for discussion and approval at a subsequent meeting.

Once the factors for data collection were identified and the criteria established, we applied the criteria to the data set and commenced the screening process for all facilities. A series of steps were designed to systematically reduce the area to be examined, until only areas that could be considered suitable remained. The first screening phase required establishing thresholds outside which a particular facility was unlikely to be sited. Such thresholds were relatively few and were predicated on minimal requirements of the facility, e.g., minimum distance to sufficient cooling water. If no suitable sites were found after finishing the entire screening, it was possible to relax thresholds and begin again, since thresholds were usually based on cost considerations and did not reflect absolute limits in a technological sense. The next key phase involved the identification of superior conditions within the thresholds. A coincidence of superior conditions for several factors identified the most suitable or optimal candidate areas from a developer's standpoint.

Of course, the facility criteria that qualified an area for potential siting did not preclude the possibility of great public concerns in the same area. Also, the potential clearly existed for the same area to be identified for many types of facilities, some of them quite incompatible with others. All this reflects the fact that conflict is nearly automatic when attempting to site large industrial facilities. Indeed, it is what fuels much of a planner's efforts. Recognizing this, the planner must begin using techniques that serve to avoid the conflict or, more likely, to focus attention on the real issues. Thus, in the Maryland work, the need to deal with potential conflicts served as an umbrella concept under which the RGH regional screening process was developed.

The planning team concentrated on three basic sources from which potential conflict could arise:

* the facility, which was assumed to be either a new development proposal or an expansion of an existing facility;
* the site, which was assumed to include the environs of a property and existing neighboring facilities, as well as the actual construction area; and
* public policy, which involved more complex assumptions because policy can either present a direct conflict with the other categories separately or indirectly work to produce conflict among them.

Using these definitions, potential conflicts were identified in terms of distinct pairs: Facility-Facility, Facility-Site, Facility-Policy and Policy-Policy. Thus,

* Facility-Facility conflicts occur when two or more facilities seek to locate on the same site, or so close to each other that a question arises whether the area can support both of them;
* Facility-Site conflicts occur when the resource requirements or environmental effects of a facility

cannot be accommodated by the site on which the facility seeks to locate;

* Facility-Policy conflicts occur when the location of a facility on a particular site would contravene a state or local policy;
* Policy-Policy conflicts occur where two or more policies, or their application in particular circumstances, are at cross-purposes.

Because policies often determine whether a facility and a potential site are in conflict, considerable overlap is possible between Facility-Site and Facility-Policy conflicts. For our purposes we found it useful to define a potentially mitigable conflict as a Facility-Site conflict. An example would be a potential erosion problem. Instances where the opportunity for mitigation is low -- as when a facility would seek to locate in a highly valued natural area -- were viewed as Facility-Policy conflicts.

In the course of the regional screening in Maryland, several potential conflicts were identified in the Facility-Facility and Facility-Policy areas, and a Potential Conflict Map was generated. Facility-Facility conflicts were derived by overlaying the most suitable candidate areas for each facility with those of every other facility, to determine the location of coincident optimal areas. With fifteen major facility types there were a total of 105 pair-wise combinations. Similarly, a Facility-Policy conflict was noted whenever an otherwise optimal candidate area was coincident with a resource of concern to the state (i.e., a potential restriction). Regarding industrial parks, for example, there were four otherwise optimal areas subject to potential restriction and therefore prevented from being designated most suitable. The map thus presented the state with a picture of where clear choices between facility location and the protection of a resource would potentially have to be made.

In the Maryland case, the process of compiling and mapping policy-affected areas had several potentially beneficial results:

* agencies with overlapping interests can establish a common ground in terms of parallel policy considerations and strategies;
* the relative effects certain policies may have on land uses throughout the region become more apparent when mapped;

* redundant, overlapping, and conflicting policies that have a spatial aspect can be identified;
* since many policy concerns have a geographic coincidence, there is a basis for beginning to establish planning priorities among areas of greatest concern;
* the state can give developers clear guidance on what issues they will especially have to address in different areas.

A further advantage to be gained by the screening process is the identification of alternative locations, always important in a conflict situation. While this may not be especially helpful when dealing with a private developer constrained by and operating within the market system, it can be most useful for public and quasi-public siting efforts.

The foregoing recitation of benefits is not meant to suggest, by the way, that the regional screening work in Maryland was without limitation. In particular, one should realize that attention was focused on developing a mechanism which, by definition, had to be geographically based. Physical phenomena are readily employed in such a system, but non-spatial or less well-defined concerns (e.g., employment, urban revitalization, energy efficiency, and so forth) are difficult or impossible to map. What's more, in the absence of a particular demand, and with ever-changing technologies, the results are constantly in need of updating. Still, it should be noted that the screening process employed is a flexible device; rather than dictating a "facility zoning" map per se, it indicates suitable areas using certain assumptions. Despite its limitations, therefore, the results of the methodology can be used to anticipate potential conflicts and help shape overall planning strategies. By coupling a clear articulation of policies -- in maps and narrative -- with a definite procedure for reviewing a proposal, the planner fulfills his or her obligation to be thorough and evenhanded. In doing this, everyone (including the planner) can readily identify trouble spots requiring more attention.

In this regard, it is also important to bear in mind that in practice, fortunately, most theoretical conflicts fail to materialize. In Maryland, as elsewhere, the state follows the general policy of 'first come, first served,' such that most of the potential conflicts that could be mapped will never arise. Those that do, however, obviously require formulation of a resolution procedure of some kind.

As shown in Figure 10.1, such a procedure consists of an ordered sequence of decisions, the purpose of which is to eliminate as many conflicts as possible and to make the best possible choice given the conflicts which cannot be eliminated. In developing such procedures in its Maryland work, RGH concluded that they can be expected to vary somewhat depending upon the kind of conflict involved, but that all are usually based on certain general assumptions.

One important notion is that market forces should be allowed to function to the greatest extent practicable. Our market economy -- subject to government regulation -- allows market forces to exert themselves regardless of what is decided in a planning exercise. Therefore, it will be preferable to work with these forces whenever possible, rather than attempt to work against them. For example, if two facilities are in competition for the same site or area, it will be preferable for them both to locate there if possible, rather than force one of them to locate elsewhere. This reflects a presumption that it is better to avoid conflicts altogether rather than have to attempt to resolve them.

A second fundamental principle for a conflict resolution procedure is that it is desirable not to make decisions before they have to be made. If all facility location decisions were made at the outset, the result would be an end-state plan and the appearance of certainty. But to do this is also to foreclose options sooner than necessary, and to lose the flexibility inherent in deferred decisions. Be reminded that in finding candidate areas for facilities, it is necessary to make certain assumptions about the requirements for each facility type. However tempting it may be to think of these requirements as fixed, though, the fact is that many can be adjusted, often within quite broad limits. By making conflict-resolving decisions only as conflicts arise, then, one is able to deal with actual prospective facilities rather than hypothetical ones (at a much lower level of specificity). This allows the planning and decision process to be much better informed, and correspondingly more meaningful, than would be the case if facility-site marriages were specified in the absence of a known demand or a more broad-based growth policy.

What results from the application of the foregoing assumptions is not so much a land use plan, but a flexible land management and planning process. It should also be emphasized that this does not preclude use of an encouragement/discouragement policy regarding facility development in particular areas. In fact, another assumption made by

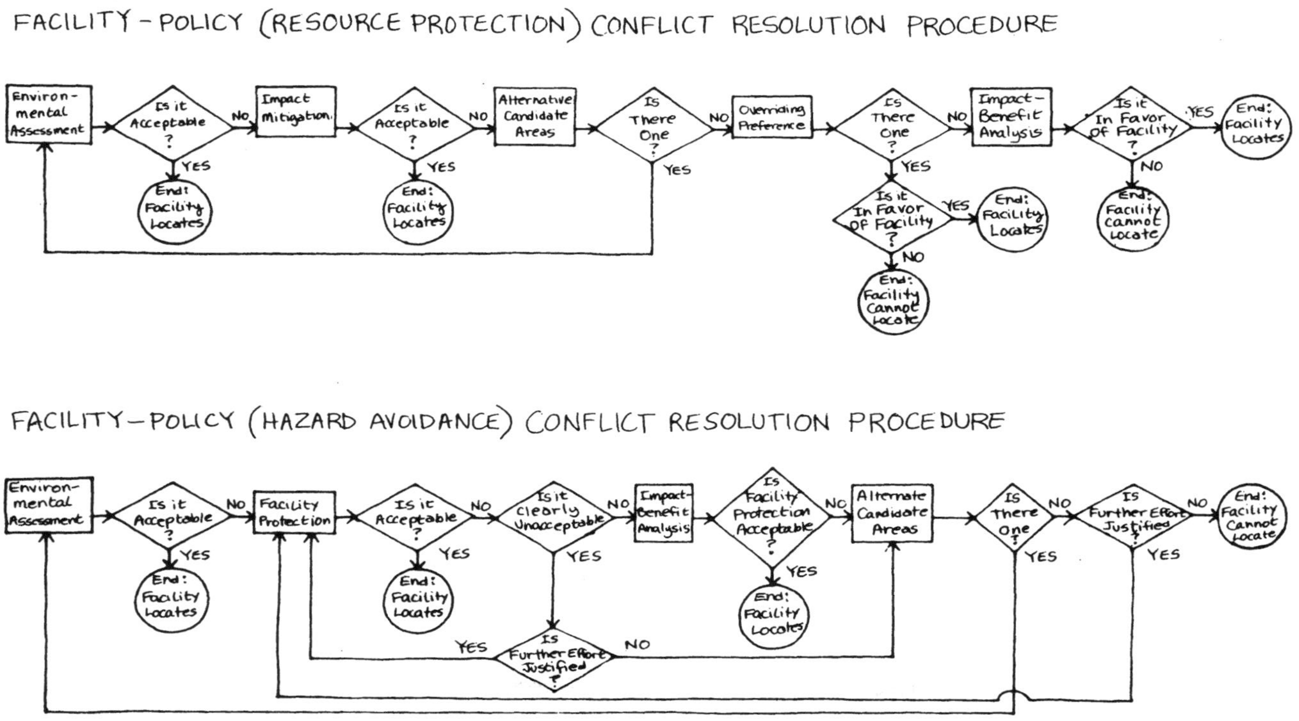

Figure 10.1 Conflict resolution procedures

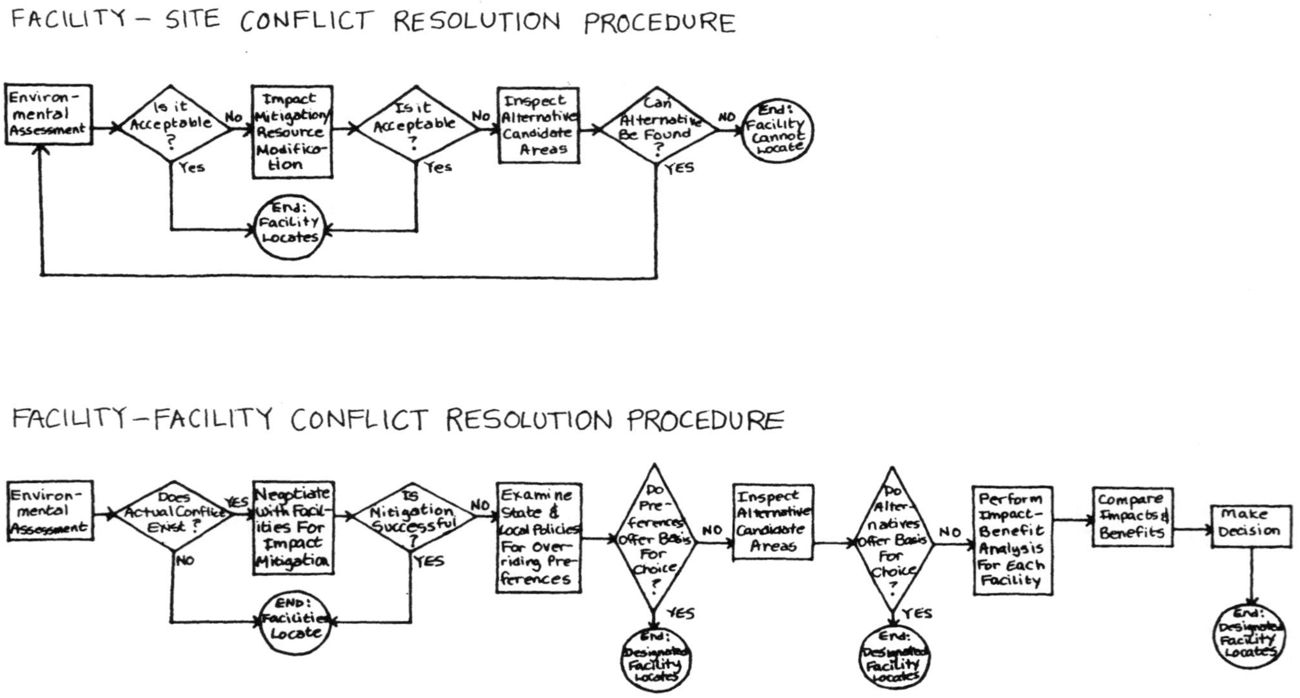

Figure 10.1 (Cont.)

RGH in the Maryland case was that -- other things being equal -- it will be desirable to reserve at least one area for each type of facility, so that the facility seeking to locate ten years hence will not find that all its candidate areas have been preempted by other types of facilities or development in general. This principle may be difficult to realize as a practical matter, as it may turn out to be impossible to hold an area for a facility in the absence of an immediate demand. But it certainly applies to power plants, in that eventual need has been anticipated and site banking has been authorized by legislation.

The availability of special legislation relative to power plant siting, of course, added an additional dimension to the study, the outcome of which is deserving of mention in concluding the Maryland case. As noted earlier, the siting procedure for this type of facility (only) was applied to the evaluation of actual sites, thus providing an opportunity to test the approach to conflict resolution in a realistic context. As it turned out, four sites were identified that satisfied the various facility and public-concern criteria. Unfortunately, the open process came under attack by the local press, which alleged that some of the sites had been selected before the study began. There is no way to deal with this sort of allegation, regardless of the number of people with whom one interacts in the course of planning. Perhaps the most effective rejoinder lies, however, in the fact that the outcome enjoyed high credibility among the groups that had been directly in involved in the participatory program.

INCORPORATING ADVOCACY IN SITING METHODS: THE CANADIAN ARCTIC CASE

The RGH methodology for energy facility siting has evolved considerably since the Maryland project, largely in response to new challenges manifest in the needs of subsequent clients. One of these was the Canadian government, who in 1982 contracted for the development of a generic procedure for siting hydrocarbon facilities along the Arctic coast. Although not explicitly designed for siting electric facilities, as was the Maryland procedure in part, the approach resulting from this effort is readily adaptable for such use. Even more apropos, as a point of departure from the Maryland work, is that for Environment Canada we were much less concerned about elaborate means of data manipulation and evaluation than we were with developing a process that would allow for meaningful involvement of affected parties in all phases of site planning.

Despite the difference in goals, there were several lessons RGH derived from the Maryland experience that found direct application in the Canadian study. Identified as important issues, for example, were the following:

* the choice of the region of interest - In the case of power plants this did not include parts of utility service areas outside of Maryland, and did include areas that were subsequently ruled politically infeasible. Subscribing to artificial jurisdictional boundaries and ignoring political realities reduced the effectiveness of the study.
* differences in facility options and outputs - The versatility of a site, i.e., its adaptability for, say, a nuclear power plant or a coal-fired power plant with fuel delivery by barge or rail, was emphasized in selecting preferred sites. This can lead to a compromise site that is not the best for any single option.
* capacity of the facility - Two of the preferred sites in the Maryland study have more recently been used in an alternate site comparison for a much smaller power plant. Differences in original requirements could have led to selection of entirely different sites.

It was found, in other words, that the initial specification of groundrules per se can introduce potentially worrisome distortion into the siting process.

Of equal importance, the Maryland experience produced some insight as to the participatory aspects of site selection. It demonstrated, to begin with, that it is difficult for participants in a siting study to agree as a group to exclude areas from further study (other than those already excluded by law or regulation). This reluctance was a major stumbling block in the original approach to regional screening, which called for using subjective thresholds of acceptability, analyzing their implications, and adjusting the thresholds systematically. Another significant finding was that, at and beyond the level of selecting potential sites, diverse points of view will usually exist and any consensus will be forced. A participant in the process will feel that his or her point of view has been ignored if at least one of the candidate sites evaluated as a viable contender does not represent that point of view. In the power plant siting phase of the Maryland study, heavy emphasis was placed on the various points of view of the watermen of Chesapeake Bay, and the final sites represented an integration of their respective inputs. In retrospect, the process would have been more satisfactory if these points of view had been separated and at least one site selected that clearly took into account each principal point of view, rather than four sites reflecting a mix of priorities.

Partly as a result of these considerations, the siting process developed for use in Canada is based to a much greater extent on an <u>advocacy</u> approach to consensus building. It brings people with a vested interest in a siting problem together at an early stage and involves them in a structured process that focuses on the development of mutually agreeable solutions. By starting with broad regional considerations and gradually narrowing down to selected sites, the process eliminates areas and sites that are clearly not acceptable, thereby avoiding potential conflicts, and focuses on areas and sites where there is a reasonable degree of consensus. This process has the additional benefit of limiting the need for detailed data and information to a small portion of the study area.

It is important to understand the philosophy behind the siting process. The following general principles apply:

* There may be more than one participant (interest group).
* Objectives and siting criteria are identified by the participants, and each participant may have different objectives and criteria that will indicate how well potential sites meet his or her objectives.
* Each participant is assured that at least one candidate (potential) site selected from the region of interest represents his or her point of view.
* The process is structured so that the sequence of decision points and steps minimizes the data requirements and the number and complexity of tradeoffs that have to be considered.
* There is provision for decisions obvious to all participants to be recognized and made directly.
* Tradeoffs and conflicts among participants are resolved in selecting a preferred site, not in selecting candidate areas or candidate sites.
* The process makes decisions as explicit as possible by favoring choices between options over less direct

approaches such as ranking of factors or importance weights.

* Conflicts are resolved by negotiation and not by the mechanics of the process.
* Decisions are made as explicit as possible and the process is structured to minimize the number of tradeoffs.

Another key feature of the siting process is that, because of difficulties in assigning valid, meaningful importance weights to various factors, the methods for making tradeoffs between factors are based on comparisons and ranking of specific situations. The rationale is that decision-makers are better prepared to indicate a preference between two specific alternatives than to provide an abstract numerical weight for each factor that would be used to indicate their preferences for all situations involving that factor. A corollary to this is that the number of tradeoffs that have to be assessed by each decision-maker must be limited.

To distinguish various roles in the Canadian siting method, specific names are given to the different actors. Persons or groups asked to provide subjective inputs regarding their objectives and preferences for various tradeoffs in siting the facility are termed participants. Persons or groups providing technical support in understanding the facility or in developing criteria are members of the siting team. Although each participant is a decision-maker in the sense of making intermediate decisions on objectives, criteria, and tradeoffs, the term decision-maker is reserved to the person or group ultimately choosing the site. The decision-maker is likely to be a participant as well. The decision-maker will typically be the facility proponent, although under certain circumstances the choice may rest with the federal or territorial government, specific governmental departments, or local interests. This will be the case when the proponent is indifferent about several options and other participants have strong feelings about which option is preferred. Other groups may also become the decision-makers because of land use control, ownership, or claims at industry's preferred sites.

In addition to the above actors, the process will require a facilitator or project manager to organize and direct the site selection effort. The project manager keeps the process on schedule and on track by coordinating the various inputs and outputs for each step and preparing schedules. The project manager will normally have a support staff, which can include experts in siting, public or group participation, and data management. Because siting may result in conflict or an impasse leading to negotiations among participants, an arbiter or referee may also be required.

The various possible roles in the siting process -- project manager, decision-maker, participant, technical staff, and arbiter -- will not all be filled necessarily. In some cases an individual will take more than one role. For example, one participant may also be project manager, or the project manager may also function as arbiter of disputes. The complexity of the facility and the process and the capabilities of the participants will determine the need for support personnel. In simple cases only the participants will be involved. In other cases most of the technical needs will have been set by previous studies and there will not be the need for a large technical staff.

Clearly, the process is based on democratic principles. It assumes that all important interest groups are represented by participants, including representatives of the public, public interest groups, industry and government. Participants in the process are also selected to balance the different social, economic, environmental, safety, and technical interests in the siting problem. The balanced representation is important to the consensus-building philosophy in that it encourages participants to negotiate solutions rather than sitting in entrenched positions. This results in an exchange of information on the objectives and criteria of each interest group, and leads to a common understanding on areas of concern.

The siting team, perhaps backed up by a technical team, brings potential adversaries into working contact with each other on regional screening/technical matters. The siting process therefore encourages the development of mutually agreeable solutions to the siting problem and reduces the potential for conflict. It should also be pointed out that the siting process downplays the use of numeric weighting as a decision-making technique in favor of more subjective approaches for ranking group preferences and assessing tradeoffs. In addition, the process allows technical inputs to be separated from policy inputs. This enables the process to deal with where a facility should be built, but not whether it should be built.

The siting process proposed for Arctic Canada is complex enough that it is useful to look at the general sequence of steps before examining specific tasks in more detail. The following is a summary of the five basic steps that characterize the process:

Step 1: Define the Siting Problem. This step establishes the framework for the siting effort. The siting problem must be recognized and understood, and procedures for addressing it must be established. This includes development of a profile describing and characterizing the facility, comparison of facility/site options, identification of potential effects and issues, establishment of a study area, and consideration of who should participate on the siting team.

Step 2: Translate Regional Screening Issues to Spatial Preferences. Each participant specifies objectives for regional screening purposes, and related criteria which can be used in map form to express spatial preferences. Data availability must be taken into consideration when selecting criteria. Inclusion/exclusion screening may be used at an early stage where strong preferences are expressed. In accordance with agreed constraints and procedures, the team then proceeds to select the best n% of the study area by applying selected criteria, combining inclusion areas, and making low-order tradeoffs. Step 2 is the most complex step in the process.

Step 3: Select Candidate Areas. From the 'best' percentage of the study area as determined through regional screening, candidate areas are selected in accordance with established rules. At this level in the process at least one area must be acceptable to each of the participants, but it need not be his/her preferred area. The rules are designed to keep all participants involved in the process and to select areas of interest to most of the participants.

Step 4: Select Candidate Sites from Candidate Areas. This step involves a re-examination of the objectives and criteria developed in Step 2, and the definition of more specific criteria and reconnaissance data requirements. The siting team then verifies the regional screening aspects of previous steps and revises the selection of candidate areas where appropriate. Using professional judgement, an agreed number of candidate sites is selected, with at least one potential site acceptable to each participant.

Step 5: Select Preferred Site(s). This step includes the development of site-specific criteria, preferably based on refinements of previously established criteria, the collection of new data on each of the sites, and the evaluation of potential effects at each site using more detailed studies. A 'decision analysis procedure' may be required to deal with probable risks that are quantifiable. Each participant or interest group then ranks the candidate sites, and determines which sites are acceptable or unacceptable. Where there are no, or an insufficient number of, sites acceptable to all participants, then negotiation procedures are employed. The final product is the selection of preferred site(s), preferably by consensus, but through arbitration where necessary.

Inevitably, decision-making becomes more difficult as the process advances through these various steps. Also, it must be recognized that despite the full-fledged involvement of the public, industry and government in the process, the final product will always by subject to political decision-making. Therefore it is important to define the final product of the process in such a way that it facilitates political decision-making.

Having now completed an overview of the siting process developed for Environment Canada, attention can now turn to the more detailed flow chart that describes the collection of tasks and decisions that comprise the 'nuts and bolts' of the process. The chart is shown in Figures 10.2 through 10.5, which correspond to each of the basic steps described above. The flow chart is based on the following conventions:

* Steps within each numbered heading indicate how the major step is accomplished.
* YES/NO decision points indicate which step follows based on the answer to a question or questions. Otherwise, the steps proceed sequentially.
* Circled letters A and B indicate that the process provides input to a step later in the process that is identified by a similar circled letter.
* Headings such as For each participant indicate that the subsequent steps are repeated as many times as is appropriate. Indentations within these bold headings indicate loops that are repeated within other loops.

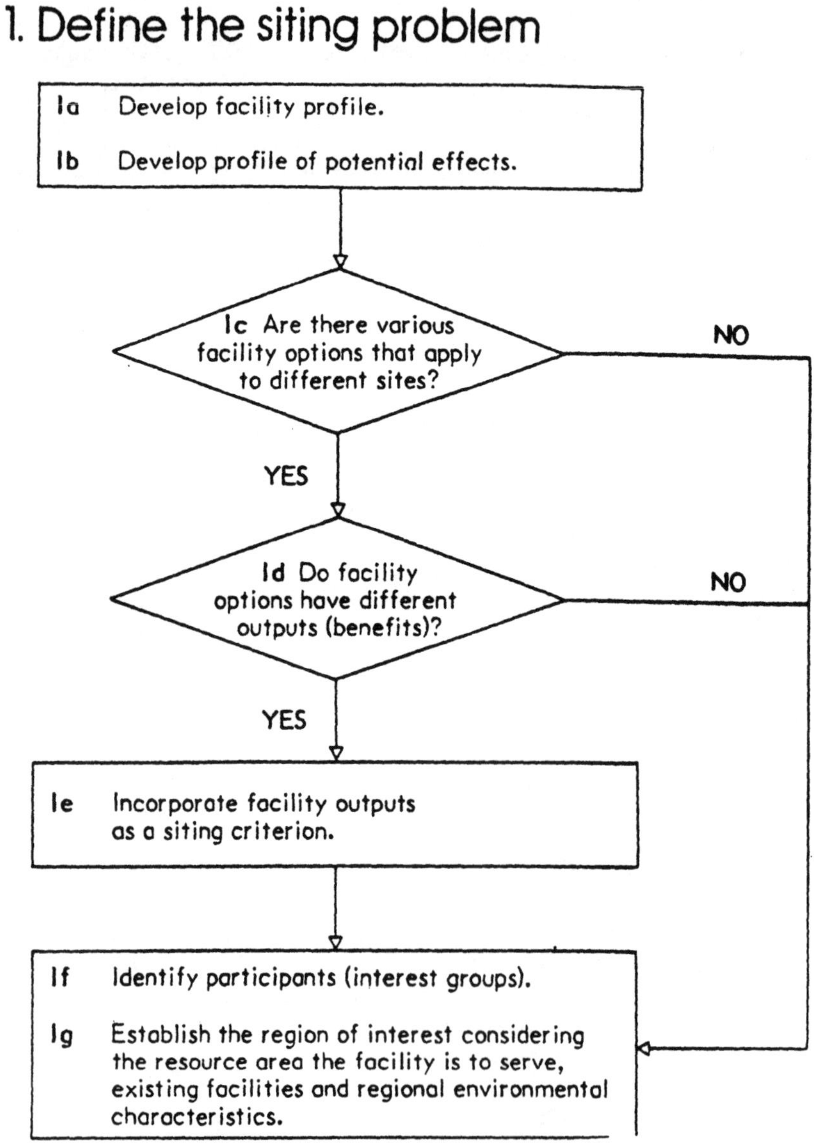

Figure 10.2 Defining the siting problem

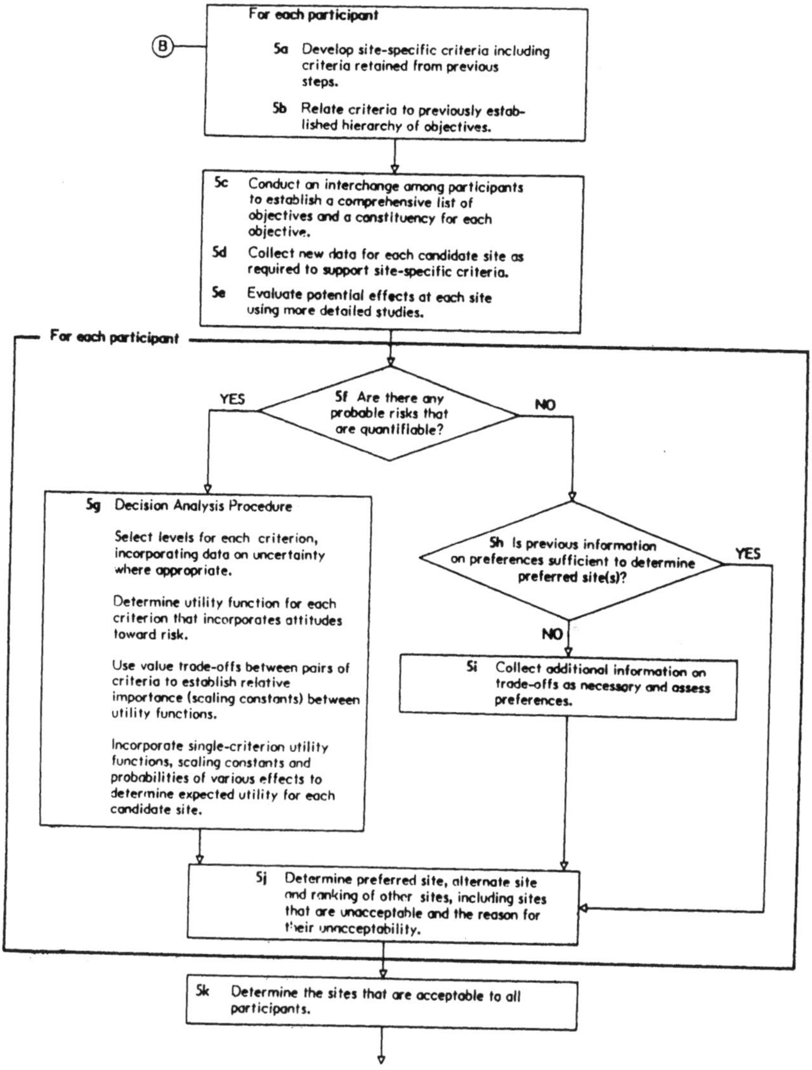

Figure 10.3 Translating regional screening results to spatial references

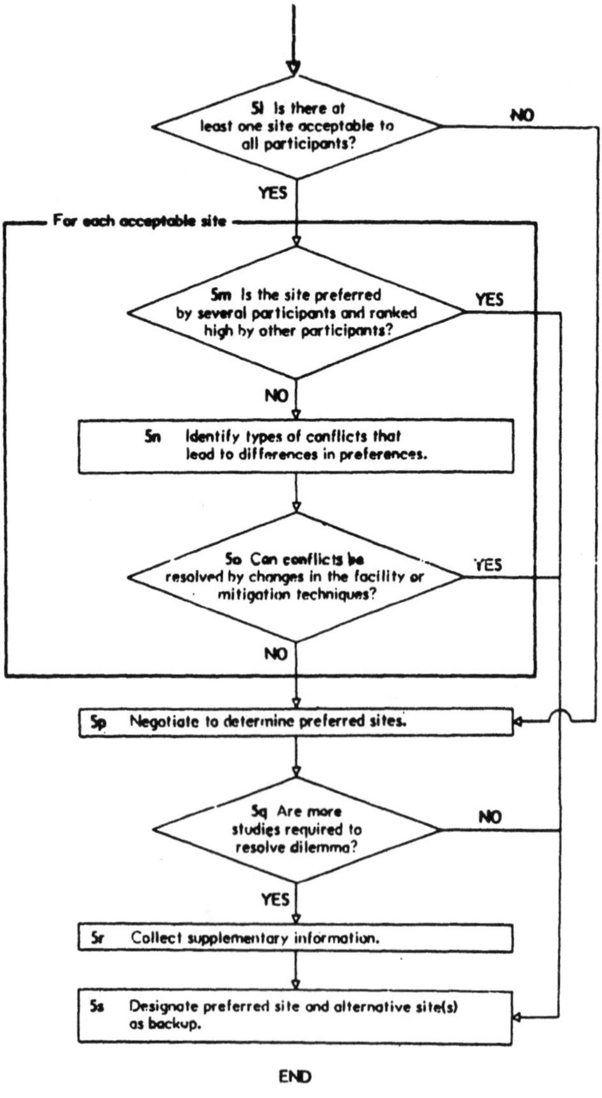

Figure 10.3 (Cont.)

3. Select candidate areas

Select candidate areas to include

- at least one representative inclusion area from each participant's inclusion areas, if any;
- at least one area from each participant's preferred n%. This area should include the most preferred combinations of criteria levels if they can be determined and should not include other participants' exclusion areas unless unavoidable;
- areas within the best n% for two or more participants where these occur; and
- a variety of sites representative of the diversity of better sites in the siting region.

4. Select candidate sites from candidate areas

(A)

4a Define and document additional criteria and reexamine criteria used to identify candidate areas.

4b Collect additional data about candidate areas from published reports, flyovers and site visits as appropriate.

4c Verify regional screening data and revise selection of candidate areas if appropriate.

4d Select candidate sites by professional judgement and the consensus of the participants, with each participant deciding candidate sites that best represent his point of view. (The number of candidate sites is also determined by consensus.)

Figure 10.4 Selecting Candidate Areas and Sites

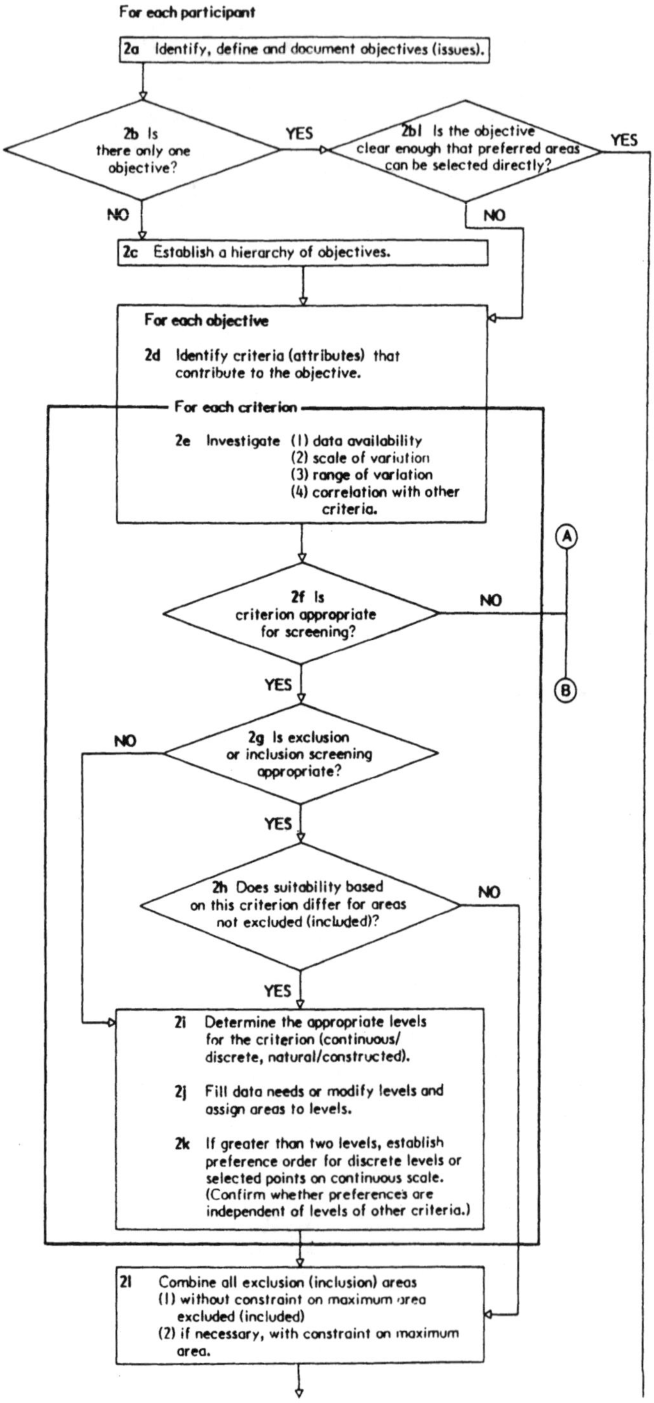

Figure 10.5 Selecting the preferred site(s)

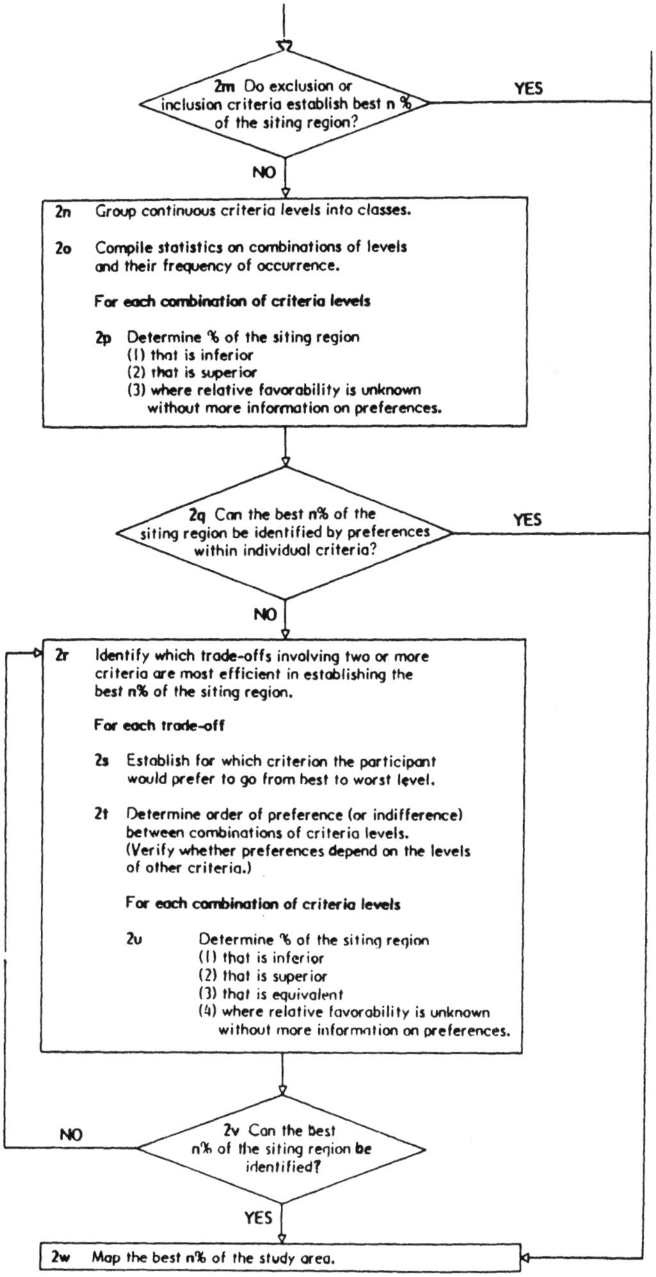

Figure 10.5 (Cont.)

* A series of steps surrounded by a box is repeated as indicated.

In all there are 55 numbered steps, of which some have several parts. Thirty-one of the steps are repeated for each participant, and of these some are further repeated for each criterion, for each combination of levels, for each site, or for each site acceptable to all participants.

Portrayed in this manner, the siting process has the appearance of being extremely complicated, especially from the viewpoint of the participant. However, in practice the degree of complexity will vary depending upon the specific siting problem: for some siting problems, some steps may either be trivial or unnecessary to perform. The application of the siting process can also be simplified, without diluting it, by assigning steps to the appropriate actors in the process, who were described previously. Each actor is only aware of the steps that provide him information and only responsible for steps that require his inputs. Only the project manager must be familiar with all of the steps and responsible for coordinating them.

As an example of steps performed by others, consider the fact that, of the five major steps, steps 1, 3, and 4 do not heavily involve the participants. Only 1g, 4a, and 4d require the efforts of the participants, and 1g and 4d only call for the participants to sanction the results of a group effort. For what seems to be the most complicated block of steps -- 2n through 2v -- only Steps 2s and 2t require input from the participants. The other steps are performed by the project manager and his supporting staff and are not apparent to the participants. Likewise, other steps that deal with data gathering, data analysis, or mapping (such as 2e, 2f, and 2w) are performed by supporting personnel, and only the results are visible.

Other possibilities for simplifying the siting process depend on circumstances. For example, it is simpler to go directly from Step 2bl or Step 2m to Step 2w, but whether it is possible or appropriate will depend on the nature of the siting problem and the participants. In effect Steps 2, 3, and 4 are a procedure for nominating sites to be evaluated. The calculated constraints placed upon the nominators and the data and resources made available to them during their selection are what distinguish the proposed method from others. Obviously, a participant could choose sites directly, either at random or based on knowledge of the region, but this would not be equivalent to Steps 2, 3, and 4. Objectives, criteria, exclusion areas, and tradeoffs, which are now considered for the second or third time in Step 5, would have to be developed from the beginning. New data would not be available to help select potential sites. The chances of finding consensus sites, which are increased in Step 3, would be reduced to coincidence. Bypassing these steps would shorten the process, but only by weakening it.

The basic point here is that, while it is necessary to be rather definitive in writing a stepwise description of a siting process, in practice the RGH model is far from rigid. It can be simplified, as noted above, and it can also be modified to operate in different situations, such as the following:

* a facility on a contiguous, compact site with most effects occurring on-site;
* a similar facility with most effects occurring off-site;
* a linear facility or corridor connecting specific endpoints, with effects occurring along its entire length;
* intermediate circumstances and combinations.

The flexibility of the method is also manifest in the fact that it allows for improvisation in response to problems arising on an ad hoc basis. For example, one common difficulty is that certain people who have not participated previously become concerned upon realizing that a site they oppose might be chosen. The siting team, though, can deal with this within the framework of the suggested process in one of two ways: possibly by clearly demonstrating that the choice was reasonable, or possibly by allowing information provided by these newly concerned people to be used in evaluating potential sites.

As a final thought, it is important to emphasize that the nature of the facility, the region of interest, the affected parties, and data availability are never the same for any two siting problems. Since each siting decision is unique, the best siting method is one that outlines a general process for reaching a decision, tutors the participants in the pitfalls of data use and misuse, suggests ways of interacting within the group, and provides a mechanism for external technical inputs when needed. Ultimately, of course, the success or failure of the participatory process will depend on the skill of the persons directing the siting study.

CONCLUDING REMARKS

In both the Canada and Maryland cases the direct contact and information flow between potentially adversarial groups during the planning and decision-making process permitted much more enlightened and sensitive decisions. Exposing all participants to the viewpoints, values, problems, technical points, and cost of the others in the siting group greatly increases awareness. It also forces each person to closely examine his or her own values and viewpoints and encourages their clear articulation.

Involvement of all affected groups in deliberations increases their familiarity with the rationales that support the siting decision. It is hoped that such involvement will lead to mutually agreeable choices of sites. On occasion it will be politically infeasible for a group to support any siting. No amount of involvement in a decision process will overcome this problem. But, if the question of whether a power plant is needed is not at issue, then a representative group working toward the goal of finding the most suitable site will be able to work through the process described here to arrive at a decision with which most (if not all) affected interests can live.

PART FOUR

Detailed Case Studies

11

Open Power Plant Siting: The Pioneering (and Successful) Experience of Northern States Power

Dennis W. Ducsik and Thomas D. Austin

At the close of the sixties, amidst the acrimonious debate and drawn out court battles with citizen activists across the Nation, virtually no one in the electric utility industry would entertain the notion that controversy over power plant sites might be avoided or reduced through advance public participation. No one, that is, except for two chief executive officers who were undeterred by trepidation -- or at least willing to set aside their fears -- over the perceived pitfalls of open planning. One of these was Earl Ewald, Chairman of the Board of the Northern States Power Company (NSP) of Minneapolis, Minnesota.* Toward the end of January, 1970, Ewald took the unprecedented step of sending invitations to roughly thirty environmental groups, asking them to assist NSP in identifying a preferred site for the company's next baseload plant. He asked these groups, many of which had previously opposed new power projects, to designate one or more individuals from their membership to serve on an Advisory Task Force which would consider a range of siting issues and recommend specific courses of action to the utility.

With the formation of this Task Force, NSP embarked on an experiment in non-conventional outreach which would ultimately lead, in matters of facility siting, to an

---

Thomas Austin holds a Ph.D. in Economics from Clark University and is currently on the staff of Energy Systems Research Group, Inc. in Boston, Massachusetts.

*The other pioneering CEO was Lee Sillin of Northeast Utilities in Hartford, Connecticut, whose citizen participation experience is reported in Chapter Twelve herein.

ongoing commitment to citizen participation as a principle of corporate operation. What follows is an account of the genesis of that commitment over the period 1970-1973, during which the company involved environmental groups in siting two major power plants and worked, in addition, with concerned citizens and government officials to design a new siting and licensing law for Minnesota (which institutionalized the concept of public involvement in state siting endeavors). By virtue of this early experience, NSP deserves to be listed among the pioneers in the field of participatory site selection, and as such their 'adventures' and the lessons to be derived therefrom are well worth recording. What is also fascinating about this story of collaborative problem-solving is that it stands in such marked contrast to the torturous pattern of events that occurred in the CPA-UPA case described in Chapter Nine. While the latter took place at a subsequent time, in another part of the state, and in a different planning and regulatory context, the underlying circumstances are sufficiently similar that the NSP case illustrates, by way of comparison, how strongly the outcome of an interaction process depends on its structural design.

BREAKING THE ICE

NSP has been badly burned in prior years in its attempts to site major facilities, and the difficulties encountered had not been confined to one type of facility or even to one class of objection. For example, in 1964 there had been strenuous public resistance to the siting of a major coal-fired facility, the Alan S. King plant, along a section of the St. Croix River with significant recreational potential (being at once relatively undeveloped and not far from the heavily populated Twin Cities area). Here, as another commentator has put it, "part of the problem was that private boat owners also liked the stretch of the river they would have to share with the plant; another part had to do with thermal pollution; and finally, there was a fiercer than usual debate over transmission lines."[1] In another troublesome case, involving a 345 kV transmission line which circled the Twin Cities region, the necessity for taking large amounts of land seemed to be the major cause of dispute.

It was the company's recent battle over the proposed 545 MW nuclear plant at Monticello, however, which had caused the most distress.[2] The fight had resulted in the state pollution control agency requiring that nuclear plants in Minnesota adhere to a limitation on radioactive discharge more stringent than that of the Atomic Energy Commission. The ensuing court battle, whatever its contribution as a precedent on the question of federal primacy in the regulation of nuclear power, was a public relations nightmare for the company. NSP was forced into the highly impolitic stance of favoring radioactive pollution while opposing states rights.

Although NSP eventually prevailed in each of these cases, like most other utilities it found the problem involved in siting facilities during the sixties to be perplexing, and occasionally incomprehensible. To make matters worse, there was every reason to believe that these difficulties would increase. Plant siting had become an increasingly expensive process, not only in terms of direct pecuniary costs resulting from litigation, delay, and uncertainty, but also in that confrontations had significantly drained the reservoir of goodwill which NSP had built up over the previous decades. For a firm in such a heavily regulated industry, and one which expected to continue expanding at an annual rate of seven to nine percent, this was a consideration of no small consequence.

As a result of its confrontational experiences, a perception surfaced within NSP that opposition to their siting attempts was at least in part attributable to the company approach to the politics of siting, rather than resting solely on objections to the specific facilities. Consider, for example, the following list of 'tactical errors' described in an in-house study of the handling of the Monticello affair, prepared by NSP's then newly-appointed director of environmental affairs -- Roland Comstock:

* spoke only in the regulatory forum for over two years;
* 'can of worms' syndrome -- assume there won't be trouble until it occurs;
* result -- always on the defensive, always behind the problem;
* misgauged issues and severity of the problem;
* talked millirems when the issue was states rights, and measured public opinion vertically instead of horizontally;
* worked with the wrong people -- regulators and politicians;
* should have worked directly with public and press, and early.

Comstock summarized the problem more succinctly in another internal memorandum in October of 1969. "We are getting hammered simply because the public will no longer concede NSP the right to make, by itself, judgments on environmental matters. Yet, that is our posture -- defending judgments we have made. Although it is these judgments which are under attack, I believe the public is really attacking the method by which these judgments were reached." He elaborated on this a couple of years later. "The basic problem lies in the fact we make internal judgments based on our evaluations of economic and operational needs -- although increasingly tempered in recent times by our interpretation of the public will."[3]

Nineteen seventy was a time of great flux in the field of plant siting, with the laws and institutions dealing with environmental protection being held up for scrutiny in many states. With this in mind, it may well be that NSP was interested in something more than simply finding a new strategy within which to approach the siting of its next plant. At the time there was no governmental organization in the state charged with the overall regulation of new generating plants or transmission lines. Indeed, surprisingly, there was not even a public utilities commission with which to work in this respect. From the utility's point of view, then, an open planning approach would have the additonal advantage of allowing the company and others who might be interested to experiment with a new approach prior to deliberations on any future siting laws. Still, it is not clear why NSP settled on this novel step. Another alternative would have been to simply attempt, as utilities in other states were doing, to approach the legislature with a siting law that was crafted in the company's interest. Given the climate then current in Minnesota, perhaps they were concerned about the potential for successful opposition to such a law on the part of environmental lobbyists.

If the advantages of opening the decision process to those outside the company were readily apparent, the disadvantages for NSP were equally obvious. Such an experiment would require that NSP publicly announce its intentions in some detail long before it could be sure that the proposal was easily defensible. More troubling still, the company would be forced to supply potential opponents with a considerable quantity of information which might prove devastating should the process break down. It was entirely possible that the environmentalists would not be willing to stick close to the matter at hand, but rather would want to investigate a wider variety of utility practices in an effort to question the need for a plant. If the company had previously been suffering from a 'can of worms' syndrome, it should hardly be surprising that some elements within NSP were dubious of the wisdom of opening the can even further.

Clearly, NSP did not harbor any naive illusions as to the risks associated with citizen participation. As Rollie Comstock later noted, "... the Chinese ideogram for 'crisis' is a combination of the symbol for 'danger' and opportunity, and this sums up a lot of things about open planning."[4] As a potential solution to its crisis in public confidence, however, NSP apparently saw more opportunity than danger in public involvement. It decided to proceed with the experiment.

The NSP invitations went out to a number of citizen groups, including such organizations as the local chapter of the Sierra Club, Zero Population Growth, the League of Women Voters, and the Minnesota Environmental Control Citizens' Association (the company's fiercest foe). In addition, NSP invited the members of the state Academy of Science and of the Minnesota Environmental Cabinet to join the Task Force. The cabinet, recently created by the governor, was made up of the heads of all state agencies concerned with environmental matters. No effort was made, however, to include such groups as industry associations or chambers of commerce; the Task Force was to have a distinctly non-business viewpoint.

The NSP proposal to form a Task force did not come as a surprise to the groups receiving invitations, for each had previously been visited by Rollie Comstock to explain the novel approach and to ask their cooperation. The initial reaction had been positive, but guardedly so because the environmentalists, too, could envision potential problems with a new and unfamiliar process. In the atmosphere of distrust, it was not hard to suspect that NSP was trying a new public relations gimmick to obtain ratification of its actions; or that, in an even more sinister scenario, the company hoped to diffuse opposition by getting the various environmental groups to take 'pot shots' at each other. If the groups had doubts about the honesty of NSP's intentions, however, they also realized their own credibility was at stake; as one group spokesmen put it, "You can't raise hell and then refuse to sit down and talk."[5] In a sense, the environmental groups saw no other option than to participate.

The men and women who agreed to take part in the Task Force were by no means average in their attainments. The group included, for example, two physicists and a biologist from the University of Minnesota, as well as a few other scientists and engineers serving as delegates from local environmental groups. And there were others who, though laypersons, had established themselves as credible and well informed voices on energy issues. Besides comprising several of NSP's most ardent critics, then, the Task Force was generously endowed with technical credentials as well. There was perhaps some concern that the group's expertise would be exploited to buttress company decisions, but the more likely scenario was that it would serve to 'keep the NSP honest'; a noisy resignation by a prominent scientist or two, after all, would make it extremely hard for the public to accept that company intentions were indeed good.

The utility made two formal requests of this newly formed Task Force. First, it wanted the group to recommend a site for a new plant, which the company had already determined would be a 680 MW coal-fired unit that would burn low-sulfur fuel, and begin operation in 1976. NSP specified, moreover, that since a six year planning horizon was necessary to meet engineering requirements, the Task Force site recommendation would be needed within a mere three months. The second request made of the Task Force was made more open-ended. After coming to a conclusion about the location of the 1976 plant, the Task Force was asked "to look at the long-range problems of plant siting (including transmission lines) in a totally unrestricted way, and hopefully to develop environmental criteria which the company could apply in the acquisition and utilization of future sites."[6]

For its part, the utility agreed to carefully consider advice received from the Task Force, although it was made clear that no recommendations could be taken as binding. In fact, the company took steps to ensure that the outcome of the selection process would be a reasonably available and acceptable site. The most important was a restriction that, ostensibly due to the length of the acquisitions process, the plant had to be located on one of four parcels of land already owned by the company. The list included two locations mentioned earlier as having resulted in environmental battles: the Alan S. King site, and a site across the river from the Monticello unit (known as the Sherburne County site). To these NSP added the Prairie Island site, where two nuclear stations were in the midst of licensing, and a new site variously known as Louisville

Swamp or Carver Rapids (the latter will be used hereinafter). For reasons of economy and operating efficiency, Carver Rapids was NSP's first choice, and it made no attempt to disguise the fact.

As illustrated by this constraint on the scope of alternative sites, NSP's groundrules were hardly calculated to relive the anxieties of Task Force participants, many of whom voiced displeasure with the restrictions. But they were willing to go along with the constraints, partly because Rollie Comstock's personal visits had gone a long way in "breaking the ice," and partly because NSP in other ways had been careful not to aggravate misgivings about participation. To begin with, the company recognized that environmentalists were concerned not only about a location for the new plant, but also about a variety of related company practices (i.e., advertising and promotion, rate structure, and environmental monitoring of plant emissions). So, NSP stated it was interested in what the environmental groups had to say on these matters. The second and probably most important action was that NSP was particularly careful to avoid placing the individual participants in a position of being unable to repudiate the findings of the Task Force, should they so choose. This is reflected in the following excerpt from Earl Ewald's letter to the president of the local Isaac Walton League.[7]

> We recognize the practical problems preventing any one person from representing your group in any official manner. We do not expect that. However, we do feel that you are in the best position to evaluate the capabilities of individuals within your group to meet the large responsibilities to be assumed by this Task Force. Further, it is expected that any such appointee will provide a continuing communication link between your group and the Task Force...

What NSP desired from both the private and governmental members, in other words, was thoughtful input rather than official commitment.

There remained the question of how, specifically, the Task Force and NSP would interact. During the early stages of the Task Force's deliberations, in particular, it would be necessary both to provide the members with quite a bit of background knowledge of utility operation and the process of plant siting. Many representatives were scientists or engineers qualified to deal with complex technical data, but even they required briefing on certain key topics. Furthermore, there was the ever-present problem of developing and maintaining some threshold level of trust, which was critical in many respects for an alliance of such unlikely composition as a utility and a panel of environmentalists to be able to function. As Comstock later recounted:[8]

> One of our big uncertainties at this time (when deciding how to implement an open planning approach) was whether NSP should conduct all these meetings or whether it would be better if some outsider were hired who had environmental credibility, i.e., a big name. After interviewing any number of such persons and firms, virtually on an international scale, we decided to do it ourselves.

Not surprisingly, Comstock, who had been so closely involved in the design of the experiment, was chosen to be the utility's chief spokesperson. Although the staffing requirements of the Task Force would be filled largely by utility people, they did make use of outsiders particularly during the early meetings. This policy served a dual purpose; not only would these outside speakers help fill informational gaps, they would also serve to lend a sense of objectivity to the proceedings.

THE INITIAL TASK FORCE EXPERIENCE

The early meetings of the Task Force were marked by considerable rancor. One is tempted to characterize these early meetings not as an emerging dialogue but as two simultaneous monologues. The utility's prime concerns were to present the various alternatives and to "bring everybody up to speed." But the environmentalists, who finally had access to a forum in which they had NSP's complete and undivided attention, were not about to pass up the opportunity to take the utility to task for a variety of past practices (in regard to rate structures, promotional campaigns to stimulate load growth, and the like). The state officials, moreover, were subjected to a considerable dose of criticism from their fellow Task Force members, who seemed to regard government as part of the problem rather than a source of possible solutions.

Roughly one month into the process, with no indication that the Task Force was ready to settle down to the business at hand, the situation looked bleak. Critics of the process within NSP took this as evidence that open planning was subject to fatal flaws, since Comstock and his staff really couldn't force the Task Force to start evaluating the candidate sites. Indeed, in the view of one Task Force member at least, the 'educational' program that had been presented to the group had backfired somewhat in this regard:[9]

> Rather too much time was spent by NSP in general background presentations. It was as if NSP was delaying focusing in on the principal issues. Some of the wide ranging criticism of utility practices was, I think, a natural reaction to this heavy dose of spoon feeding by NSP, however well intended background presentations may have been.

At this point it seemed clear that one of Murphy's laws -- "if anything can go wrong, it will" -- was definitely at work.

But, as Rollie Comstock would later recount, "along about the sixth week, something very significant happened. The group itself turned upon one of the more outspoken critics and in effect told him to shut up. They felt they had a chance to do something constructive and that the constant hostility expressed by this one man was getting in the way. From that point on, group dynamics began to change."[10] One member recalled that the Task Force was well aware of the same change, and began to take the attitude, "now that we've got the donkey's attention, it's time to stop hitting him over the head."[11] One reason for this change is seen in another remark of a Task Force member, who said that during a heated discussion of advertising he felt that the time had come to "get down to the meat of helping NSP determine where their next plant was going to be located . . . I said, 'NSP is going to have to make a decision by a certain date, and if we haven't joined them by that date, they're going to make the decision and go on, and we'll be where we were two years ago."[12]

Concurrent with this major shift in the focus of discussions, two other changes of significance to the operation of the Task Force occurred. First, the government officials with direct regulatory power withdrew to form their own separate group, amidst rumors that they had actually been recalled by the state governor. Whether their departure was precipitated by the abuse forthcoming from others on the Task Force, or whether they left in order to remain impartial when the time came to review the chosen site in an official capacity, the result was that of forcing NSP to work with two groups simultaneously. In a second important development, the Task Force decided to change the locale of their meetings from the NSP offices to a more neutral ground, the University of Minnestoa, where several Task Force members were employed. Apparently the Task Force was increasingly suspicious that NSP was already committed to the Carver Rapids site, and it was not comfortable with the basic methodology and criteria for comparison being advocated by the company. This involved a quantitative model for evaluating environmental impacts, which the Task Force rejected as an example of the "engineering approach" that it felt had led to environmental abuse in the past.

With its composition stabilized and the change of venue effected, the Task Force needed only a few weeks to reach a conclusion as to its preferred site, as well as to make suggestions as to plant design and operation. These were contained in a report that commented on the four candidate sites in order of increasing suitability. The group considered the installation of another unit at the King site to be the least desirable alternative since the area was still relatively undeveloped and possessed significant recreational potential. For similar reasons the Task Force felt the Carver Rapids site to be unsuitable, despite the fact that it was the utility's first choice and strongly favored by local officials because of tax base and other revenue benefits. The report noted that this area had been proposed as a state park and that it contained features of geologic and historic significance as well as an interesting variety of flora and fauna. Then, in a choice which was much less obvious, the Task Force decided against the Prairie Island site because construction work on the new plant would have to begin before the nuclear units, then under construction, were completed. The Task Force worked to avoid "the confusion and environmental impact of two large construction projects on the same site"; and it felt that, with only nuclear generators under development at the site, it was desirable to monitor the effects of this plant type alone for further study, rather than mix the effects of two different plants.

Thus the Task Force arrived at a determination that Sherburne County would be, using the words of the group report, the "least objectionable" site. This location had none of the potential land use conflicts that were associated with the King or Carver Rapids sites, nor any of the complications of a dual construction project since, unlike the situation at Prairie Island, the new coal-fired plant would be across the river from the partly-built Monticello unit. This candidate site was, on the other hand, only thirty-four miles from the Twin Cities, and -- more importantly -- both upstream and upwind. Thus, in giving this site priority status the Task Force was (somewhat reluctantly) relying on the use of pollution abatement equipment to keep air and water emissions within tolerable levels.

Perhaps what eased the Task Force's discomfort about Sherburne County was that it also made a series of generic recommendations (i.e., for any plant at any site) as to mitigation measures. These were as follows:

* closed-cycle cooling should be used (it was not determined whether cooling towers or ponds would be preferable);
* emission of all air pollutants -- in particular sulfur oxides -- should be reduced to the minimum level technologically feasible, and adequate space should be provided for future pollution control equipment;
* mechanical rather than chemical scrubbers should be used in the condenser water system to avoid introducing dangerous chemicals into waterways;
* under no circumstances should coal be delivered by barge (in keeping with NSP's stated intention);
* contracts with coal suppliers should require that all state codes concerning mine operation be strictly adhered to;
* associated transmission lines should be as attractive as possible and designed so as to have a minimal impact on scenery;
* the company should follow through with its intention to consider the aesthetic aspects of the plant and should strive to keep noise levels at a minimum.

Also included was a strong exhortation on behalf of comprehensive monitoring of air and water quality in the vicinity of the plant, to indicate whether further pollution control measures might be required in the future.

There is no evidence that NSP took issue with any of these generic recommendations, but the question of how to respond to the Task Force's site findings turned out to pose some difficulty. What complicated the decision was that a week after the group had made public its preference for Sherburne County, the officials from the Governor's Environmental Cabinet (who earlier had left the Task Force) issued a report which came to a different conclusion. In a terse letter to NSP, these officials proposed that the plant be built at Carver Rapids, NSP's original preference. The logic behind this recommendation, however, was presented only in the most sketchy of terms. Basically, they agreed with the Task Force on the unsuitability of the King site, but opposed Sherburne on the grounds that it was upstream of the metropolitan area. It should be noted that the municipal water intakes for Minneapolis and St. Paul are above the point where the Minnesota River joins the Mississippi, and hence downstream from Monticello (on the Mississippi) but not from Carver Rapids (on the Minnesota).

Not surprisingly, this point of departure in the advice it was getting created much confusion at NSP. It was obvious that the citizens group and the government officials had used different criteria in evaluating the alternative locations. Whereas the Task Force was willing to trade off some risk of pollution impact in order to protect land use values, the Environmental Cabinet found it difficult to strike a similar balance. Indeed, there was some conjecture at the time that the political advantages of strengthening local tax bases at Carver Rapids weighed heavily (though not explicitly) in the thinking of the state officials. Whatever the reasons for the variant recommendation, however, it gave NSP executives a non-trivial question of corporate strategy to ponder: should the company go ahead with its own first choice, which had now in some sense been ratified by government, or should it accept the rationale of the Task Force, a group which included some of its most vocal critics?

Ultimately, after a great deal of internal agonizing, the company elected to follow the recommendation of the Task Force. As Comstock later commented, "while the reason for that decision was complex, it basically stemmed from the feeling that the rationale presented by the citizens' group was more sound."[13] Apparently this proved to be the case, because the Sherburne plant subsequently proceeded through the state licensing process without difficulty. Interestingly, the final location was not opposite the Monticello installation but on a new tract about three miles further upstream, acquired in a very short span of time after NSP discovered geological flaws at the original site. Although the rapidity of this adjustment seemed to contradict NSP's rationale for having the Task Force consider only sites currently in hand, no serious objections were forthcoming when the Minnesota Pollution Control Agency gave notice of the final project plan early in 1971. To the contrary, individual Task Force members went on record as saying the plans were generally good and predicting the plant would be among the cleanest in the country. The unit went on line in 1976, as per the stated schedule.

THE TRANSITION TO ANOTHER EXPERIMENT

That the Task Force had devoted the bulk of its time to the siting problem did not mean that other concerns about electricity planning had been set aside permanently. Thus, after the site evaluation report had been submitted, the group continued to meet for several months to consider a number of broader issues in the areas of energy demand, site identification, plant development, and environmental monitoring. Insofar as these matters had traditionally fallen under exclusive management control, a somewhat reluctant NSP decided to relinquish the role of administrator and the Task Force consequently elected one of its members to chair the additional meetings. What's more, the group chose to divide into four committees to come up with specific recommendations and reports to be adopted by the Task Force as a whole.

The deliberations which ensued produced a modicum of concrete results. For example, the plant development committee met on technical engineering policies concerning Sherburne, gathering information and aiding NSP in producing and evaluating an environmental impact report. And the site identification committee ultimately proposed a location methodology to be applied to new generating facilities, called for a comprehensive site search and related surveys, and produced a job description for a consultant to assist in the process. Apparently, NSP subsequently followed these guidelines and hired the outside firm of Commonwealth Associates to carry out the requested reconnaissance. Among the issues which NSP specifically wanted to be addressed in this study were the following:

* environmental factors such as the ability to accommodate closed cycle cooling (ponds were preferred to towers);
* the feasibility of incorporating ash disposal and water re-cycling systems;
* the desirability of sites remote from urban centers and preferably downwind of them;
* the impact of a potential plant on surface and ground water, with attention to seasonal flow problems;

* existing zoning along with community plans and recreational usage;
* plant aesthetics;
* impacts on recreation, industrial development, and other potential uses.

Another Task Force product which elicited some (albeit indirect) response from NSP was a report entitled, "Establishment, Operation, and Evaluation of Environmental Monitoring Programs," which elaborated the important concept of follow-up included in the earlier site report. In this document the Task Force mentioned that NSP had an obligation not just to live within the letter of the law but also, because of its commitment to the public welfare and its technological knowledge, to go beyond that. Thus the company was urged to:

* make every effort to obtain early warning of potentially hazardous conditions;
* sponsor research programs directed toward a better understanding of the environmental effects of generating plants;
* set up an internal panel to review monitoring programs and environmental research;
* support the creation of an outside panel as a counterpart to that of NSP, under the sponsorship of an independent organization like the American Association for the Advancement of Science;
* support improvement in the Minnesota Pollution Control Agency's capability to evaluate monitoring and research activities (and thus decrease reliance on a citizen panel for this function).

As it turns out, NSP never delivered a statement to the Task Force on these suggestions, although the company did respond indirectly by incorporating many of the specific ideas in the report into its corporate planning policies.

The only other report issued by the Task Force -- and that which received the most tepid response -- concerned advertising and other means of promoting electricity consumption. This included some rather pointed suggestions, most of which were blatantly contradictory of the companies past practices. To wit, the report said:

* The company should refrain from advertising designed to promote the increased use of electricity.

* NSP should take steps to promote efficient and non-wasteful use of electricity. These should include education projects, direct conversations with major industrial users, and specific suggestions on how to conserve electricity. In addition, NSP should place particular emphasis on reducing the summer and winter peaks.
* The purpose of these measures would be not only to lessen the company's environmental impact, but also to demonstrate a knowledge of and concern for the environment.

NSP's reaction to all this was hardly enthusiastic, although it did agree that trimming the seasonal peaks was desirable. Otherwise, the following excerpt of a statement made by company chairman Ewald is representative:

> "NSP can and will exercise its corporate responsibility of informing our customers how to use energy wisely, if they so choose. Equally as important . . . is the challenge of developing new means to generate cleanly produced, ever abundant supplies of energy."

Clearly, the company was not prepared to seriously entertain this notion of demand management, which at the time was still viewed as a radical departure from the norms of long-range planning.

There is no evidence that NSP's ambivalence toward what the Task Force had to say about need for power was a serious detriment to the company's improving relationship with the environmental community. But it was perhaps no coincidence that, as 1971 wore on, the Task Force began to lose a clear sense of purpose and direction, and attendance waned. This may have been attributable in part to a lack of stimulus and quality data from the utility, but it was also clear that some people were simply beginning to lose interest. Participation of regular members also decreased due to scheduling difficulties, the press of other community-interest business, and a perceived lack of prospect for additional results. By the end of the year, the Task Force had not met for six months, and for all intents and purposes had ceased operation entirely.

The concept of open planning, however, was still very much alive in the minds of NSP and a number of former Task Force members. By April of 1972 the company had carried out a baseline siting study and was ready to find a location for another coal-fired baseload plant, this one sized at 800 MW and scheduled for completion in 1979-1980. Impressed with the overall results of the first participatory experiment, and having gained practical insight into the mechanics of the interaction process, NSP decided to convene a second task force. But in designing this next experiment, several important changes were made in both the organizational arrangements and the agenda for the advisory group.

In particular, NSP chose to abandon the original approach of asking regulatory officials and private citizens to serve as co-equals on an unofficial entity, one entirely of the company's own making. It was important to avoid a recurrence of the dilemma precipitated by the withdrawal of agency representation in the first experiment, and the best way seemed to have Task Force II operate more directly under the auspices of state government. Conferring semi-official status on the group would not only add to its sense of legitimacy, but also provide access to a more neutral staff and thereby diminish reliance on the utility and its consultants for critical information. Finally, to the extent NSP saw its initiatives as a prelude to the development of a new siting law, an opportunity was presented to test an alternative model for citizen participation which might later be institutionalized by the legislature.

Against this background, the utility turned to a new unit of state government -- the Environmental Quality Council (EQC) -- with a request that it "take an active role in the evaluation and selection of sites for further NSP power plants."[14] The EQC had recently been created by executive order of Minnesota's governor, Wendell Anderson, and consisted of the Governor, the commissioners of highways and natural resources, and the directors of planning and pollution control. The Council agreed to take on the review function and then promptly created its own Citizens Power Plant Siting Task Force, to study candidate locations for the next facility and make appropriate recommendations to the Council within approximately six months (by December 15, 1972). The EQC, in turn, would consider the Task Force advice and then make its own report to the utility. The new approach thus segmented the roles of citizens and the government in such a way as to ensure, at least in principle, that coordinated input would be forthcoming from each.

Associated with the modified format was (predictably) a new composition for Task Force II, which comprised twenty-three members from communities throughout the state. Unlike its predecessor, which included designees from environmental groups almost exclusively, the new group was broadly representative of the constituency spectrum (including ethnic minorities, the farming sector, and organized labor). It is not clear how the individual members were chosen, or what influence NSP might have had over the process. But it is obvious that a principle of comprehensiveness was at work, presumably because the EQC -- a product of the political system -- saw the necessity for equality of representation, in much the same way as party conventions select candidates who are perceived to lend an appropriate 'balance to the ticket.'

Another significant point of departure from the previous open planning effort was that, because of the longer review period, Task Force II was in a position to be more encompassing in its approach to the siting problem. It could explore, for example, such generic issues as the method used to estimate future loads, the potential role for conservation, the desirability of purchasing rather than generating electricity, the relationship between facility location and statewide land use planning, and the tax effects of putting several plants in one town or county. What's more, it was entirely possible to expand the range of siting options to be considered, especially because the results were in from NSP's year-long study of potential sites for fossil-fueled units in the southern Minnesota. Indeed, the consulting firm NSP had hired for this purpose had identified no fewer than sixty-eight potential sites for further analysis.

NSP was willing to have the EQC and its task force consider all of the aforementioned sites, if they so desired, but pointed out that it would be rather unwieldy for the group to do so. Thus, the company went through an additional screening process and reduced the list to what it maintained were the seven "most promising" candidates. In presenting this list the utility argued that its screening approach gave equal weight to environmental and cost factors, and that it was based on the following elements:

* Site Development -- site preparation (grading, clearing, etc.); access, both road and rail; necessary relocations; and potential cooling factors.
* Air Quality -- remoteness; prevailing wind patterns; stack height; and the potential for problems with fog or inversion.
* Water Quality -- withdrawal, injection and stream flow; seepage from the reservoir; reservoir fluctuation; and weed and algae growth.

* Land Use -- extent and manner of effect on existing uses; land use and zoning restrictions; possible effect on land form and physical and biological aspects; aesthetic beauty, view and screening; and the existence of historical, recreational, or cultural factors.

NSP went on to say that it was open to any other suggestions as to plant location that any of the participants would care to offer. This left the door open for Task Force II to follow the same avenue which the first group had taken, i.e., to propose that the next plant be located at the site of an existing plant.

Clearly, NSP was making an effort to correct the problem which had rankled many on the original Task Force, that of having only a limited number of sites from which to choose. In fact, to make the process more meaningful in this regard, the company went one step further: it announced that no property (or even options thereon) would be acquired pending the outcome of the advisory review process. This left the company vulnerable not only to speculative price increases, but also perhaps to the political disadvantage of having to resort to the blunt legal instrument of eminent domain to obtain the necessary rights to land (if a new site was designated). While neither of these possible consequences would have been especially debilitating in the larger scheme of things, still, the company was making a gesture that was important as evidence of its sincerity in the matter.

THE SECOND TASK FORCE EXPERIENCE

As expected, the final report from Task Force II to the EQC applied a holistic perspective to the questions of utility expansion, focusing at the outset on the underlying rationale of growth in power consumption. At the most fundamental level, for example, the group made a point to distinguish between the concepts of demand and need for electricity:[15]

> Demand, as we understand it, represents desire and preference for a particular item under a given situation . . . 'Need', on the other hand contains within it a consideration of values. For example, although there might be a certain 'demand' for electricity to operate air conditioners, not all would agree that

> there is a 'need' for this electricity given the circumstances which prevail in Minnesota. A determination of 'need' is subjective, and it requires value judgments regarding which there may be a wide diversity of opinion. Some criterion for determining electricity needs must be established before needs can possibly be analyzed or projected.

The Task Force did not go on to outline the conservation methods which might be appropriate to reconcile any divergence between demand and need, but simply suggested that conservation methods be given highest priority by the EQC.

The group did find it within its ken, however, to call into question the process employed by NSP to project future demands (whether or not such was a true representation of need). The company had made a key assumption that peak demand would continue to follow a pattern similar to that which had prevailed historically, although an adjustment was made for the apparent saturation of the air conditioning market. Hence, peak demand was expected to grow by 9 percent in 1973 and then to taper off by 1/4 percent per year, so that by 1980 the annual growth rate would be just over 7 percent. The Task Force, however, took issue forcefully with such reliance on a simple time-trend approach to load forecasting, citing a recent government report on the state-of-the-art which indicated more accurate methods (e.g., econometrics) were available.[16] On this basis, the group suggested that NSP re-evaluate its forecasts to take better account of such factors as rising fuel prices, population growth rates, and price elasticities.

Interestingly, Task Force II did not translate its qualms about NSP forecasting practices into a recommendation to defer the decision to build a plant. It did take note of the fact that NSP had been in the process of negotiating with Manitoba Hydro (in Canada) for the importation of 10 MW of 'firm' power, which would certainly help ensure the state would have access to sufficient capacity in the event the 1979-1980 unit was delayed. Beyond this, however, discussion centered on the strategy NSP should follow if the plant came on line and demand levels were subsequently found to be lower than anticipated. Of the various possible courses of action NSP could take in response, two -- the sale of power to other systems and aggressive promotion of electricity demand -- were felt to be "highly undesirable," while the options of early retirement of older plants or the delaying of construction of additional peaking capacity were recommended.

Although not convinced of the need for the plant, the second task force, like the first, was willing to address the question of the most desirable site. In this regard the group decided to consider not only the seven "most promising" sites which NSP had suggested, but also the seven sites at which plants were currently on-line or under construction. In addition, perhaps motivated by a desire to be comprehensive and/or to assert its independence, the Task Force included the Carver Rapids site to which the first task force had objected. There was even some sentiment to expand the list still further, in light of a statement from a University of Minnesota geographer that some good locations might well be found in the northern part of the state. This was an area of relatively sparse population and, unlike that which Commonwealth Associates had surveyed, was not prime agricultural land. Ultimately, Task Force II did not pursue this suggestion, although it is not clear whether the reason was related to time pressures or because a majority of the members felt that this would represent a departure from NSP's request and the EQC's charge to the group.

Like its predecessor, Task Force II attempted to find some way to strike a balance between land use considerations on the one hand, and air and water quality degradation on the other. The real question, in its collective mind, was how to compare the disadvantage of devoting new land to the production of electric power against the environmental and health impacts which might occur when several large plants are operating in proximity to populated areas. From the available documentation it is difficult to learn much of the actual group deliberation on this issue, although we do know it felt compelled to seek opinions from the public at large via a series of informal hearings in the vicinities of the candidate sites. And it is easy to surmise from the final report how the Task Force ultimately resolved the dilemma.

Basically, instead of attempting to make an unequivocal choice, the Task Force reduced the list to two 'first' choices along with a third option (known as Jordan County) ranked slightly below the other two. The leading candidates were the Sherburne County site, where the 1976 plant which the first task force had labored over, was presently under construction; and one of the sites which Commonwealth Associates had identified, known as the Henderson site, about 60 miles southwest of Minneapolis. The difference between them was, in essence, that Sherburne was a site with a relatively high level of air quality impact, and one that had property tax implications (i.e., the effect of putting several generating units within one locality); whereas Henderson was a site which involved displacement of people, the withdrawal from use of highly productive agricultural land, and the creation of a large reservoir for make-up water (which further aggravated the need for usurping farm land). Being so dissimilar the alternatives obviously represented a real choice, but it was a choice the Task Force did not feel that it -- nor anyone else -- was yet in a position to make intelligently.

One realization which apparently bothered the task force members was that this sort of tradeoff had always been made without the benefit of a state land use plan, and was thus unduly susceptible to what many considered to be spurious considerations. A prime example in their minds, although it was not discussed at length in the final report, was the distorting effect of the property tax windfall that normally accrues to the municipality where a plant is to be located. In fact, when the first task force was deliberating, a delegation of local officials from Sherburne County had come to ask that the plant be sited there for just that reason. Still another concern was that pollution control regulations might produce poor site choices, as can be seen from the following excerpt from the final report of Task Force II:[17]

> The Task Force is concerned about permitting air quality considerations to be the overriding factor in determining land use. While ambient air quality is a boundary condition, the air quality standards were never meant to be considered a land-use plan in and of themselves and they should not be used as such. A multitude of factors must be incorporated into land-use decisions. Thus while Sherburne clearly causes a marginally heavier pollution load to fall on a larger number of people, the Task Force is not convinced that the degradation in air quality which would result from additional generating units at the Sherburne site, or for that matter at the Henderson site, would cause a violation of the ambient air quality standards in the Twin Cities. If the Task Force were convinced that any site would result in such an effect on the Twin Cities' air quality, then we would feel compelled to reject that site.
>
> The Task Force understands that in considering the issuance of air discharge permits, the MPCA assumes

that the worst possible set of circumstances of record will prevail without regard for the probability of occurrence of these circumstances. In considering the permit for additional generating capacity at Sherburne County, this would entail assuming the worst possible meteorological conditions of record occurring at the same time as the peak demand for electric power and both prevailing for at least a 24-hour period. The Task Force is concerned that an adequate evaluation of the probability of such an air pollution episode had not been considered and recommends that more complete calculations be undertaken to evaluate the frequency and severity of such episodes at the sites under consideration.

With consideration to these implications, the Task Force recommends that the Henderson site be favored for power station development if, and only if:

(a) the Environmental Quality Council, working with its member agencies, establishes that the Sherburne County site would lead to substantially greater air quality impact than would the Henderson site;
(b) the Environmental Quality Council publicly announces its commitment, together with a timetable to fulfill the commitment, to propose to the Minnesota Legislature a specific and comprehensive land-use plan to guide development in Minnesota and protect the state's natural resources;
(c) the Environmental Quality Council publicly announces its commitment to actively support meaningful reform of property tax-distribution laws -- for power plants at the very least -- so that taxes do not accrue exclusively to local taxing units of government.

Thus, the whole tone of the Task Force II report was one of preference for the Sherburne site and of concern that there was no formal procedure for considering land use questions on the same basis as air or water quality aspects.

Task Force II was in some sense following in the footsteps of the original task force by giving the benefit of the doubt to protection of land-use values, and as it turned out the basic reasoning was again at variance with that of the state environmental bureaucracy. The first indication of disagreement came in a separate report to the

EQC from the Minnesota State Planning Agency, which -- despite the fact that agencies of this type are normally sensitive to a broad range of quality-of-life issues -- seemed concerned primarily with problems of air and water resource impact. Against the criteria employed in these areas, another plant at Sherburne County was deemed undesirable because:

* it would be a major consumptive use of the Mississippi River, potentially critically reducing water flow at the Pigs Eye secondary treatment plant;
* it would increase the level of non-carbonate hardness in the water supplies of both Minneapolis and St. Paul;
* it would cause significantly higher concentrations of sulfur dioxide in Minneapolis, St. Paul, and St. Cloud (a community of 40,000 people located 50 miles northwest of the Twin Cities).

On this basis the Planning Agency concluded not only that the Henderson site was preferable, but also that the site in Jordan County (the third choice of Task Force II) was second best.

Ultimately, in a written resolution and report to NSP, the Environmental Quality Council concurred with the Planning Agency -- Henderson was the first choice, Jordan County the second. Rollie Comstock, in his later testimony to the Joint Committee on Atomic Energy, felt that this recommendation came only after "a lot of jockeying -- essentially political in nature -- among the officials who sat on the EQC."[18] According to him, property tax effects played a particularly important role in arriving at the decision in the end.[19] But such was hardly evident in the resolution itself, which was more interesting for what it did not say than for what did. Simply put, it ignored the points made by Task Force II and completely skirted the land use aspects of the decisions. No mention was even made of the Sherburne site, or of the "if and only if" conditions which the advisory group had specified in connection with a decision to go ahead with Henderson.

Thus, the rationale behind the governmental position in this case was no less obscure than it had been in the first, and NSP was left -- once again -- in the predicament of having to make a site choice in the absence of consensus between a citizens task force and a collection of agency officials. This time, however, the divergent opinions had arisen in a very different sort of institutional context.

The EQC was now a formal and prominent part of the executive branch, and it would be most imprudent to discount such an unequivocal signal as the Council's manifest aversion -- however poorly rationalized -- to the Sherburne site. What's more, because Task Force II was set up as advisory to the EQC, not NSP directly, the onus of rejecting the citizens' advice was also in some sense shifted away from the company. With these factors in mind it is hardly surprising that NSP, with very little comment, announced its acceptance of the EQC recommendation and proceeded to prepare the necessary applications for a new plant at Henderson.

Ironically, the seemingly conservative decision to go with Henderson eventually became problemmatic, and was never implemented. Part of the reason was that a subsequent engineering study identified difficulties with soil conditions as well as with a parameter that had generated concern over the Sherburne site -- water availability and flow. More important, though, was the fact that when NSP started to make the necessary purchases of land, the farming community began a political campaign to stop the project.[20] An organization known as SAVE (Save America's Vital Energy) was formed, grew quickly to a membership of over 500 farmers and local residents, distributed information critical of the need for and environmental impacts of the proposed facility, and sponsored mass demonstrations which received extensive media coverage. Despite NSP attempts to counter this resistance and mobilize support in other sectors of the community, the SAVE lobby ultimately prevailed. By the fall of 1974 the company was ready to drop its plans for Henderson, and by early 1975 the EQC had begun to waffle on its opposition to a second unit at Sherburne. The end result was an agreement to proceed with the building of 'Sherco II', which SAVE applauded while reiterating -- as Task Force II had -- its concern about need for power.

Can all of this be taken as vindication of Task Force II's original judgment as to the most appropriate course of action? Before jumping to conclusions, one should realize that no significant local opposition had surfaced at the public hearings the group held in the Henderson area, nor even when the report naming that site as one of the first choices was published. One possible explanation is that at both times options were still relatively open, but another important factor may well have been the knowledge that the Task force was inclined to specify very demanding conditions under which the site should be favored. In other words, the fact that political resistance had not yet crystallized at Henderson may be an indication that the Task Force was unusually perceptive -- more so than those to whom it reported -- in recognizing the importance of land-related values in rural Minnesota, and in anticipating the problems that would stem from the intrusion of uncoordinated development on the agricultural way of life.

If the question of Task Force II's acumen regarding Henderson must remain something of an open question, one thing is certain -- the alternative site which it and its predecessor found to be 'least objectionable' has proven useful to Northern States Power. Later in the 70s Sherco II came on line there, and at this writing a third unit has recently received certification of need and approval of air and water discharges from the appropriate regulatory agencies in the state.

A REVIEW OF POSITIVE RESULTS

Based on the preceding narrative, it seems fair to say that NSP's novel attempts to bring outside interests into the siting process were largely successful. This is true even if one measures success by the narrow criterion of project implementation, since two important facilities were spared from prolonged and dogged opposition and were thus able to come into service as scheduled. In the second case, of course, the work of the task force did not preclude controversy from developing over the Henderson site, but then again there is really no reason to expect that it should have. For one thing, it is unrealistic to think that all opposition from local parochial interests can be eliminated via open planning, especially in a case where the primary purpose is to obtain input from environmentalists and others whose perspective is essentially a regional one. It should also be noted that Task Force II had been very leery about dedicating new land to power facilities, precisely that aspect of the Henderson proposal which gave rise to public resistance. Yet NSP chose to defer to the contrary recommendation of the EQC and paid little attention to the conditions the task force had specified as necessary before support for the Henderson project could be forthcoming.

The community relations nightmare at Henderson was relatively inconsequential in the end, however, insofar as the switch to Sherburne was made with apparent ease. Nor did the difficulty encountered with the farming community seem to offset in any significant way the progress the company had made toward reducing the level of skepticism the public at large had come to manifest toward its plans and policies. One member of Task Force II put it this way:[21]

> I think it is beyond question that bona fide participation by a broad public segment will enhance acceptance of the end result. Perhaps it will not overcome all difficulties, but it cannot help but make acceptance easier. Many, many times in our experience people would come up and say something to the effect of: 'Thank you. I don't know what the result will be, but I believe you are sincere (and) are looking for an honest answer. I've had my say, and I'll feel good about your answer.' What else can we hope for?

Quite apart from such positive results in the area of general public relations, perhaps the most important outcome of the NSP experience with open planning was that the perceptual lines which had previously separated concerned citizens and utility people into distinct camps had begun to blur. In talking to a local reporter, for instance, the acting chairman of the first task force -- William Cunningham -- made this comment: "I certainly tend to view NSP now as less of a monolith than as composed of individuals."[22] Undoubtedly there was a similar improvement in NSP's image in the eyes of other environmentalists, making it easier to consider the company's interests and points of view on a variety of policy matters.

Along the same lines, it is equally significant to note that NSP's view of citizen activists also changed for the better as a result of the participatory process. Conventional wisdom in the industry, based as it was on myriad stereotypes of environmentalist behavior,* had predicted that the prospects for constructive interaction would be slim; yet the problems most often cited as leading to a breakdown of collaboration either did not occur or proved to be surmountable. To begin with, there was no evidence that the people on the task forces were in any way intransigent, irresponsible, self-serving, or otherwise unreasonable. Early in the first set of deliberations, for example, the participants themselves gave overt indication that obstructionist tactics would not be tolerated; they simply

---

* A detailed discussion of utility concerns in this respect is presented in Chapter Two, <u>supra</u>.

told one of their number to 'put up or shut up.' What's more, having dissuaded extremist behavior, both advisory groups demonstrated a clear willingness to engage in a balancing process and to trade off one environmental objective for another. The second task force in particular worked long and hard in this respect, trying to weigh air quality considerations against land use preferences.

Another area where skeptics might have expected the process to falter is in the willingness of the citizen groups to work within constraints, but here again significant difficulty did not materialize. First, although the need for power was examined in both cases, participants did not become pre-occupied with long-range planning issues nor did they insist that policy disagreements in that domain be resolved as a prerequisite to consideration of siting issues. They were willing to say, in effect: 'Well, we're really not quite convinced that you need a new plant such as this one. Demand may not grow as quickly as you expect, or you may be able to buy the power elsewhere. But if you are going to build, this is our suggestion.' Secondly, while both groups were frustrated by realities that served to limit or distort the site selection process, each was able to get down to business and decide on a preferred site rather than grumble ad infinitum about imperfect circumstances. The first task force, for example, was not deterred by the fact that it had only four alternative locations to choose from and a mere three months to complete an evaluation. And the second task force, faced with certain legal and political influences having to do with property taxes, responded not with inaction on the siting question but by advocating legislative reforms that would prevent local revenue considerations from having an undue influence on decisions of regional consequence.

The NSP experience also contradicted allegations to the effect that citizen participants are typically unable to deal with complex technical information. While it was incumbent upon the company to provide task force members with some background on the siting process, this was not a troublesome procedure and the group was able to assimilate the material with relative ease (indeed, some complained about being 'spoon-fed'). What's more, the respective task force reports indicate that each group was able to deliberate intelligently on complicated issues involving, for example, the capabilities of modern stack scrubbing devices and the meteorological assumptions underlying ambient air quality standards and impacts. Particularly impressive was the report of the second task force, which reflected a high level of sophistication not only in considering air pollution matters in detail but also by taking a fresh look at the methodology of load forecasting. Specifically, a review of state-of-the-art techniques led the group to call into question NSP's use of time-trend analysis to generate demand projections, at a time when most of the utility industry still accepted time-trending without question.

In the available documentation, by the way, one also finds evidence that the task forces made important contributions to the planning effort in the area of non-quantifiable social impacts. The second group, for example, was sensitive to the fact that development of one potential site would have disturbed Indian burial mounds; another would have removed several families from land homesteaded by their grandfathers or great-grandfathers; and two others would have disrupted closely-knit communities drawn together by nearby churches.[23] While no site was rejected for such a reason only, this kind of local understanding of land use implications did enter into the task force decision process and eventually proved to be at least as influential as concerns about effluent levels and the like. In this regard one should recall that NSP, in its post-mortem on the controversy over the nuclear unit at Monticello, realized that it had "misguaged the issues" underlying the opposition. Thus the task forces had shown themselves to be a useful vehicle not only for incorporating information that might otherwise be missed, but especially for providing a different perspective as to the significance to be attached to such information. There was, after all, no evidence that the government officials -- the other group that might have served to heighten the company's awareness of land-related impacts -- were either inclined to pursue the question or 'in tune' with grass roots sentiment to the same extent as task force members.

Finally, the process of "rubbing elbows" with citizen groups had one benefit that was totally unexpected by NSP, and that not even a non-utility proponent of participation would anticipate. To wit, the opportunity to interact with outsiders apparently had a salutary effect on the morale of certain elements of the company staff. Again from Rollie Comstock comes a description of what took place:[24]

> Sometimes we'd have twenty, thirty, forty NSP people (at the task force meetings), hanging around the audience. . . . Somebody would ask a question, or get into a line of inquiry I was incapable of handling,

> and (these NSP employees) would kind of tenatively raise their hands, and the next thing I knew they were up at the podium and they were talking . . . And pretty soon they were getting involved . . . You'd have engineers, economists, generalists of various kinds . . . meeting in the company . . . saying, 'How can I help?', and 'Gee, do you need this kind of information from me?'. . . And this was all volunteer time. So that there was a staffing cadre that evolved informally. . .

This commitment of personal time, Comstock went on to say, was a reflection of:

> . . . a new vibrancy within the corporation, particularly among the younger engineers. . . Many began to see corporate problems with a broadened understanding and a larger perspective.

The open planing initiatives, in other words, had served to effect in the minds of some company personnel a redefinition of an old public service doctrine, to which utilities had traditionally subscribed but which had become more difficult to relate to in an era of acute antagonism and dispute.

What all this adds up to is a strong testament to the effect that advance citizen participation in power plant siting can work, and that it can be beneficial to a utility in reducing conflict over individual projects, in creating a more temperate political climate, and even in some unexpected ways like improving the company self-image. Perhaps the most significant insight to be derived from the case, however, is that it suggests there is reason to have faith in human rationality and to allow at least certain kinds of decisions to be made through pluralistic means. Rollie Comstock put it this way:[25]

> There is an inherent wisdom in the body politic that on balance has served this country well in many troubled times . . . I become impatient with the militant assertion of 'lets tell our story'. It assumes a basic disparity between utility goals and public goals.

These comments should take on particular significance because they are based, not on wishful thinking, but on hard fact.

WHY THE SUCCESS?

To this point our reflections on the NSP case have demonstrated that the potential for effective public involvement in site selection is indeed real, but nothing has been said about what it takes to realize such potential on a recurring basis. One encounters an immediate problem in this respect, however, in that it is difficult if not impossible to ascertain to what extent outcomes can be attributed to how the process was designed, as opposed to a variety of other important factors (e.g., personalities of the individuals involved, dynamics of group behavior, etc.). Would an otherwise identical experiment generate similar results if carried out by a different utility in a different context? Obviously, there can be no clear-cut answer; there is no such thing as a set of necessary and sufficient conditions for successful open planning. Rather than try to generalize about cause-effect relationships, therefore, it is now perhaps most instructive simply to identify both the idiosyncratic circumstances and the more-or-less conscious actions by NSP which together -- in some mysterious combination whose intricacies no amount of analysis would be able to reconstruct accurately -- seemed to have a bearing on the scenario which unfolded.

First, one must consider the broad contextual parameters of time and place. Several observers, including Rollie Comstock, have pointed out that conditions in Minnesota during the early '70s were in certain ways conducive to the sort of cooperative venture NSP sought to initiate. Not the least of the relevant factors in this regard was the rather unique demographic situation, which if nothing else afforded a major logistical advantage. One task force participant described it thusly:[26]

> More than half the population in the state lives in the Twin Cities metropolitan area. This means that the state government, the only really large utility, the university, and the headquarters of all the environmental groups are here in close proximity. It would have been very difficult for us to have even met regularly, let alone come to any decision, if members had been scattered over a large geographical region. After the NSP task force, several of the environmental groups tried to form an umbrella organization to coordinate activities on other fronts. We wanted to include people from northern Minnesota so we agreed to meet at Sandstone, about 80 miles from the Cities and

> half way to the Iron Range Cities. It was so difficult to make that trip, especially in the winter, that our coalition never really accomplished anything.
>
> . . . Another factor that was important was the presence of the university. Most of the committee members who stuck through to the end were from the University of Minnesota. It provides a pool of people with technical knowledge, time, and an interest in this sort of process.

It is perhaps also significant that while the Twin Cities region is heavily developed, it is not populated in the same sprawling way as, for example, the Bos-Wash corridor or the eastern seaboard generally. Within a half hour's drive from Minneapolis, the city -- and with it the special intensity of urban life -- can be left completely behind. There is simply less congestion, and partly because of this a somewhat more relaxed atmosphere exists than what is characteristic of metropolitan areas elsewhere in the country.

A second and probably even more consequential factor has to do with the social climate in Minnesota. For one thing, as a relatively new society it is not difficult for one whose parents or grandparents farmed virgin prairie, or helped build the economy of the Twin Cities, to feel that it is still possible to have a personal impact on the institutions that affect one's life. And as William Cunningham notes, this perception had ethnic roots as well:[27]

> This is the midwest of sturdy yeoman Scandinavian or German farmers. There is a tradition here of respect for authorities, progressive politics, open government, and feeling that one ought to contribute to society. That was especially true in the early '70s. Even if one felt personally cynical, there was pressure to represent the general public who would probably expect us to get down to work and to do the best that we could. That climate surely doesn't exist other places. It may not exist here any more but it influenced our task force.

Rollie Comstock had similar things to say when asked why the first task force had been willing to undertake an evaluation at all.[28] He answered that, compared to other areas in 1970, there was considerably less cynicism and anger amidst the general populace of the upper Mid-West, and a greater willingness to work within the establishment system. Even the earlier battles NSP had fought with environmental groups had fostered an attitude of grudging respect on both sides, rather than one of total alienation.

One can take this argument just so far, however, because the fact is that both task forces deliberated in an atmosphere of skepticism as to the company's motives and a general lack of trust on both sides. This was particularly evident during the early meetings of the first task force, but it continued to be the case (albeit to a lesser extent) throughout all subsequent proceedings. However pervasive it was, though, this distrust never grew to a magnitude that would have precluded good faith bargaining and make the process, as a result, an abortive failure. In trying to understand why, one must turn to the subject of process design and, specifically, to the question of whether there were identifiable steps on the part of NSP which helped maintain a threshold level of trust.

What makes this a particularly intriguing question is that in certain respects the company's original proposal for an interaction process could easily have had the effect of heightening environmentalist fears, not mollifying them. First of all, while NSP explicitly stated a desire for input on a wide variety of company actions and policies, it was also insistent that a project had to be sited in a period whose brevity (three months) would not allow time for prior discussion of the long-range plans underlying that project (especially the determination of need for power). Secondly, the project itself was highly circumscribed in that key features of the plant were presented as a given (i.e., coal-fired, closed-cycle cooling), and only four company-owned sites were provided for evaluation. The lack of options regarding fuel and cooling method were not likely to cause the task force much distress of course; it is hard to believe they might have preferred a nuclear unit with once-through cooling. But the potential participants were not at all comfortable with the other critical constraints, either of which might well have been interpreted as a sign of insincerity and prevented the experiment from ever getting off the ground. NSP may have felt it had no choice but to specify restrictive groundrules, but the risk of doing so was great nonetheless.

Fortunately for all concerned, this major disincentive to participation was apparently outweighed by a perception within the citizen groups that they could not afford to pass up the opportunity to make "some contribution toward siting and developing the 1976 plant."[29] One could explain this attitude as being consistent with the prevailing mileau of good citizenship described above, but it may also have been based on a concern that constant opposition to utility projects could have a damaging effect on the credibility of the environmental movement. Whatever the reasons for the commitment to playing a constructive role, however, without it the task force might never have gone beyond the diatribe over promotion of electric power which marked its first month or so of discussions. Certainly NSP was not in a position to shift the focus to site specific issues; at best, the tight schedule it had laid down probably helped push the group toward making the transition, i.e., when they decided enough time had been devoted to general criticism of NSP and castigated one of their own members. Still, had it not been for the basically positive spirit of the task force, the time constraint might have only aggravated the situation.

Notwithstanding the fact that NSP was the beneficiary of a cooperative predisposition on the part of those it invited to serve on the task force, it is important to realize that the company also made a concerted effort to avoid polarization and to earn some degree of confidence from the group. In order to reinforce the point that there was no hidden agenda, for example, the company made no secret as to its preferred site; made a point of using outside experts to provide certain kinds of information; and was conscientious about collecting all the data which the task force desired to peruse. In short, the company simply laid its cards on the table and asked for a response. What's more, anticipating the need to afford the task force some room to maneuver, NSP was careful not to ask the participants to represent their respective groups in any official sense. If they had, it is quite possible that several individuals would simply have declined to become involved, and those who remained may have been forced to take more critical positions and be less flexible in considering various courses of action.

Further evidence of NSP's sound judgement -- or at least good intuition -- is to be found in the fact that it decided against hiring neutral third parties to work directly with the task force. It was originally thought this could help overcome the problem of mistrust, but a different rationale eventually prevailed. Rollie Comstock put it this way:[30]

> I think it is essential to successful communication that the real parties in interest are face to face. I am convinced that the presence of an intermediary gets in the way.

Undoubtedly this lack of outsiders to act as a buffer caused the utility some occasional discomfort, particularly at the early meetings, and it required a heavy commitment of time by company personnel. But in the end the decision appears to have been critical to the success of the program; Comstock, in fact, called it "one of the wisest we made."

As a final comment on how NSP structured its first experiment, and one that relates closely to the decision to interact directly with the task force, it is important to highlight the part played by Rollie Comstock himself. While a discussion of his role is largely absent from the available documentation (much of which he authored), it is not difficult to imagine the qualities necessary to fulfill it. To begin with, one must have a balanced perspective on both the actions of the utility and the concerns of the environmentalists. And one must be able to function comfortably in a charged atmosphere, to provide direction and insight and be able to accept criticism without losing sensitivity. Apparently, Comstock was abundantly possessed of such qualities, especially the latter. Consider this testament by William Cunningham.[31]

> Your suggestion that Rollie Comstock had a lot to do with the outcome of our experiment was certainly correct. The fact that he was a part of the company and yet was willing to stand up and take abuse without blowing up was very important. He was what Tom Wolfe calls a 'flakcatcher.' It was fun to take pot-shots at the company for the first few meetings, but then we began to sympathize with Rollie. If the meetings had been run by a third party we might have continued for a long time to castigate the company in absentia. . . if a company wanted to repeat the experience, I would strongly recommend that they assign a smart, personable, clever flakcatcher for the public to vent their wrath on.

The lesson to be learned can be stated succinctly and by using Comstock's own words: "selection of the lead person is critically important."[32]

At this point it is easy to fathom how the temper of the times, generally, and NSP's style of working with the task force, specifically, had a lot to do with making the first interactive experience a satisfying one overall. But in terms of getting over the hurdles of mistrust and laying the groundwork for future cooperative endeavors, these factors combined were probably less influential than one other: NSP's acceptance of the task force recommendation to build at Sherburne. What made this so difficult, one should recall, is that it meant rejecting a counter suggestion from regulatory officials that the company use the site it had originally preferred. Fortunately for NSP, this choice did not seem to prejudice the licensing process for the Sherburne facility, and it certainly paid dividends in the area of goodwill. As the task force chairman said, it "indicates that NSP really did have an open mind and that they were listening to us."[33]

The company's mind was not quite so open relative to what the task force had to say about demand management, of course, but at the time this was perhaps seen as forgivable (at least temporarily). The most important thing was that chief executive Ewald had essentially been true to his word when he had said, at the outset, that NSP would give serious consideration to task force opinions on the question of siting new plants. What's more, the second task force experience undoubtedly reinforced this perception of responsiveness on the part of the utility. In providing more time and a much larger inventory of potential sites, for example, it demonstrated an ability to learn from past mistakes and to be sensitive to environmentalist perspectives. Perhaps the latter was not so much in evidence when the Henderson site was finally chosen, but then again Task Force II had waffled on the question of a preferred location and the signal coming from government officials was much more forceful than it had been in the previous case. Even so, it is significant that in the end NSP was flexible enough to change its plans in favor of another unit at Sherburne, rather than carry on the battle with the opposition at Henderson.

One can see, then, a pattern of flexibility in the company's actions which, if not manifest across-the-board, was certainly sufficient to demonstrate good faith and thus make the interaction process workable on the whole. Such flexibility, we believe, was the real key to success in this one particular episode of collaborative problem-solving.

THE LEARNING PROCESS CONTINUES

Having now described and evaluated NSP's initial ventures in citizen participation, there is but one additional part of the story to recount: that of how the experiences of 1970-1972 served to shape subsequent events in the field of power plant siting in Minnesota. To provide a detailed longitudinal analysis of the subject is well beyond this paper's intent, but an important source of perspective would be lost if the evolutionary nature of the situation were not highlighted in at least a broad way. The fact is that, however satisfying the task force experiments had been to both NSP and the environmental community, there was a feeling in both camps that the basic advisory model employed to that point -- wherein NSP received input from concerned citizens and agency officials on an ad hoc, voluntary basis -- did not constitute an acceptable approach to decision-making in the long-run. In other words, the more fundamental issue of defining an effective utility-citizen-government relationship had yet to be resolved.

What made the experiments less than an unqualified success in this regard had nothing to do, really, with any doubts about the constructiveness of public involvement. Rather, the problem emanated from a source that was least expected -- from the government personnel who took part in the open planning process. These were the people with primary oversight responsibility and ultimate approval authority over utility projects, and yet, in the company's view at least, they had not exhibited the same degree of commitment to the interactive concept as had NSP staff and the respective task force members. This seems to have been the result of several factors, not the least of which was that the agency heads serving on the first task force had been subjected to quite a bit of verbal abuse during its early meetings. Nor did anything happen thereafter, it appears, to offset this inauspicious start and make the process less awkward for these officials. They were, after all, under pressure to voice opinions and even take positions in advance of formal regulatory proceedings, which could well be seen as prejudicial to their ability to make objective decisions on the merits of the case as ultimately presented. And they were being forced in some sense to share the function of evaluation with an independent group of citizens who might at some point challenge, directly or indirectly, their professional judgement. Given this potential for infringement of role, it is easy to see how the bureaucracy might have felt threatened by the presence of the NSP-initiated task forces, and resentful of the perturbations open planning had introduced to its traditional operating climate.

Apart from such difficulties of attitude, another source of disappointment for NSP was what it considered to be the sub-par performance of the state agencies in substantive matters. The agency heads, for their part, seemed unduly responsive to political considerations that were arguably extraneous in terms of the broader public interest (e.g., property tax benefits); yet at the same time, they had been palpably inflexible when it came to balancing the mandated concerns of their special-purpose agencies (e.g., air quality) against competing objectives of comparable social value (e.g., agricultural land preservation). Nor were the company's expectations met with respect to the quality (and quantity) of contributions from government at the staff level. As Comstock put it: "I recognize that most state agencies are under-staffed and over-worked, but so is everyone else."[34]

Thus, despite NSP's concerted efforts to establish a link between advance citizen participation and the machinery of state government, neither of the two (very different) approaches it had tried resulted in a truly integrated process of site evaluation. Even in the second experiment, which took place under the auspices of the Environmental Quality Council, there had been a marked reluctance on the part of agency personnel to work closely with Task Force II. That such an arm's-length relationship developed was particularly distressing to Comstock, the strong advocate of elbow interaction, whose testimony to the Joint Committee on Atomic Energy was prefaced by the following statement:[35]

> Despite these ringing testimonials to voluntary and open planning, I nonetheless am uneasy about it. As I describe the process, you will note that it essentially involves an end-run around government. All the potential adversaries gather together, negotiate a solution, and present it as a gift-wrapped package to government for approval -- and usually get it. In a way, an analogy might be to a criminal case, in which the attorneys out of court decide the essential question of guilt or innocence and expect the jury to concur.

The possibility that the jury might not concur was very real to Comstock, of course, insofar as the EQC site recommendations had in fact differed from that of the respective task forces on both occasions. While such divergence eventually proved inconsequential, the prospect of repeatedly having to cope with a 'damned if you do, damned if you don't' situation was most assuredly not an acceptable one to the company. To avoid it, though, something more would have to be forthcoming from state government in the way of support and leadership of future open planning endeavors.

Interestingly, the environmental community also had reasons to be dissatisfied with the relatively subordinate role in which state officials had been cast in the initial participatory projects. Basically, these center on the idea that in the long run it is government and government alone -- not a consortium of utilities and citizen activists -- that should be expected to represent the environmental interests of the public at large. Consider, for example, the following sentiments expressed by William Cunningham:[36]

> In looking back at the experience . . . . I'm not sure that I endorse the process as a way of making decisions. There are several reasons for this. One of them is that I'm not sure that private citizens, no matter how well-meaning, can make balanced decisions in a complex case like this. In the case of the first NSP task force, we considered land use questions almost exclusively because none of us knew much about air pollution. In the intervening years I have often wondered whether we made a serious mistake in recommending a site upwind of the metropolitan area. This is especially true since it has turned out that NSP will now use that site for three large (600-700 MW) plants. If we had known more about the contents of the plume and where it would fall out (right on Minneapolis-St. Paul) we might not have been as comfortable with the Sherburne County site. . . . Ideally, these questions should be considered by competent impartial public servants . . . .
>
> Another reason that I am suspicious of any open planning or mediated negotiations is that I see a strong potential for co-opting the environmentalists without much chance of making fundamental changes in the corporation. While it is true that some of the younger engineers and lawyers at NSP may have become more interested in and conversant with environmental issues during our meetings, I am convinced that they have

> little impact on company policy. When you come to the bottom line, the interests of the company and the stockholders take precedence over any moral or ethical qualms of the junior staff.

From this sort of statement it is evident that, whatever progress had been made toward a more understanding and informed relationship between utilities and public interest advocates, the latter were not so naive as to think of themselves as capable in the long run of acting as de facto regulators and effecting appropriate decisions on behalf of the general welfare.

Against this background, it should be no surprise that NSP and Task Force II alike came away from the siting project of 1972 with a nagging sense of unfinished business. As a result the latter, like its predecessors, continued to meet after its recommendation had been submitted, with the intent of devising a new set of institutional arrangements to govern power system planning, site selection, and facility licensing. To make a long story short, the task force issued a report describing the key elements it thought should be incorporated into a statute, and then the legislature enacted most of the group's basic recommendations in the form of the Power Plant Siting Act of 1973. After a lengthy dialogue the bill had come to the legislature with the support of both utilities and environmental groups (albeit with varying degrees of enthusiasm), and this support was apparently sufficient to overcome the exclusive opposition of the state EQC, which (to quote Comstock again) "saw the whole effort as a dilution of its present activities."[37]

How the law could have been construed in this way by the EQC is somewhat difficult to fathom, since if anything the Council's jurisdiction was greatly expanded. In considering the problem of how to upgrade and add coherence to the state's role in electricity planning generally, Task Force II (and NSP) had reasoned that the task of site identification (among others) had to become a formal part of the mandate of a single agency, and the legislature decided the EQC should assume this responsibility.[38] Specifically, the Act directed the Council to assemble and publish an inventory of potential locations for large power plants and transmission lines throughout the state. Utilities seeking to build new facilities are allowed to propose a site not in the inventory, so long as the license application states the reasons and includes an evaluation using criteria established by the EQC. However, the final choice is made by the Agency -- not the company -- which compares the merits of available candidates and then designates whichever site it concludes is most suitable.

As if all this were not unusual enough (only one other state -- Maryland -- has taken on such direct involvement in siting), the new law went on to provide explicitly for maximum practicable input from the public. Among the innovative provisions in this regard were as follows:[39]

* The Council shall promptly initiate a public planning process where all interested persons can participate in developing the criteria and standards to be used by the Council in preparing an inventory . . . and to guide the site suitability evaluation and selection process.
* The Council shall appoint one or more advisory committees to assist it in carrying out its duties. Committees . . . . shall include a majority of public representatives . . . .
* The council shall adopt broad spectrum citizen participation as a principle of operation. The form of public participation shall not be limited to public hearings and advisory committees . .

Thus the Minnesota legislature, impressed by the workability of the concept, had decided to institutionalize it in a way that went even further in pluralizing the siting process than NSP had in either of its task force experiments.

At this point, to follow the Minnesota story further would be to branch off into an account of implementation experience under the Siting Act, which would no doubt be fascinating and yield a wealth of additional insight on the subject of citizen participation in power plant siting. But this task is too ambitious for present purposes and better left for a follow-up study. Suffice it to say that there is reason to believe, at least from the utility standpoint, that the "new, improved" model of industry-government-citizen interaction which the statute embodied has again fallen short of expectations, despite some 'mid-course corrections' made in a 1977 amendment. One indication of this is the CPA/UPA powerline difficulties described elsewhere in this volume,* although that particularly delicate case was one of the first to be handled

---

* See Chapter Nine, supra.

according to the new procedures and did not, therefore, benefit from the existence of any learning curve. Since then, however, even Rollie Comstock has opined that the system is not working very well and has expressed misgivings about "leaving public involvement in the hands of government."[40] Whatever arguments might be advanced on this point, pro and con, the important fact is that Northern States Power has concluded -- much as United Power Association has -- that the existence of a participatory governmental process has not obviated the need for direct communication with citizen groups, outside of the regulatory framework.

As might be expected, then, the company's style of open planning in the later '70s followed a distinctly evolutionary path in response to lessons of the past and new circumstances the future brought. For example, in its one major attempt to site a baseload plant since the enactment of the law, NSP decided to employ a more decentralized program of outreach than was characteristic of the earlier task force experiments. In this case, the company identified three broad candidate areas to screen for potential sites, and then began to communicate on an individualized basis with local interest groups. This seemed an appropriate course of action in view of the fact that strong local opposition to the 1979 plant had developed at the Henderson site, the potential for which the public meetings held by Task Force II in the area had failed to discern. Such would be avoided, presumably, by a strategy of interaction focusing on the grass-roots level, where input from individuals with a personal knowledge of the community would be valuable in assessing socio-economic as well as environmental effects. Another potential benefit, strictly from the public relations standpoint, was that this type of approach would give the utility a chance to build a supportive constituency for the project early in the game.

Such was not to be, however; as things turned out, significant political resistance developed in each of the regions involved, at which point NSP turned for the third time to its Sherburne County site (this unit, as noted earlier, had been licensed but not yet built). According to Rollie Comstock, the company has made a crucial mistake in that it had delegated the responsibility for involving the public to its regional field offices.[41] While the regional staff did use newspaper releases, newsletters, public meetings, and storefront offices to encourage citizen input, Comstock felt that they never fully understood the degree to which public participation had to be mobilized. In particular, he felt that the regional offices had devoted a disproportionate amount of time and energy to dealing with municipal officials, to the exclusion of other important groups.

While NSP has not had occasion to apply this lesson in siting another power plant, it did apparently prove instructive in the planning of a major transmission facility. In July of 1976, the company signed a contract with Manitoba Hydro to purchase 500 MW of firm power for thirteen summers, beginning in July 1980, and to construct a 500 kV line from the Twin Cities area north to the Canadian border, where it would connect to the Manitoba grid. For this project, responsibility for interaction with area residents and activist groups was retained by the head office in Minneapolis. There Comstock and his staff came up with an initial proposal, stressed that it was tentative and open to change, and set about soliciting proposals, suggestions and opinions prior to placing the matter before the Environmental Quality Board (the EQC, renamed), as required by the 1973 Siting Act. The end result was an approved route which, although roughly three-fourths had been relocated from the position identified in the company's original proposal, was eventually constructed without running into any severe political problems.[42]

Under normal circumstances, there would be no reason to hail NSP's success in developing this major new route as a singular achievement, with or without public involvement. However, what makes it especially significant -- and a fitting capstone for this case study -- is that it occurred just as the controversy over the CPA/UPA line was approaching the height of its intensity in a neighboring area of Minnesota. To be sure, unlike the CPA/UPA route to the west of the Twin Cities, the NSP route was to the north where there was little prime agricultural land and therefore no well-organized farming community whose hackles would be raised. Still, the fact that conflict had been so virulent and so close, that it had generated such acute animosity toward a sister utility, and that it had brought so much visibility to the alleged adverse effects of high-voltage transmission lines (especially health effects), certainly did not bode well for the NSP endeavor. Yet the project was relatively smooth in implemention, making it a dramatic counterpoint to the CPA/UPA experience and suggesting that the big difference in outcomes might be attributable to some key variation in the structure of the respective planning processes. There are probably many factors here, but it seems reasonable to infer that a large part of the explanation lies with the fact that the NSP's modus operandi represented a fundamental departure from the traditional decide-announce-defend approach, whereas that of CPA/UPA did not.

This points to a concluding observation as to the efficacy of NSP's initial decision to 'go out on a limb' and experiment with advance citizen participation. Clearly, its outreach efforts have produced mixed results in specific cases and have by no means led to comprehensive public acceptance of preferred actions; and its program designs have gone through several iterations without arriving at a formula that seems close to ensuring constructive interaction on a regular basis. But all this seems rather beside the point, because the company appears to have reached a higher goal -- that of making a transition away from confrontational, adversarial relationships with those affected by its decisions, and toward a more relaxed political climate. While the atmosphere could hardly be characterized as one of complete trust, the company does enjoy a reputation of being willing to listen to criticism of its proposals and to alternative suggestions, and consequently does not face automatic opposition to its plans. The lines of communication are now open, in other words, and on the bottom line that is what really counts.

NOTES

1. J. Holdren and P. Herrera, Energy, p. 208, Sierra Club (1971).

2. R.W. Comstock, "Memorandum Re Facilities Siting and Development," submitted to NSP vice-presidents of public relations and engineering (October 27, 1969).

3. R.W. Comstock, "Text of Remarks to P.I.P. Thinkshop Conference," Scottsdale, Arizona, p. 13 (October 1, 1970).

4. R.W. Comstock, "Northern States Power Company and Open Planning," remarks presented at Atomic Industrial Forum Meeting on Nuclear Power and the Public, New Orleans, Louisiana (February 26, 1973).

5. Remark attributed to Lawrence Cohen of the Minnesota Environmental Control Citizens' Association, in Northern States Power Company, Plant Siting Task Force: A Documentary Narrative, NSP presentation for an Edison Award, p. 13 (April, 1972).

6. R.W. Comstock, "Open Planning and Power Plant Siting," statement to the Joint Committee on Atomic Energy, United States Congress, p. 9 (March 20, 1974).

7. E. Ewald, letter to Mr. Raymond Haik, president, Isaac Walton League, Minneapolis, Minnesota (January 30, 1970).

8. Comstock, op. cit. note 6 *supra*, at 8.

9. Personal communication with William Zimmerman, Jr., School of Physics and Astronomy, University of Minnesota (July 18, 1983).

10. Comstock, op. cit. note 6 *supra*, at 9.

11. *Ibid*, p. 20.

12. Remark attributed to John T. Rose of the Minnesota Conservation Foundation, in M. Harwood, *Getting Together*, unpublished manuscript, Washington, Connecticut, p. 240 (1974).

13. Comstock, op. cit. note 6 *supra*, at 11.

14. "NSP Plant Siting Study and the Minnesota Environmental Quality Council," anonymous paper, p. 1, (1972).

15. Task Force on Power Plant Siting, Governor's Environmental Quality Council, "Findings and Recommendations on the Need and the Site for the Proposed NSP 1979-80 Electric Generating Station," p. IV-7 (December, 1972).

16. U.S. House of Representatives, Committee on Interior and Insular Affairs, *Energy Demand Studies: An Analysis and Appraisal*, Government Printing Office (September, 1972).

17. Task Force, op. cit. note 15 *supra*, at V-3, 4, 7, 8.

18. Comstock, op. cit. note 6 *supra*, at 14.

19. Personal communication with R.W. Comstock (Fall, 1978).

20. See "Fighting Back in Minnesota," in *The Powerline*, monthly newsletter of the Environmental Action Foundation, Washington, D.C., vol. 1, no. 6 (December, 1975).

21. Letter to D.W. Ducsik from William P. Cunningham, Department of Genetics and Cell Biology, University of Minnesota (August 1, 1983).

22. J. Blade, "NSP Gains in Eyes of its Citizen Advisors," *Minneapolis Star*, p. 4a (May 30, 1970).

23. Task Force on Power Plant Siting, Governor's Environmental Quality Council, "Future Power Plant Siting in Minnesota" (January, 1973).

24. Harwood, op. cit. note 12 supra, at 241-242.

25. Comstock, op. cit. note 6 supra, at 16-17.

26. Cunningham, op. cit. note 21 *supra*.

27. *Ibid*.

28. Comstock, op. cit. note 19 *supra*.

29. Northern States Power Company, op. cit. note 5 *supra*.

30. _Ibid_, p. 8.
31. Cunningham, op. cit. note 22 _supra_.
32. Comstock, op. cit. note 6 _supra_.
33. Blade, op. cit. note 23 _supra_.
34. Comstock, op. cit. note 6 _supra_, at 18.
35. _Ibid_, p. 3.
36. Cunningham, op. cit. note 21 _supra_.
37. Comstock, op. cit. note 6 _supra_, at 20.
38. Task Force II had recommended creation of an independent siting authority instead of reliance on the EQC, on the grounds that governmental agency heads each have their statutory obligations, and it is difficult to expect them to compromise those obligations to some competing considerations. For fiscal reasons, however, the legislature decided otherwise. Letter to D.W. Ducsik from Joseph F. Grinnel, Task Force II chairman (January 31, 1983).
39. See Southern Interstate Nuclear Board, _Power Plant Siting in the United States_, Atlanta, Georgia, pp. 188-193 (1974).
40. R.W. Comstock, remarks presented at the Edison Electric Institute Workshop on Utility Experience with Advance Public Participation in Planning, Washington, D.C. (March 4-5, 1982).
41. Comstock, op. cit. note 19, _supra_.
42. _Ibid_.

12

Open Power Plant Siting: The Pioneering (and Unsuccessful) Experience of Northeast Utilities

Dennis W. Ducsik

In 1968 Lelan F. Sillin left the Central Hudson Gas and Electric Corporation, where he had risen to the office of president and chief executive officer, to assume the presidency of what was then the twentieth largest electric utility in the United States and the biggest in New England -- Northeast Utilities (based in Hartford, Connecticut). A few days after Sillin arrived at NU, a prospective news release crossed his desk which made it clear that the company soon would be facing an energy-environment dilemma of the kind that was becoming commonplace throughout the power industry. The announcement awaiting his approval was that Northeast would investigate building one of the world's largest (at 1,000-2,000 MW) hydroelectric generating facilities at Schenob Brook in (mainly) the town of Sheffield, deep in the Berkshire hills of southwestern Massachusetts. The plant was needed to meet anticipated peaking requirements of the 80s, and with New England utilities committed to a major program of nuclear baseload expansion, economic and operating advantage pointed to pumped-storage as the hydro technology of choice.

But to attempt such a project in the bucolic Berkshires also had serious ramifications from a public interest standpoint, as Sillin himself well knew. Later he recalled having the following reaction to the draft press release:

For assistance in the preparation of an early draft of this case study, the author is thankful to Dr. Thomas D. Austin, a graduate student at Clark during the research phase and currently a staff economist with Energy Systems Research Group in Boston, MA.

> It was like somebody dropped a load of bricks on my back. Because I happened to know that area. I had hiked over it with my family, just the other side of the mountain. We'd camped there, and I knew what a lovely area it was.

Nor was Sillin's appreciation for the environmental trade-offs associated with the project simply that of a weekend outdoorsman. As the founder of regional planning in the mid-Hudson Valley, as a member of the National Industrial Pollution Control Council, and as a Presidential appointee to the National Citizen's Committee on Environmental Quality, he had far more than personal knowledge of the area to draw upon as a frame of reference.

If Sillin was less than thrilled with the burden of siting this new plant because of personal environmental values, he had another reason as well. Specifically, he had no desire for his first important project at NU to become embroiled in the sort of intense controversy which its neighboring utility to the south -- Consolidated Edison -- had been experiencing since 1962. This was the famed conflict over a hydroelectric plant proposed for Storm King Mountain along the scenic Hudson River, a conflict which ultimately spanned nearly two decades and which many consider to have spawned the environmental movement. Sillin, it turns out, had had a front row seat as the CEO at Central Hudson, which was actually the utility that had first conceived of the project. When it became apparent that the Storm King facility would be too large and expensive for its needs, Central Hudson suggested the site to Con Ed, agreed to purchase some of the power produced, and offered to help in securing the approval of local officials. Sillin's utility was thus spared much of the agony Con Ed had to endure in trying to secure a license from the Federal Power Commission (FPC), but its top management was still too close to the case not to be deeply affected by the political turmoil it engendered.

The recent and nearby debacle in New York, it should be noted, had also made a lasting impression on the people of the southern Berkshires. It would be inaccurate to describe the region as a hotbed of environmentalism at the time, but there is no doubt that opinion leaders and other public-spirited citizens had been sensitized to the potentially negative effects of hydropower development. What's more, there was an element of the populace that was likely to feel particularly threatened by any suggestion that a similar plant, of even greater size, might be located on their 'turf.' Dubbed the 'newly-landed' class by some, this was the well-educated, upper-income, ex-urban elite, among them a sizable contingent of the most successful publishing and advertising professionals in New York City. Their affluence had allowed them to discover and capture the special ambience of the region's splendid natural resources, and that same affluence -- along with the drive and expertise that had brought them to the top of their fields -- was available to help fight any technological progress perceived to be at the expense of the cherished landscape. If ever there was a time and place a utility might expect to hear the determined cry of "Don't Put It Here," it was in the Berkshire hills in the early 1970s.

The problem of political resistance, it should be noted, had been essentially unknown during the '60s to the several electric companies which would eventually be combined to form Northeast Utilities. In fact, the only difficulty in facility planning that had been encountered concerned land availability at a site on Northfield Mountain, along the Connecticut River near Turners Falls, where the Western Massachusetts Electric Company (WMECO) wanted to build a 1000 MW pumped-storage plant. The problem -- which for a while appeared insurmountable -- lay in the fact that the reservoir would have to be built in an area under the control of the state Department of Natural Resources (DNR), which had embarked on a popular program to expand the state's park and recreation system. The problem was resolved in 1965, however, in a rather lengthy bargaining process with DNR Commissioner Charles H.W. (Hank) Foster, which resulted in a land swap and a WMECO commitment to develop recreational facilities and to allow the Northfield facility to be used, if necessary in the future, to divert Connecticut River water in to the Quabbin Reservoir (the principal water supply of Boston). Groundbreaking began in 1968, with the compromise having been accepted as beneficial to NU, the interested state agencies, and the local populations (which welcomed the tax windfall and were generally unconcerned about water withdrawals from the already-impounded lower Connecticut). Indeed, Hank Foster was so favorably impressed by the exercise in collaborative problem-solving that he published a detailed account of it, entitled "The Northfield Mountain Pumped Storage Project: Counterpoint to ConEd."*

---

* The article appeared as Harvard Forest Paper No. 19, Harvard Forest, Petersham, Massachusetts (1970).

Lee Sillin, then, was stepping into a company that had not yet experienced heated controversy, and wanting to keep it that way (at least for the time being) he decided to postpone issuance of the press release on the Sheffield study. Putting the project on the back burner made it possible, first of all, to broaden NU's capability for environmental impact assessment. This occurred with the creation of a new unit, the Regional and Environmental Planning Department, to be headed by a man whom Sillin had previously worked with closely in regional planning for the mid-Hudson Valley -- Peter Stern, then chief of regional studies at the Tennessee Valley Authority. Secondly, the deferral also allowed time to consider other potential sites in southwestern Massachusetts, as well as those across the state line in the northwestern Connecticut region of the Berkshires. In hand for this purpose was a site reconnaissance study prepared by a consultant during the mid-60s, a study which had identified some 28 potential sites and then screened out four (including Sheffield) as the best candidates.

Considering the size of the proposed plant and the fact that no nearby river had water flows approaching that of the Connecticut, Northeast needed to find a large site that could accommodate both a lower and an upper reservoir. As it turned out, the most reasonable alternative to Sheffield was situated just ten miles to the south -- at Canaan Mountain, near Falls Village, Connecticut. At this stage, there didn't seem to be any reason to prefer one location over the other. Either would apparently satisfy NU's power supply objectives, yet neither could be built without considerable disruption to the surrounding communities and to the aesthetic qualities of the valleys involved. The Canaan Mountain site seemed to require somewhat less land, whereas the Sheffield (Schenob) site might allow for some recreational usage -- a potentially important compensatory factor. But there was no doubt that for both sites the environmental price of development would be high.

By the time the Canaan Mountain site had been identified it was well into 1969, and Sillin was beginning to feel pressure to put the planning process into high gear. Company engineers, pointing to the need to have a new plant on line by 1978 or 1979 and voicing concern over the potential for delay such as the Storm King project had experienced, were saying that even a ten year lead time was cutting things uncomfortably close. Under a best-case scenario, they felt it would take 2-3 years to get through the FPC licensing process (including the treacherous public hearing) and another five years to build the facility, leaving 2-3 years at most for exploratory study, land acquisition, and preparation of the highly detailed FPC application. The latter should be ready for submittal no later than 1972, Sillin was reminded repeatedly, so the sooner he could make up his mind as to the preferred site, the better.

The decision was hardly a routine or straightforward one, however, and Sillin recognized that there were no pragmatic criteria to apply as far as the environmental issues of choice were concerned. As he had once said: "How we effectively meet the public interest -- that is, how we provide power while protecting the environment -- is in the end a matter of judgement...."* In Sillin's view, moreover, it was not for a utility or anyone else to exercise such judgement unilaterally; rather, it should be done on a collective basis. He put it this way:

> What I'm really trying to say is that part of my own ethic and my own belief was...the imperative need (for) people being able to talk with one another... hopefully, as unemotionally as possible, about the issues with which they are faced, and on the basis of the kinds of intelligence and expertise that each of them has to make the highest quality of input and hopefully arrive at that...better consensus and that better judgement as to where they wanted to go for the future.

It was this sort of reasoning which led Sillin to conclude that the site evaluation process should be opened up to receive direct input from the public whose interests might be affected by the NU hydroelectric project.

What follows now is a condensed description and analysis of the events which flowed from this decision, historic in the sense that it -- along with that taken by NSP's Earl Ewald at virtually the same time** -- was essentially without precedent in the electric utility industry. Fortunately, like its sister experiment at NSP but unlike most of those carried out by other companies in later years, the NU experience with open planning has been well

---

* Quoted from J. Holdren and P. Herrara, Energy: A Crisis in Power, Sierra Club Books, p. 240 (1971).
** The citizen participation experiences of Northern States Power are described in detail in Chapter Eleven, supra.

documented. In fact, it has been chronicled in rich detail in a manuscript entitled Getting Together,* prepared by free-lance author Michael Harwood of Washington, Connecticut. The descriptive portions of this chapter are based on the material contained in that manuscript, and unless otherwise noted all direct quotations are drawn therefrom.

DESIGNING THE EXPERIMENT

Having become committed to the idea of citizen involvement, Sillin and his environmental chief, Peter Stern, next faced the task of conceptualizing a participatory program. The desired outcome was relatively clear in their minds: ideally, NU should be the recipient of some sort of recommendation as to which of the two candidate sites should be developed, or even if the plant should be located outside the lower Berkshires. Far less obvious, however, were the particulars of who should participate, how the process should be structured, and so forth. After some deliberation a judgement was made that the political atmosphere of acute distrust made it infeasible to run a participation program as an extension of in-house planning efforts, by establishing (for example) a company-financed citizen advisory committee. This in turn led to a crucial decision: to seek the active assistance of a third party and monetary support from an independent source. As Peter Stern later put it, succinctly: "NU from the outset wanted to have no part in the organization of the citizen effort."**

Ultimately, Stern suggested that the problem be laid before Hank Foster, the former DNR Commissioner with whom NU had successfully negotiated the Northfield land swap. Foster had recently returned from Massachusetts after completing a doctorate in geography and environmental engin-

---

* The book is appropriately subtitled "Can the Electric Utilities and the Environmentalists Learn to Communicate Sensibly, Thus Saving Us All a Lot of Trouble?". Completed in 1974 but never formally published, it was prepared under a grant from the Fund for the Preservation of Wildlife and Natural Areas in Boston, Massachusetts. The present authors extend sincere thanks to Michael Harwood for making the document available as the primary source for this paper.

** Personal communication with Peter Stern (September, 1983).

eering at Johns Hopkins University, where his dissertation had sought to discover reasons for the lack of success of water resource planning in New England. His findings had given rise to a strong interest in non-governmental approaches to planning problems, as the following statement indicates:

> The primary conclusion I came to...was the fact that New England has this governmental machinery for water resources planning, but...lacked...any non-governmental machinery to do battle with the feds or the state people or the professionals, to help shape the policies and programs and recommendations and then to go to bat for them, or oppose them if they were wrong; and that what New England needed was this kind of constituency which would be sophisticated and professional enough to work with the public resource people..., and yet be in a position to modify and criticize and then mobilize support or opposition.

Feeling an obligation to follow through with this germ of an idea, Foster proceeded in 1969 to set up the New England Natural Resources Center, a non-profit organization to be funded by donations from private foundations, conservation groups, and business corporations. Peter Stern had been one of the people Foster had approached for support of this new entity, and it had occurred to Stern that such a Center might well provide the institutional base that NU was seeking for its open planning experiment.

Foster came up with the basic concept that a citizens' group be set up and provided with funds for staff and consultants, an ill-defined idea which evolved slowly -- after discussion with Sillin, Stern, and people connected with the Natural Resources Center -- into a specific proposal for what Foster began to call Project CEVAL (for "citizens evaluation"). This would be carried out under the auspices of the Natural Resources Center, which would seek a two-year, $180,000 grant from the philanthropic community (more specifically, the Ford Foundation in New York). The money would be used in part for a project director and staff at the Center in Boston, but mostly to support the grass-roots evaluation effort to take place in the Berkshires. In this regard, Foster envisioned that the Center would:

> furnish funds to the Housatonic River Watershed Association (a private citizens group) so they could hire a project director and establish an office to serve as

> an informational clearinghouse on the siting proposals; and finance a set of citizen task forces put together in conjunction with advocacy and planning groups in the region, to review the social, environmental and economic aspects of the available alternatives.

The money available to the task forces could be used to hire outside consultants and/or to contract with a professional resource agency (Foster proposed the New England River Basins Commission, a federal-state entity) for staffing services. As an adjunct to this, the Center would also invite various governmental agencies, such as the Federal Power Commission, to contribute to the deliberations.

At this point it is important to realize that Foster's model met the NU requirement that responsibility for getting citizens involved should rest squarely on his shoulders. This understanding on the part of the company was articulated in no uncertain terms by Peter Stern, who wrote the following in a letter to the Ford Foundation:

> In our talks with you, we explained why we think that Hank Foster's leadership in organizing and molding a participatory planning program is the key to this innovative approach in resource management...Hank's personal contribution would lie principally in the organization and guidance of citizen groups, including professional advice to an organization such as the Housatonic River Watershed Association, which would spearhead the evaluation process.

This made for a neat division of labor, in which NU's principal role was simply to carry forward with its siting studies and make them available to the CEVAL Project for purposes of evaluation. In fact, there needed to be only two (relatively minor) departures from business-as-usual. First, as a sign of good faith, the company would agree not to buy or option any land until one of the two sites had been eliminated. And second, as a gesture on behalf of a regional approach to the siting problem, NU would urge sister utilities in Maine and Vermont to seriously investigate the possibility of building a comparably-sized pumped-storage plant, to which NU would have access, in their own service territories.

With an outline of the proposed project now in better focus, Foster set about the task of making it a reality. Beginning in early November of 1969 he labored over a draft proposal for a Ford Foundation grant, and in mid-December he and Peter Stern made their first contact with a person they hoped would be a key figure at the local level -- Joan Flood, president of the Housatonic River Watershed Association and long an active member of the League of Women Voters. By her own admission, Flood's initial reaction on first hearing that NU was interested in the Sheffield site, not far from the town of Lenox where she resided, had been: "...over my dead body -- that's my valley." Apparently she also had reservations about working closely with Foster, whom she respected but also felt had a tendency to run a closely-knit, Boston-oriented organization. Yet she was willing to keep an open mind, realizing the significance of the utility initiative and the advantages of public involvement in planning. As she put it:

> I had been and was doing at that point a bunch of citizen involvement conferences throughout the United States for the League and for the National Air Quality Administration, and I was very high on this kind of approach to difficult resource problems.

Thus, at meeting's end, Flood was guardedly positive. She agreed to bring the proposal to the board of the Watershed Association, and to stay in touch with Foster and Stern.

Over the next two months, Flood broached the matter with her own board and with the executive directors and/or officers of other groups in the area, particularly the Berkshire Natural Resources Council, the Berkshire County Regional Planning Commission, and the Berkshire Panel on the Environment. Among those groups there was quite a bit of interest and even some rivalry, not surprising given that Foster had mentioned a figure of $50,000 as available for transfer to local organizations. By contrast, things were more quiet on the Connecticut side of the line, partly because there were fewer environmental and planning groups and partly because Flood and Foster, both Massachusetts residents, had fewer contacts there. Peter Stern, however, did meet with a group known as the Housatonic Valley Planning Association to try to interest them in a joint venture with Flood's Association. At this stage, the apparent hope was that at least one established regional organization from each state could be induced to share the lead role in Project CEVAL.

While the initial responses Foster and Stern received from these various citizen groups had been quite positive and encouraging, the same could not be said of the feedback on the draft proposal to the Ford Foundation. Lee Sillin had explored the idea informally with friends there and had found the initial reaction to be unenthusiastic. And Hank Foster, who had attempted to get the Conservation Foundation in Washington to lend its prestige to the application, had been advised that feelings were mixed and that advisors were concerned that the project might disrupt the environmental community (citing the famous split in the California Sierra Club precipitated by negotiations with Pacific Gas and Electric Company). Further complicating the matter was the fact that Hank Foster had had personal disagreements with one of the Ford staffers. So when the formal proposal was finally submitted in late January, 1970, the prospects for acceptance were uncertain at best.

The uncertainty didn't last long, however, for on February 9 NU was formally notified that the project would not be funded. The following excerpt from a letter to Peter Stern from Ned Ames of the Ford Foundation recites the reasons:

> There are two fundamental problems. The first has to do with the nature of the choice which is being offered to the citizen groups. As I pointed out to you and Lee Sillin at our recent meeting, the choice between two sites for a pumped-storage plant is a very narrow one and the apparent alternative of locating sites elsewhere in the region may not in fact be a real one since it would seem to be clearly beyond the capability of the citizens groups to make such a choice. We are a long way down the road at this point, and while your willingness to invite citizen participation before purchasing any land is laudable, I am afraid we must go back to a more fundamental level...
>
> The second problem that we have with your proposal is that it seems to put the citizen groups in the position of being responsible for site selection, which is a role that I am not sure they would either be very comfortable with or able to play out very successfully. In fact, the citizen groups do not have the answers you need to solve your dilemma. All they know is that site selection has been made without consideration to the environment in the past and that they must react to force the industry and the government to take environmental values into account in the future.

These must have been very distressing words for Sillin and company to read. It was one thing not to be receiving financial support, but in effect they were also being told that the project was ill-conceived in the most fundamental respect.

One could infer from the Ames letter that the Ford Foundation was advocating that the NU experiment be cancelled altogether, but Sillin was apparently in no frame of mind to entertain such a notion. He was philosophically committed to the project; considerable interest and tentative support had been cultivated in the environmental community; and the planned date of public announcement was fast approaching (in about two weeks). So it must have seemed that the point of no return had been reached, with nothing to do but make the best of a bad situation as far as the funding was concerned, which meant financing Project CEVAL with company money. As it turned out there was a way this might be done so as to, as Hank Foster put it, "give credence to the company's good intentions" and avoid having his Center appear as "a consultant employed by the company to promote its own point of view, which we are definitely not." Recognizing the possibility that NU might be forced to bankroll the citizen evaluation effort, Foster in December had approached a small Boston organization known as the Fund for the Preservation of Wildlife and Natural Areas, to see if it might be willing to act as a conduit through which some $180,000 could flow from Northeast -- with no strings attached -- to the Natural Resources Center and the citizen evaluators in Berkshire and Litchfield counties. The reaction had been favorable, so when the Ford grant fell through Stern and Foster began to negotiate in earnest with the Fund to make the necessary arrangements.

What proved more difficult to secure, however, was an endorsement of this alternative method of financing from the environmental groups that had previously indicated enthusiastic support of the open planning experiment. Some felt that the use of an intermediate financial recipient was sufficient to protect the CEVAL project from perceived conflict of interest, but others feared it would be seen as merely a cosmetic scheme to 'launder' money that would inevitably exert a coopting influence on the participants. The existence of this split was clearly established in a meeting Foster had with some directors of the Berkshire National Resources Council and the Housatonic River Watershed Association, less than a week before the scheduled announcement. At the close of the meeting, in fact, it was painfully obvious that the rift was deep enough to prevent either group from participating, as an organization, in CEVAL.

Lacking the built-in constituency that established groups would provide for the evaluation process, and in light of the reservations which the Ford Foundation had expressed about the program design itself, it would not have been surprising if Lee Sillin had decided to postpone further the media event that would launch Project CEVAL, then only days away. But there were weighty factors to consider on the other side of the ledger as well, both internal and external to the company. Within the NU ranks there was acute concern over timetable slippage already, and amidst the body politic there was a growing danger of false speculation and misinformation (the unavoidable product of 'leaks' from off-the-record discussions with government and citizen leaders). Indeed, the company had just been publicly accused of plotting a "land grab" via eminent domain, by a person of no less stature than the Republican floor leader of the Massachusetts House of Representatives -- a Sheffield resident who apparently exploited the confidentiality of a private briefing he had received from NU representatives.

Thus, February 25, 1970 found Lee Sillin at a rural Connecticut Inn, presiding over the long-awaited news conference. Surrounded by an array of maps, charts, photographs, and background reports, he officially announced that Northeast Utilities was considering two sites as candidates for a major pumped-storage plant, stressing that further planning would follow an unusual -- and progressive -- course. For one thing, NU studies would be conceived as "part of a regional effort" in which "utilities elsewhere will examine other promising sites so that a final decision can be reached on a regional basis." And most importantly, for its part the company wanted both government agencies and a citizens evaluation group (organized by Foster's Center) to "join us in essential studies and to assist in the decision-making process." The company, he went on to say, hoping to forestall the "crisis atmosphere" so commonly associated with utility construction projects, was "committed to a completely open evaluation" of the merits of a Berkshire-based plant, well in advance of the ultimate corporate decision.

Notwithstanding the fact that Sillin went to considerable length to stress the novelty of his approach to the siting problem, it is important to realize that his statements in this regard were highly general in nature. Nothing was said, for example, about how locations identified in other service territories might be compared to the NU sites on an inter-utility basis. Such was an unknown practice, then as now, although the company had attended one inter-utility meeting (of the New England Interim Planning Committee) at which site-related data was exchanged with two other companies that had identified potential pumped hydro locations. On the subject of public involvement, moreover, Sillin could not specify the composition of the central advisory group since it had yet to be assembled. Nor could he divulge the intermediate fiscal role envisioned for the Fund for the Preservation of Wildlife and Natural Areas, since no final agreement had been reached. However problemmatic he knew such "loose ends" to be, he apparently had decided it was necessary to rely on the ability of Foster, Stern, Flood et al. to iron out the myriad details in a satisfactory fashion.

Locally, the two principal newspapers in the area -- the Berkshire Eagle (Massachusetts) and the Lakeville Journal (Connecticut) -- reacted favorably to Sillin's presentation. News stories in each highlighted the citizen participation aspects, and both ran editorials which reserved judgement on the facility but applauded the open style of work which NU seemed to be adopting. The only commentary that could be considered equivocal in this regard appeared in the Eagle, which wondered out loud about how the money source (NU) would affect the credibility of the studies to result from CEVAL:

> ...no matter how objectively the work were done, it would obviously suffer from the widespread (and not unreasonable) public suspicion that surveys have a built-in tendency to tell those who underwrite them what they want to hear. From a community relations viewpoint, the appearance of collusion can be almost as harmful as collusion itself.

Given the ambiguity which existed at that point in just who would do the evaluation and how the funds would be handled, such a statement could only allow to grow the seeds of doubt which had already been planted in the minds of many environmentalists.

There were other reasons for the local activist community to be skeptical, as well. Despite Sillin's avowed desire for a broad-based examination of how to meet New England's peaking needs into the '80s -- which could mean importing Canadian power or building fossil-fueled units or hydro plants in other parts of New England -- the predominant interpretation was that Northeast wanted to build a pumped-storage plant at either Schenob Brook or Canaan Mountain. Sillin himself had probably contributed to this impression at his press conference, where a slick pamphlet on the Northfield facility was distributed and where the discussion centered on pumped-storage technology (prompting some media folk in attendance to variously describe the event as "oily," "arrogant," and a "sales pitch"). And if NU's objective boiled down to just getting the public to accept one or the other site, the reasoning went, wasn't the suggestion of citizen involvement really a smokescreen? The choice, one should realize, was hardly calculated to satisfy an environmentalist's desire to deal with a range of options. The Sheffield site was in an area highly valued for (among other things) the presence of a rare and much-studied type of wetland, whereas the Canaan Mountain area was noted for its scenic attributes. This, coupled with the spectre of massive construction impacts at whichever site was developed, made it difficult for many to view the situation in anything but zero-sum terms.

Thus, while area environmentalists were not necessarily hostile to the open planning part of Sillin's announcement (ambivalent is probably the better word), they were nearly unanimous in their opposition to what was perceived to be the NU 'plan' to site a plant in the Berkshires. This distressing reality was driven home to NU literally the very next evening after the Sillin announcement, at a meeting of the environmental council of the Housatonic Audubon Society, to which company representatives had been invited. There, Peter Stern and the head of NU's public relations department, Charles Bragg, were subjected to a two-and-a-half hour barrage of questions prepared in advance (and in writing) by a series of committee chairpersons, who had obviously made good use of the advance warning provided by the company's exploratory discussions with Flood and others. Reporting on the session to Hank Foster the next day, Stern wrote as follows:

> What they were in fact trying to test were all the soft points in our intentions and the process outlined in the white paper (i.e., the original nine-page company announcement).....In a newspaper article the night before, the Housatonic Audubon Society had stated that they would conduct their own studies of our project and would not associate themselves with you or accept any funds. Both Charles and I concluded that this group will be playing along with us and

> using the open planning process to document every possible negative aspect of either siting proposal.....

Stern went on to say that the meeting had been useful in that now that he knew what engineering investigations would be necessary "to respond to the types of environmental inquiries that we're now getting from the more sophisticated public." But he also knew that the same sophistication might allow the citizen participation process to be used against the company, and his fears to this effect had hardly been allayed.

Coming when and as it did, then, Lee Sillin's press conference had played the ball into the local citizens' court, and what they had done with it was hardly positive from the company's perspective. Indeed, the adverse public reaction had led to trouble in other quarters: only two weeks after the announcement, Sillin complained to Foster that he had been "buffeted on all sides -- by stockholders, by investment specialists, (and) by regulatory agencies" as well as by conservationists. He probably didn't give serious thought to cancelling the experiment altogether, since to do so would be to betray his own hopeful and idealistic instincts, to suffer embarassment before his peers in the industry, and to amplify the cynicism of environmentalists (and with it their propensity for legal and political resistance). But it must have been unsettling to feel compelled to forge ahead without being sanguine about the prospects for success. It was far from an auspicious beginning for NU's -- and the eastern branch of the utility industry's -- maiden voyage into the uncharted waters of open planning.

THE CEVAL CONCEPT TAKES SHAPE

In the months following Sillin's press conference, the situation would get worse before it got better. In fact, hardly two weeks had passed when a new activist group began to organize, a group with the express purpose of opposing the plant and whose negative feelings toward the NU endeavor were succinctly summarized in its own acronym: BLECC (for Berkshire-Litchfield Environmental Conservancy Council). The initial group comprised Connecticut residents mostly, but it soon grew to embrace concerned citizens in the Massachusetts portion of the Berkshires as well. At any rate, there is no small measure of irony in the fact that a new resistance group should have formed with such apparent ease, while the vision of a group with which the utility could work cooperatively remained just that, despite months of labor by Foster, Stern, and Flood.

If there was ever any doubt about BLECC's intentions, it was soon removed when Lee Sillin hosted a face-to-face meeting requested by two of its founders -- Donald Warner and H. Lincoln Foster. At the meeting Sillin had hoped to establish a spirit of cooperation, but Warner and Foster emphasized their group would do everything possible to prevent construction of a pumped-storage plant in the Berkshires. The door was left open for future talks, however, and the next time it was Sillin who travelled -- in late May -- to a Sunday luncheon at the Foster residence, at the picturesque foot of Canaan Mountain. Perhaps sensing the not-so-subtle hint this choice of venue represented, Sillin was visibly uneasy during the event, although he was somewhat encouraged by an express desire on the part of Linc Foster to understand the problems of the power industry and the role of pumped storage in long-range electricity planning. But the prospects for having any meaningful discussion along these lines soon faded, because a short time later BLECC published a highly inflammatory pamphlet which damned both the project and the company. Already sensitive about the way the environmental community had "prejudged" the process, Sillin could only express dismay in a letter of response to Foster:

> From our point of view the brochure completely misrepresents our efforts and, without any evidence of study, expresses opposition, in the strongest terms, to any further activity on our part. This is indeed disheartening in view of the effort we are making to establish an affirmative working relationship among conservationists, consumers, government, and the utility.

For all intents and purposes, the letter marked the end of informal interaction between NU and BLECC, and the beginning of a more traditional adversarial relationship.

While all this was going on, the new citizen group which was to be the centerpiece of Project CEVAL, and with which Sillin had high hopes for constructive interaction, had finally come into being. It did so toward the beginning of May, and had taken the form of a private, nonprofit corporation in Massachusetts. The reasons for becoming a formal legal entity are not entirely clear, although this may have been necessary to facilitate the transfer of funds for the evaluation process. In any event, its organizational status required that CEVAL be run by a board of directors, complete with a president, vice-president, secretary, treasurer, and other officers as necessary. From the comments made earlier as to the NU expectation of a leadership role for Hank Foster, it would seem that the responsibility for setting up the Board would fall to him. But this did not happen. Instead, it was Joan Flood who talked to individuals in the area and eventually came up with ten able and interested people (all men) to join her on the Board.

The majority of board members on the final roster were Massachusetts residents (as required to incorporate in that state), and a total of seven out of the eleven turned out to be either artists or writers -- including Harrison Salisbury, a senior editor of the New York Times. Rounding out the group were a retired teacher, the owner-operator of a ski slope, a forester, and career volunteer Flood (as president). Apparently she relied on her own judgement in assembling this group, although she had one explicit criterion: that persons serving on the Board agree to withhold judgement on the question of whether a plant should be built in the Berkshires at all. This is not to say they did not share a general antipathy to the idea, for most (especially one whose house bordered Schenob Brook) would have opposed the plant were it not for a sense of commitment to the ideals of citizen participation. In fact, two of the CEVAL board members -- the vice-president and treasurer -- also belonged to BLECC, and a third was actually the chairman of that (highly contentious) organization. However strange this interlocking of directorships may seem, the fact is CEVAL could not hope to enjoy much credibility in the area if it failed to have board members who were skeptical of Northeast Utilities. And what better credentials could one have in this regard than affiliation with BLECC?

Speaking of Northeast, it is important to note that the CEVAL Board which emerged was not the type of group that Sillin and Stern had envisioned. Having decided not to exercise any control over the structure of the evaluation process, they could only express an opinion, and what they had favored expressly in this case was a broad-based group. In the words of Peter Stern:

> We had in mind that...the filling station operator would be part of this; that the beauty parlor operator -- there'd be a spot for her too; that the

> local official would be in on it...and retired people, people of wealth, people of poorer means; that this would be a cross-section of the communities.

What the company witnessed, instead, was the formation of a board in no way representative of the larger population, but a seemingly self-selected group with a uniformly preservationist perspective. Charlie Bragg put it more succinctly: "The CEVAL Board was, very frankly, elitist."

Joan Flood didn't see it that way at all, of course.* To her it seemed that it was terribly naive of Sillin to think that "the local hairdresser or Joe six-pack would spend time on a committee;" and that with representation of groups like the Berkshire Natural Resources Council (2 members) and local conservation/planning agencies (2 members) in addition to BLECC, the Board was far from homogeneous in outlook. What's more, she took particular offense to the "cynical attitude that men of intelligence and integrity cannot be objective." In this regard she was quick to point out that, while no one on the Board was very sanguine about the prospect of a large generating plant in their backyards, as a group they were "internally very concerned about our responsibility to ensure that there was adequate power for people and industry." In Flood's mind, then, the Board may not have been exactly as NU would have composed it, but at least it wasn't packed with 'hyper-environmentalists' either.

Whatever misgivings NU might have had about the make-up of the CEVAL Board, they were not serious enough to jeopardize the company's pledge of financial support. At about the same time that Board membership was finalized, $180,000 was given to the Fund for the Preservation of Wildlife and Natural Areas, with the understanding that "NU would retain no control whatsoever over this grant and that disbursements would be made entirely at the discretion of the Fund." Five-sixths of the money was allocated for use by the CEVAL Board, and the remaining $30,000 was set aside for Hank Foster's Center to provide "personal and other services of a liaison nature between CEVAL, Northeast Utilities, federal, state and regional agencies and organizations, and the Fund." The scope of these "other" liaison services was not spelled out, especially as to whether the Center would have any influence in budgetary matters.

---

* The following citations are drawn from personal communication with Joan Flood (August, 1983).

But it seemed clear that the role of the Foster organization would be more than perfunctory; as he saw it, the Center's responsibility was to handle "the mechanics of how the CEVAL operation would take place."

With cash soon to be on hand, the CEVAL Board turned to its first important order of business: the hiring of an executive director. The prime candidate had originally been George Wislocki, executive director of the Berkshire Natural Resources Council, but upon further consideration the Board decided against this choice because it was judged impolitic to draw yet another key person from the Massachusetts side of the border, especially one whose existing (albeit casual) acquaintance with Lee Sillin could prove to be an embarrassment. Another candidate soon appeared, however, in the person of Roy Lerman, a twenty-six year old lawyer in Washington, D.C., who had previously written to the Connecticut River Watershed Council offering to serve as an attorney in some land purchases. Someone at the CRWC showed Lerman's letter to Flood, who arranged to meet with him when she was in Washington on other business. Lerman told Flood during their meeting that he was interested in environmental questions but knew very little about them and less about environmental law, having worked primarily in a public defender program. Still, Flood liked him and suggested that he be given an eighteen month contract (starting in June), a suggestion with which the Board concurred without feeling the need for a full interview before the group.

The Board also voted to hire a second staff person, 47-year-old Addison Cate, a trained geologist who had worked for several oil companies, for the Pennsylvania Geological Survey, and as a university lecturer. Cate's role would be to monitor the engineering aspects of Northeast's studies until the Fall, when he would return to the University of Michigan where he was pursuing an advanced degree in resource planning. Cate was interested in the social aspects of the open planning experiment as well as the technical, and even proposed that he study how various political institutions might actively contribute to the process and what their relationships might be with NU, the CEVAL Board, and each other. Joan Flood preferred that he stick to engineering, however, and he accepted the job on this condition despite being unsure how important this function would be, given that Northeast's technical work was still in the formative stages.

In another respect, though, it was fortunate that NU's plans were still highly tentative: it allowed some time for the CEVAL Board and its staff to 'gear up' for the evaluation task that lay ahead. First on the agenda was the question of what basic role the group should play, about which those who had originally conceptualized the project had notions that were not only vague but also contradictory. Lee Sillin had expected the citizens group associated with CEVAL to suggest some course of action ultimately, based on the members' collective perception of costs and benefits and with careful consideration given to patterns of opinion within the community at large. But the Ford Foundation had warned that such an advisory role could be troubling to participating citizens, and Hank Foster had apparently taken this warning to heart. In his early pronouncements on the subject, he portrayed CEVAL as primarily a clearinghouse for information which would "(establish) communications multilaterally, with all the different parties, ... (and then) render a report to the region on these project proposals, with facts and alternatives, to help the region itself come to a decision...." This role could well be described as that of a non-partisan catalyst for discussion and debate within the general populace, where the bulk of the evaluation process would (somehow) take place.

Foster never said exactly how the nebulous goal of crystallizing the public will could be achieved (does one take a poll, hold a referendum, or what?), and it is doubtful the CEVAL Board had a clear vision of implementation either. In fact, the ambiguities were such that the group debated at length whether the basically reactive approach Foster had advocated was really superior to the alternative of assuming more direct responsibility, along the lines Lee Sillin had envisioned. This would mean promising to make a recommendation (with minority views included, if necessary), and letting the credibility of that recommendation rise or fall on the legitimacy of the process and the power of the reasoning which together stood behind it. Taking the bull by the horns in this manner would certainly be simpler than attempting to help the public make up its collective 'mind'; and as long as the public's help was sought as the Board made up its own, the process could be just as democratic as well.

For a variety of reasons, however, the Board grew uncomfortable with the notion of committing itself to taking a stand in the end. For one thing, with most of the group not belonging to what might be termed "problem-solving professions," some members were less than confident in their ability to work effectively with technical information and reach judicious decisions. Somewhat related to this, and even more compelling in the eyes of most Board members, was a perception that a failure to reach internal consensus would be fatal to the entire effort, given how politically 'loaded' the siting issue had turned out to be. Unanimity might easily be forthcoming with the appearance of some clear cut reason to reject the facility at either location; but could the Board ever agree to a finding in favor of one or the other candidate site? This seemed highly unlikely given the instinctive aversion several members felt toward the pumped-storage proposal, to say nothing of everyone's reluctance to be accused further of 'selling out' by BLECC and others for whom CEVAL's acceptance of NU funds was already grounds for intense distrust.

According to this line of logic -- much different from that which Sillin, Stern, and Foster had followed originally -- the only way for the Board to be effective and remain faithful to its pledge to keep an open mind was to become a dispassionate collector and disseminator of information, at least for the time being. Thus the Board chose for itself a mission similar in many respects to that of the League of Women Voters, with which Joan Flood had so long been associated. It was Flood, in fact, who later provided this capsule description of the CEVAL position:*

> (We) steadfastly refused to exercise the All-American prerogative for being an advocate, and traded that right for the stance of an umpire-like participant in an emotionally-charged controversy. Ours was an effort to disavow narrow protectionist viewpoints and at the same time -- though it seemed inherently contradictory -- to establish 'grass-roots' support for an undefined consensus.

Notwithstanding this statement, however, it is important to realize that the Board did not explicitly foreswear in advance the possibility of coming to a conclusion on its own. It simply waffled on the matter, thus leaving open the option of joining the ranks of the opposition if warranted by circumstances in the future.

Having thus resolved (at least temporarily) the question of defining a role, the Board turned over to executive director Lerman the unenviable job of inducing the public to participate in a consensus-building program of evalua-

---

* CEVAL Reports, newsletter no. 13 (September, 1972).

tion. New to the area and given responsibility for a project that was still unstructured in many respects, his was a difficult situation. The Board, to be sure, had assigned some specific tasks like building files on the power industry, developing a mailing list, producing brochures, lining up members for citizen task forces, compiling lists of questions to be submitted for NU, and so forth. But Lerman was more or less on his own when it came to preparing an overall plan of action, and in coming to grips with he larger questions about CEVAL's strategy of operation. Would it be productive to get involved with the ongoing NU study of the two sites, or should the evaluation commence after the company issues a report on the in-house findings? How could the issue of need for power -- seemingly outside of the CEVAL charter but of great concern to opposition groups -- be addressed in the evaluation process? The answers had to be found rather soon.

To get his bearings and develop a sense of direction, Lerman spent most of his time that summer meeting with opinion leaders and other people in the area. He sensed that to be effective CEVAL would have to be credible to a range of constituencies, and to him the best way to build the organization's credibility was to begin developing his own. Lerman also conferred at length with Robert August, whom Foster had recently hired to handle the Center's end of the project. This association began in mid-July, at a meeting which August described to Foster in the following words:

> Roy indicated that up to now, the organizational format that he had to deal with (i.e., State activity, NU activity, Center role (River Basins Commission) was all very nebulous.... Basically, he had no plans to organize the task forces at this time. More importantly, he had not yet determined what CEVAL should do. I suggested to him that his Board probably wouldn't help him conceptualize (the) CEVAL program and he would have to suggest things to them. He agreed!

Over the next several weeks Lerman continued to feel he wasn't getting much help from his Board, and consequently spent a fair amount of time with August discussing the pros and cons of CEVAL's strategic options.

The CEVAL Board was initially patient with Lerman's gearing up endeavors, but by the early Fall it had begun to worry about his performance and general suitability for the job. Having devoted the bulk of his energies to personal outreach on behalf of credibility, in three months he had produced only a brochure on CEVAL's purpose and a few letters to local residents, and had yet to map out an overall approach and timetable for the project. Another source of disenchantment was apparently Lerman's general demeanor. Some on the Board felt he was too often abrupt and argumentative, while others were unimpressed by the way he conducted himself at a first round of meetings held -- in early October -- to brief interested citizens on the CEVAl effort. Then, later the same month, he committed the unpardonable sin of sending a financial statement to the Natural Resources Center which not only was inaccurate, but also had never been seen (let alone approved) by the Board. This prompted the Board to ask for Lerman's resignation on October 20, 1970. When he refused, he was fired.

As Lerman's successor the Board turned to Joan Flood, who agreed to serve but refused compensation except for reimbursement of out-of-pocket expense. It would be up to her to establish CEVAL in the 'League of Women Voters' mode with which she was so familiar, and to deal with what she knew to be the number one obstacle to effectiveness in this regard -- that of apathy among the public at large. A 1969 poll, after all, had shown that 84% of all Berkshire residents were unaware of environmental problems, and there was no reason to expect a higher degree of consciousness of power plant siting issues. The disclosure of Northeast's pumped-hydro proposal had evoked some reaction, but with the preferred location not yet designated and with several months having elapsed since the original announcement, interest on the part of the average person on the street had all but evaporated.

To counter this state of lethargy, Flood wasted little time in taking the first steps to attract attention and disseminate information to the community. In November, for example, she convened the first in a series of "study group" meetings, the purpose of which was to foster discussion of substantive topics of interest (fish and wildlife, transmission lines, planning and zoning practices, etc.), which in turn would hopefully lead to the formation of task forces to focus on specific issues of concern. At about this time she began to issue two mimeographed publications: <u>CEVAL Reports</u>, a one-page bulletin mailed to all area boxholders; and <u>Facts and Issues</u>, a longer and more detailed piece sent to those on the mailing list CEVAL had managed to compile by then. The latter consisted entirely of written answers from NU to questions that had been posed during the summer, and was somewhat difficult for a lay reader to understand. Still, it was a manifestation that the CEVAL Project was beginning to overcome its initial paralysis.

Flood envisioned that both Reports and Facts and Issues would be published regularly throughout CEVAL's lifetime, providing readers with a list of upcoming meetings, status reports on the general progress of the utility proposal and its evaluation, and so forth. Unable herself to assume the editorial task, she soon hired a sophomore in a co-op program at Beloit College -- Gloria Freund -- as a sort of 'press secretary' for CEVAL. It was not an easy task for such a young person to be responsible for, that of communicating with the general public. But in Flood's mind, Freund's inexperience was offset by the fact that she was articulate, hard-working, and a native -- her father ran a local dairy farm -- whose contacts and viewpoints could be very helpful. Besides, Flood suspected that a more 'professional' approach to the outreach function might actually be less credible; in her words, "slick Madison Avenue stuff would not have been accepted locally."*

In addition to building the CEVAL staff, Flood also worked actively during the winter of 1970-1971 to line up consultants to help critique the Northeast studies, and to investigate whatever issues the Board felt deserving of independent scrutiny. Ultimately, four subcontractors were commissioned. The research staff of the Massachusetts Audubon Society would study the environmental aspects of Northeast's upcoming report. Hill and Lind, Inc., a firm comprised of two retired Army Corps of Engineers people, would investigate the engineering data. A group from Yale University Divinity School, Connecticut Citizen Evaluators, would survey local knowledge of and attitudes toward the proposed plants. And Sosnoff, Cooper, and Whitney, a public interest law firm in New Haven (Conn.), would investigate the influence of rate structure on environmental quality. Public announcement of these choices began toward the end of January, 1971.

Approaching the first anniversary of Lee Sillin's announcement of an open planning experiment, then, Project CEVAL had finally developed some momentum. The Board had defined its goals and a plan of action, stabilized the staff situation, hired consultants, issued its first newsletters, and conducted several study group meetings. In logistical terms, at least, the organization seemed ready to tackle its principal objective -- to mobilize a con-

---

* Personal communication with Joan Flood (August 21, 1983).

structive and coherent public response to the NU site evaluation report, due at any time. This was to be the real touchstone for the participatory process, and it would also put to the test CEVAL's ability to serve as an effective liaison between NU and the community of affected interests. And the test, for reasons that will now be described, would in all likelihood be a very severe one.

THE WEB OF (STRAINED) RELATIONSHIPS

The question of whether the CEVAL group was really in a position to fulfill its self-chosen role as a catalyst for public input is, of course, not so much an administrative one as a diplomatic one at root. Those with experience as a go-between of any sort know that to be a successful broker one must enjoy the confidence of all involved parties. Yet to build a web of positive relationships is a delicate matter at best, especially in the presence of a rancorous dispute where the proximate question is "whose side are you on, anyway?" This brings to the fore an issue which must be addressed before the second major phase of this open planning chronicle can be reported: the status of 'diplomatic relations' between CEVAL and the two key protagonists on the scene, i.e. Northeast Utilities and BLECC. This in turn leads to a review of another set of events played out during 1970 and early 1971, events having to do not with internal preparations but with the evolution of CEVAL's standing within its broader external context.

This thread of the CEVAL story begins with a problem that surfaced early and often: the inability of the financial arrangements to remove the 'taint' of using utility money. When the Board was first assembled several persons had refused to join, in part because NU was the sole source of funding; and those who did join soon became acutely aware of the distrust it had engendered among their peers. Consider, for example, this anecdote told by Board member Roger Drury, lay reader in an Episcopal Church in Sheffield and the parish clerk for twenty years:

> By the middle of the summer (of 1970), when the thing was heating up -- you won't believe this; I still don't believe it, looking back -- someone whom I assume is a pretty good friend of mine in Sheffield asked my wife...if it was true that I had been offered $10,000 by Northeast Utilities not to take sides.

One of the reasons such rumors spread easily, it appears, was that the press did not always take pains to clarify the funding situation. Articles in the Berkshire Eagle, for example, never failed to refer to NU as the original donor but from time to time neglected to mention that the Fund for Wildlife and Natural Areas was also involved to ensure the utility exercised no control over CEVAL's grant.

The Board tried to be scrupulous, therefore, in avoiding any further appearance that local support was simply being 'bought' for a pre-ordained project, a goal which could only be realized through total financial independence for CEVAL. Joan Flood's understanding of this 'clean money' principle, expressed in her mid-May request to Hank Foster for a disbursement from the Fund for the Preservation of Wildlife and Natural Areas, was that the project "will remain completely under the control of the Project CEVAL Committee...(which will) be free of any direction from the Natural Resources Center..." Foster had already told the Board as of April that it would be "entirely independent", and in June he had seemingly applied the notion to pecuniary matters when he told the group that it was free to spend some $80,000 for general purposes over the first two years.

Thus, it was initially unclear to the Board that the degree of autonomy Foster envisioned for the group was actually somewhat less than what they had inferred from his statements. As noted previously, NU had always assumed Foster would assume a position of active leadership, and both he and James Mosely (director of the Fund) had interpreted this as extending to control over the budget as well as other project features. As Mosely put it, the Center "was going to be the watchdog, to make sure the money was spent right, because I, from the Foundation's point of view, didn't want to disperse the money if (Project CEVAL) was starting to go bad, or wasn't doing what it was supposed to." And Foster, for his part, had no intention of abdicating this oversight responsibility. To the contrary, he intended to solicit regular financial statements on expenditures from the first payment to CEVAL; and he planned to release the second installment of $70,000 -- targeted mostly to meet the important need for consultants -- only upon mutual agreement among his Center, the Fund, and the CEVAL Board.

Mosely and Foster obviously felt their organizations should act in a fiduciary capacity vis-a-vis the NU grant, and given the amount involved this was not unreasonable from the standpoint of prudent financial management.

Still, it was less than sensitive to the Board's perception that they too were subject to a trust, a sort of public trust, and that the CEVAL image in this regard would suffer if the organization were to become closely aligned with Foster and his Center, either financially or programmatically. At any rate, with the Board not fully aware of what was occurring at the staff level, the Center had proceeded to play an active role in CEVAL affairs during the summer and early Fall of 1970. Bob August talked often with Roy Lerman, sent suggestions to him and pressed him for a financial statement, and even worked out -- at Lerman's request -- a timetable for evaluating the NU site report when it was issued.

It was only a matter of time until the Board would realize that the relationship with the Center was developing in a way counter to its wishes, and the occasion came with the firing of Roy Lerman. Dismayed at the revelation of what had been going on thus far, the group met with Foster and August in November to voice their displeasure. The level of feeling approached that of outrage, as seen in this reflection on the event by Board member Robert Osborn:

> The one reason that quite a few of us went on (the Board) was the idea, okay, we'd get some money, we'd get use of the money -- and we specified time and time again that there must be absolutely no strings attached....I was really trembling with rage, and.... saying that, 'Can't you see how you expose us, you make us look not very clean....if you people keep associating yourself with us?'....

Marked by such unpleasantness and lengthy argument over both procedural and financial issues, the meeting produced an uneasy compromise: CEVAL would report only to the Fund on financial matters, but could not receive special funds for contract work without prior consultation with both the Fund and the Center as to "priority of need." The latter point still rankled many Board members, however, and it was not the last time they would clash with the Center over its attempts to share decision responsibility when it came to the hiring of consultants.

CEVAL's determination to keep Foster's Center at a distance was, of course, a derivative of its determination to be entirely free of any real or perceived influence from Northeast Utilities. Interestingly, though, it appears that the people at Northeast (like those at the Center) were slow to appreciate how pronounced was the desire of the Board to establish its independence from the company. This in turn gave rise, at an early stage, to a certain amount of tension between NU representatives and CEVAL, especially with Roy Lerman being obsessed at that point with the problem of developing credibility. At a June meeting between the Board, Peter Stern, and others from Northeast, Lerman had been "astounded" that the company had attempted to lay out more or less "what they wanted to group to do"; and that they had offered to share drafts of the utility siting report as it developed, prior to public release. A further source of irritation was NU's refusal, a few weeks later, to follow a similar policy of disclosure relative to BLECC. More specifically, when A. Mitchell Finlay of BLECC requested direct access to NU reports, publicist Bragg responded as follows:

> Very frankly, Mitch, we don't know quite how to respond. We understand that CEVAL wishes to act as a public information conduit, and, of course, we would want to respect that position....With BLECC representation on the CEVAL Board, perhaps the whole thing can be resolved between you....As you know, our purpose is to have the reports publicly evaluated. We will abide by whatever mechanism is thought best.

In short the company was taking the position that, as Bragg would later put it, "open planning...meant dealing with CEVAL."

Roy Lerman, however, did not share this view as to how the two sides should interact. Given the delicacy of setting up CEVAl as a neutral facilitator, he felt it was inappropriate for the organization to develop a close working relationship with the company or to be treated differently than others in the area. Indeed, he believed such a role would preclude any chance a success for open planning, and he thus made every effort to counter the impression that CEVAL was anything but a completely autonomous entity. For example, he was openly critical of NU (and even Hank Foster) at the June meeting between Stern and the CEVAL Board. Later, after complaining to Bob August that Northeast "seems to think CEVAL is their citizen group and is favoring them with information prior to release," he let it be known that he did not condone nor would he support the practice. In a July letter to BLECC, he explained that the preferential treatment "was not solicited nor was it welcomed by us," and promised the immediate release of "any information of public interest" provided by NU.

Thus, if it can fairly be said that the open planning experiment had gotten off on the wrong foot with Sillin's press conference in January, by the summer of 1970 it was even further out of step as far as the utility was concerned. The Board's composition and especially its interconnection with the upper echelons of BLECC were unanticipated and troubling, as was its policy of automatic disclosure of even preliminary planning data. What's more, the Board had hired an executive director who was clearly intent on keeping the company at arm's length; his antagonism, in fact, was such as to prompt Peter Stern to remark that he appeared to be "BLECCed." The situation was hardly conducive to the development of a frank and mutually satisfying dialogue, yet under the circumstances Northeast was not in a position to do much about it. Even to voice dissatisfaction with the structure of the Board or to complain about Lerman's modus operandi, it seemed, would only risk further degradation of a relationship that had already become cool and aloof at best.

There was yet another, probably more significant reason for Northeast to grow increasingly cautious in its interaction with CEVAL. Back in February, applications had been submitted to the FPC for preliminary permits at the candidate sites, ostensibly to reserve them for study by NU and to get some feedback from the agency. In response, BLECC -- on advice from a participant in the Storm King resistance movement -- had been hard at work during the spring getting residents of Berkshire and Litchfield counties to write letters of protest. Even more worrisome was that the group had decided to formally oppose the applications, and the FPC had extended the deadline for filing a protest in the case from June 8 to July 8. The lawyers at Northeast were up in arms -- worrying that information released for purposes of open planning would somehow be used as ammunition by BLECC -- and consequently insisted that all direct communications with CEVAL be routed through the legal department. This hardly made it any easier for Sillin to implement his notion of bringing people together; apparently it even precluded him from meeting personally with the Board (an event that would not occur until CEVAL's final phase of operation, two years down the road).

Ironically, NU's perception of CEVAL and BLECC as being kindred spirits if not outright allies -- probably the greatest impediment to fruitful communication at that point -- was not an accurate reflection of the true situation. The fact was, the Board's relations with BLECC were no less uncomfortable than those with Northeast; indeed,

BLECC had virtually no desire to cooperate with CEVAL. In its eyes open planning was a public relations ploy designed to procure acceptance by coopting potential opponents, and was therefore a dangerous threat to the BLECC cause. Linc Foster's wife elaborated on the nature of this threat in the following terms:

> Certainly BLECC's official feeling was not that CEVAL was pro-utility, though we mistrusted Hank Foster and consequently to some extent Joan Flood; rather, (it was) that CEVAL was not really capable of doing the job it had been asked to do and that even if they came out against pumped storage their finding could and would be ignored by NU, (whereas) if they found in favor of NU their findings would be used to quash any opposition in front of the FPC.

Quite obviously, BLECC was concerned about the possibility CEVAL would eventually side with the utility, a fear no doubt aggravated by the fact that the CEVAL Board had not unambiguously ruled out making a site recommendation. Neither the fact that (if anything) most Board members were privately predisposed toward opposition, nor that the group had worked so hard to disassociate itself from NU, were sufficient to overcome BLECC's acute anxiety in this regard.

In this hypersensitive context, several things the Board did simply in the interests of maintaining objectivity vis-a-vis NU were interpreted by BLECC as evidence that CEVAL was leaning toward acceptance of a plant. For example, while the Board didn't hesitate to attack industry-wide policies and intended to criticize the pumped-storage project (as necessary), it was not openly critical of Northeast per se; as one member put it, "it didn't seem very ethical of us (to criticize our funding source)." And then there were Joan Flood's attempts to widen the range of views represented at the public meetings, which she felt had become one-sided due to the disproportionate presence of opposition forces. To BLECC, however, such actions were signs not of impartiality but of bias. Hal Borland, BLECC's chairman who also sat on the CEVAL Board, put it this way:

> It seemed to me that CEVAl worked right along with the friends of the pumped storage plant. Falls Village was pretty well split, and the larger faction of it was in favor of it and CEVAL, it seemed to me, was

> working with that faction right along, and against the other faction... ...encouraging those (who were in favor) to bring in their friends to the meetings and to state their position. It was just, to me, a... sense of bias right straight through.

By December of 1970, Borland had resigned from the Board. Besides organizational inaction, the principal reason he cited was that "the CEVAL viewpoint, when articulated at all, has struck me as apologetic for Northeast Utilities at too many points."

To Joan Flood, BLECC had completely misinterpreted what CEVAL was trying to do and was guilty of outrageous and unconscionable behavior. "Most of the members of BLECC I personally liked and respected," she would later say, "but collectively they permitted themselves to run a totally dishonest campaign against CEVAL and the utility." In that remark she was referring to the fact that the resistance BLECC had mounted with publication of its anti-pumped storage pamphlet, and which had been carried before the FPC as well as into a series of informational meetings NU has convened in the area, was later applied with equal fervor to the public meetings convened under CEVAL sponsorship. The audience at nearly every event was dominated by those who opposed the plant, and they did not shrink from the use of disruptive, confrontational tactics. There were even attacks on the integrity of the CEVAL Board, as at one public meeting in the spring of 1971 where BLECC-inspired questions about the origin of CEVAL's funding were raised for the sake of embarrassment only.

How strange and unexpected was the predicament that CEVAL was in! Far from establishing itself as a dispassionate body, the Board had come to be viewed by both sides as being aligned with the other. It was in a classic no-win situation, in which every action seems to have had an erosional impact on its legitimacy with some sector of the community. For example, the group chose not to follow a suggestion that the seat vacated by Borland be filled by a person from the Falls Village Chamber of Commerce; to replace him with someone with diametrically opposing views would simply lend credence to his charges of bias. Yet to residents who supported development of the Canaan Mountain site, the action was taken to mean that the Board only wanted members who were dyed-in-the-wool conservationists. Thus when a pro-utility group was eventually formed -- Area Development Ecology/Electricity (ADEE) -- it chose to ignore CEVAL altogether and to proceed on its own.

The great irony here is that the politically sensitive circumstances which had made it seem so imperative for the Board to adopt an umpire-like role were also making that role impossible to carry out. In the capacity of a broker CEVAL couldn't be 'for' any one position or perspective, and yet the operative criterion was 'if you're not for us, you're against us.' Thus the unavoidable consequence of neutrality was to be seen as inimical to all special interests -- hardly the image that is required of a catalyst for reasoned debate. Call it Catch-22, or being caught between a rock and a hard place, or having to walk a tightrope blindfolded. Whatever the metaphor one prefers, the people affiliated with CEVAL seemed to be in a most precarious -- and lonely -- position.

Nor was it within the capability of the Natural Resources Center, by the way, to dissipate the atmosphere of estrangement surrounding the CEVAL project. For one thing, it seems improbable that the Board would welcome efforts to do so, considering its general antipathy toward allowing Hank Foster or his staff to exercise any operational influence. Secondly, the Center itself was loathe to intervene (at least directly) in this domain, as indicated by these comments by Bob August:

> ...(we saw our role as that of) a third party....that was able to talk in a credible way with CEVAL people, with Northeast Utilities people, with government people, whatever.....and in order to maintain that relationship with those organizations.....I think we felt there was a need for some confidentiality. Things we talked about with the utility we shouldn't really have felt free to...go blabbering around the countryside. In the same vein, there were things that we knew from talking to BLECC people and CEVAL people or other people (that we didn't divulge to the company).

Foster and August did use their insider's knowledge to bring about an exchange of ideas on an anonymous basis; as August said, "I think we probably did a fair amount of that." But it wasn't enough, apparently, to repair the lines of communication between CEVAL, NU, and BLECC, which were essentially 'down' by the close of 1970.

In sum, with the approach of its first birthday -- and on the eve of receiving the NU site studies -- CEVAL's woes encompassed far more than mere sluggishness in gearing up for the ambitious evaluation process over which it hoped soon to preside. As an organization it was devoid of a grassroots constituency of significant proportion, a reality of ironic contrast to the fact that membership in BLECC had grown to the vicinity of 1000. Next, CEVAL's ability to interact with Northeast Utilities had declined markedly; they were talking to each other, but little of real consequence was being said. And even within itself there had been increasing signs of polarization, insofar as not all Board members had made a good faith effort to withhold judgment on the merits of a possible project. One wonders, in short, if there was really any hope that the organization would be able to serve as an effective link between NU and the community at large, i.e., to achieve the purpose to which it owed its very existence.

Despite all this, however, there was still considerable feeling within CEVAL that its situation was not wholly untenable, and that as the concrete task of reviewing NU studies got underway the Board's plan of action would prove to be workable. Thus it remained for future events to determine whether the rocky road CEVAL had travelled so far would turn out to be smooth up ahead.

THE ABORTIVE EVALUATION

The dawning of 1971 found Northeast Utilities somewhat behind schedule in the preparation of a site comparison report, but enough information was in hand to convince company managers that Canaan Mountain (in Falls Village, Connecticut) should be the preferred alternative. At the beginning of the in-house study the other location -- Schenob Brook in Sheffield, Massachusetts -- had seemed cheaper and more attractive in terms of opportunities for multiple use (especially public recreation), but as the investigation proceeded new data indicated otherwise. Test borings there had found less-than-optimal conditions in the rock, reinforcement of which would make construction costs comparable to those at the Connecticut site. Secondly, in order to realize the recreational potential of the project, the water level of the lower reservoir had to be controlled; yet no amount of engineering design could avoid significant change in the appearance of the valley and considerable disruption of the surrounding communities (including, in one scenario, extensive flooding of homes, roads, schools, and the like). Peter Stern described the situation this way:

> ...in the landscape in which it lay,...every improvement in public utilization, multi-purpose use, enhancement of shore (meant) you enlarge the area of the project. It's inevitable. By that very fact, I thought this was so much more difficult an environmental intrusion that we should rule it out....I think the decisive consideration was that from a rational public acceptance point of view we were just creating a monster.

The man who dealt with the public on a day-to-day basis, Charlie Bragg, agreed: "To me, it was just too damned obvious that we shouldn't go into Schenob."

To Hank Foster at the Natural Resources Center, however, it was premature for NU to announce a decision to eliminate Schenob from further consideration. Foster and project manager Bob August had received draft copies of the forthcoming report summarizing NU's findings and conclusions, and both objected to the idea of narrowing the range of options before CEVAL could make the issues of choice known to the community and assimilate whatever feedback it had to offer. On Foster's part, in addition, there was apparently some distress over the fact that it was the Massachusetts site -- for which he personally saw exciting possibilities for regional resource management -- that was to be dropped. But he maintained that his principal argument "was that...the company making a unilateral choice... tended to short-circuit the open planning process, and that what they should do (was) to leave both of them open and report the facts that relate to both, but not make the choice between the two at this point, until CEVAL had a chance to contribute some suggestions of its own." Foster went on to say that he felt the company "was making some judgements based on its own circumstances -- reading the tea leaves for the Massachusetts portion -- without really having put it to the test."

It was quite understandable, of course, as to why NU might have been fearful of any further 'test' of public acceptance at the Sheffield site. On the one hand there was little reason to believe the CEVAL organization was ready and able to collate community preferences in a systematic fashion, even if this might have revealed greater support than the company expected once the severity of impacts came to light. On the other hand there was BLECC, always ready to fan the flames of opposition and always hungry for fuel to use in the process. Thus the benefits of holding fast to a key principle of citizen participation -- if any existed in fact -- were clearly outweighed in the company's mind by the benefits of removing one major source of political resistance (at least that of the 'not-in-my-backyard' variety). By this time NU was simply feeling shell-shocked from the war that BLECC and others had been waging, and if a retreat from the ideals of public involvement was necessary to 'cut its losses,' so be it!

Another manifestation that NU was on the verge of backing away from its commitment to open planning lay in the fact that neither Joan Flood nor the CEVAL Board were consulted on the question of whether a preferred site should be identified by the utility. While an awareness of the situation and an opportunity to discuss it with the company might well have brought some focus to the Board effort to define an evaluation strategy, relations between the two organizations were still cool at best; and it seemed probable, to NU, that information transmitted directly to CEVAL would be relayed to BLECC and thereafter distorted in the rumor mill. Thus consultation occurred only with Foster and his staff, although Peter Stern would later purport that he assumed the Center was in close contact with CEVAL. Perhaps he assumed as well, therefore, that the Center's advice was a reflection of the CEVAL position. At any rate, the Center in fact did not feel free to divulge the contents of the draft report, for reasons noted previously. Far from being open, in other words, the channels of communication were still hopelessly clogged, and CEVAL -- to use an expression that is aptly symbolic of just how little sunshine was actually falling on the planning process at this juncture -- remained completely in the dark.

One Foster criticism of the draft report to which NU did respond was that it was incomplete and poorly written, and the subsequent process of revision (which comprised at least four drafts) delayed the release date until April 6, 1971. The final document was 37-pages long, and was accompanied by a thick appendix containing technical reports on engineering feasibility and cost, geology and seismicity, natural history and fishery resources, and socio-economic impact. The nub of the conclusion NU had arrived at, however, was articulated in a foreword signed by Lee Sillin (and written with considerable editorial help from Foster and August). The key portion read as follows:

> Our findings lead us to believe that a facility at Schenob Brook would have a greater environmental

> impact than would a project at Canaan Mountain. We conclude, therefore, that the Canaan Mountain site should be considered better suited for further regional examination, although we are well aware that there are major environmental questions at this site which require more precise definition. However, we believe that our studies of both sites should be reviewed by concerned citizens and governmental agencies.

That last sentence may well have been a bow to Foster's wish to leave the Sheffield door open, at least a crack, but it was evident from the implied cessation of studies there (among other things) that NU considered the option foreclosed. As CEVAL editor Gloria Freund put it in her next newsletter, "the company (deems) it unlikely that an alternative conclusion as to the merits of the two sites would emerge via sincere study, and has indicated that it has no further interest in the eliminated site as a pumped storage location."*

Another thing Sillin was abundantly clear about, it is important to realize, was that because Schenob Brook was 'out' did not mean that Canaan mountain was 'in.' For one thing, he indicated that before making a commitment to any project the company still needed to refine its forecasts for peak power demand and confirm that pumped storage was the most appropriate supply technology; a report on the latter, in fact, was promised by the summer. What's more, he pointed to the fact that other utilities as well as the New England River Basins Commission were still evaluating the potential for hydro-plants of the pumped-storage type elsewhere in the region. And last but not least, both the summary report and the material appended thereto (despite its bulk) revealed that only exploratory work had been completed at Canaan Mountain -- no in-depth analysis. Peter Stern amplified on this, much later, with the following remark:

> ...(it was) a preliminary study, in the hope that that preliminary study would then lend itself to some kind of consideration by somebody, that could then allow us to do the final study, hopefully with the input of somebody else. It was (sort of) a bastard as a re-

---

* CEVAL Reports, number 5 (April, 1971).

> sult. That preliminary report satisfied nobody -- certainly nobody in here. It was certainly totally insufficient for an application to the Federal Power Commission, from an environmental, engineering, technical, economic, every point of view. The only thing that study did for us that we wouldn't have to re-do was the geology, the drilling, the assurance that the rock was good at Canaan and pretty poor in the other place.

Lee Sillin had put it very mildly, in other words, when he spoke in his foreword about the need for "more precise definition" of major environmental questions!

In a general sense the fact that NU was exposing its logic at an intermediate stage was not inconsistent with the principles of open planning, but under the circumstances it had a most damaging effect on Project CEVAL. A lot of momentum had been lost in the three months the Board had had to wait for the company's report, and the success of the entire effort hinged on the expectation that it would provide a strong stimulus to further action. But what was there to sink one's teeth into? With only one site left to consider there was nothing to evaluate in the way of locational tradeoffs, nor was there anything concrete to respond to vis-a-vis demand estimation or alternative supply strategies (which for the most part CEVAL felt to be beyond its scope anyway). Further study at Canaan Mountain offered the possibility of influencing the impact assessment process, to be sure, but most of the issues of public concern had already been raised in rather specific terms; BLECC had seen to that. What CEVAL needed to operate on -- desperately needed -- were some tentative answers that could be broadcast to a wider audience and used as a focal point for vigorous discussion and debate. But the highly preliminary NU report provided no such grist for the CEVAL mill.

Far from entering a new phase of vigorous activity in the spring of 1971, then, the CEVAL Board could do little but continue to bide its time and hope that further analysis by its consultants and especially by the utility would stir the long-awaited evaluation process. But it was not to be, for the record shows the appearance of NU's report to be the last significant event in the CEVAL story. Although the Project did not actually come to a close for another year and a half, the company provided essentially nothing new during this period in the way of planning data.

It never put out the promised report on alternative methods of peak power production, for example, nor any further assessments of impacts at Canaan Mountain. In fact, monthly briefing meetings CEVAL had set up with Stern and Bragg had to be discontinued for lack of anything substantive to report, and a major question-and-answer session with both CEVAL and BLECC was repeatedly postponed. In short, NU proceeded to pull the rug of information out from under CEVAL's feet entirely.

Why was the flow of additional information curtailed so dramatically? The principal obstacle to further disclosure in the short run was still BLECC, whose executive director (a lawyer) and a consulting engineer had prepared a long list of difficult questions designed to provide 'ammunition' for their program of legal intervention to block construction of a plant. Interrogatories are a fact of life in adjudicatory proceedings, of course, but it is poor strategy to give one's adversaries a head start in preparing their rebuttals. As Peter Stern eventually admitted to Joan Flood, "we're going to be in a law case with BLECC, and we can't give them the information they need."

From a public relations standpoint, one might think such reluctance to cooperate with subsequent evaluation initiatives on the part of CEVAL would have been most awkward for NU. Eventually, though, there was another consequential development which provided a way out of the dilemma: electricity consumption had begun to slacken, and the company thus revised downward its growth projections for peak demand. This in turn called for slippage in the timetable for the pumped storage plant, so there was no longer any pressure to reach a decision on Canaan Mountain and, consequently, no reason to incur the unnecessary expense of additional site-related research.

With NU marking time for the remainder of 1971 and throughout 1972, it was understandably difficult for CEVAL to carry on in any significant way with its fundamental mission. In particular, there could be no productive dealings with the public at large, which had continued to show disinterest in the siting isue despite Flood's publicity efforts and good coverage of CEVAL's early activities in the local press. At its peak the audience at the 'study group' meetings had only reached 70 or so, and only a handful of those were typically not from BLECC or otherwise part of the anti-plant movement. When the last of the series of public meetings was over, in March of 1971, it was clear that it would not be easy to recruit participants for the task forces that supposedly would continue the deliberative effort. And as it subsequently became clear that NU's report was too preliminary to evoke serious public discussion, the whole idea -- another of the original features of the Sillin/Stern/Foster model of participation -- was simply dropped by the CEVAL staff.

The Board's activities were thus quickly reduced to fact-finding via the independent analyses its consultants had undertaken, which if well done and sufficiently provocative held some promise as a touchstone for community deliberation. Unfortunately, though, the studies that had already been commissioned trickled in slowly and weren't much help when they did arrive, in no small measure because so little material of substance had been forthcoming from NU (in both the preliminary report and in response to subsequent queries by the consultants). At any rate, the documents were essentially greeted with a yawn by the company as well as the public at large. Undaunted by the lack of any response to the work it had contracted for, the CEVAL Board later issued grants for socio-economic contingency planning (to the Falls Village Planning Board) and for a water availability study of Canaan Mountain (to the Yale School of Forestry). These were central issues of concern, but they were all but moot by the time the results were in (for reasons to become clear momentarily).

The pattern of atrophy extended to CEVAL's information dissemination efforts as well. Gloria Freund continued to turn out newsletters and a two-part series of Facts and Issues, but the latter focused on regional energy issues for lack of any fresh copy on the home front. The publications program thus did little to arouse the sleeping public, or even to create in its mind an accurate perception of what the CEVAL Project was all about. Disappointing news to this effect came in late summer of 1971, when the Board learned the results of the poll it had commissioned to identify local attitudes toward the proposed plant and the various organizations with a stake in the planning process. The pollsters found that only half of the people who had heard of CEVAL realized the primary role of the organization was to be a clearinghouse for information. Indeed, a full 25 percent thought that CEVAL was an advocacy group, although there was confusion as to just which side it was on: nine percent said pro, and sixteen percent said con.

CEVAL even found it difficult during this period to sustain a critical mass of people willing to stay 'up to speed' on the issues and help keep public discussion alive. Addision Cate returned from the University of Michigan and took over as executive director, and along with Joan Flood served on a Consulting Committee of the New England River Basins Commission, which was doing a shoestring study of the role of pumped storage technology in the region (with some attention to the merits of fourteen alternative sites). But to widen the small circle of literate (yet uncommitted) activists was a lost cause. In the winter of 1971-1972, Cate formed a "consultative board" composed of influential residents of the Falls Village area, who would attend monthly dinner presentations by guest speakers so that they, in turn, could discuss the siting problem intelligently with friends and neighbors. These ended during the summer, however, as attendance by the 'influentials' was more often than not lower than that of the CEVAL Board and staff.

Thus, as it was beginning its third year, CEVAL still found itself struggling to define a sense of direction. But all it could come up with was to follow through with the studies then being supported, make arrangements for the preparation of a final report of some sort, and look for a way to extend its life until NU decided whether to go ahead with Canaan Mountain. But the decision had been put off until 1974 at the earliest, and the Board had no enthusiasm for requesting more money from the Fund for the Preservation of Wildlife and Natural Areas (and certainly not from NU). By now the group was totally disenchanted with the company and with Lee Sillin, who left his first and only meeting with them (in May of 1972) with the feeling that no one wanted to talk with him or even, for that matter, be seen with him.

This ignominious episode, by the way, was neither the first nor the most consequential way the interactive approach had backfired on its one-time champion. It turns out that while the CEVAL Project was underway, Sillin had attempted to engage Northeast in yet another endeavor billed as open planning, involving a decision as to the placement of transmission lines within certain broad corridors to the southeast of Falls Village. Affected citizens were supposed to participate in this routing process, but for some reason the NU subsidiary in charge of the project employed the traditional 'decide-announce-defend' approach, and in so doing alienated both the local citizenry and their elected officials. The furor was so great, in fact, that it led to the formation of a lobbying group whose efforts by mid-1971 had contributed to the passage -- over NU's objections -- of a new state law governing siting and licensing of power facilities in Connecticut.* As irony would have it, the company's ability to develop a constructive working relationship with citizens outside of regulatory channels had resulted in even greater exposure to the vagaries of public participation within them.

In view of all of this, it would be very much of an understatement to say that, as of mid-1972, Northeast Utilities' experimentation with open planning had left quite a bad taste in its corporate mouth. A far better recapitulation, one that could hardly be improved upon, is that given by CEVAL's historian Michael Harwood:

> From the company's point of view, its attempt to involve citizens in the decision-making process hadn't done much more than saddle it with impossibly demanding regulation, give the environmentalists a leg up, put off the solution to peaking power and transmission questions, cause dissension in the company, and give NU a black eye in the press. So the non-institutional approach to the public was virtually a dead letter at Northeast Utilities -- ruined by "prejudgement," then co-opted and transmogrified by the state of Connecticut. Sillin was evidently still committed to following through on the process in terms of Falls Village; he was a man of his word, after all. But as for elaborating the approach or trying to improve it for future use, he felt he had been sandbagged enough already and wasn't very interested.

Thus it should come as no surprise that by the fall of 1972, with its bank account depleted, its role more poorly defined than ever, and its relationship completely sour with the executive whose initiative had brought it into being, Project CEVAL gave up the ghost. With the question unresolved as to whether a pumped-storage plant would still be needed, though, BLECC managed to keep its doors open for another year or so. But that group eventually followed suit in early 1974, when NU deferred its decision once again -- this time, it turned out, permanently.

---

* The statute, entitled the "Public Utilities Environmental Standards Act," was passed into Connecticut law on June 23, 1971.

WHY THE FAILURE?

Notwithstanding the great deal of anguish produced in the course of Project CEVAL, in one sense it might be argued that the outcome was indirectly quite positive, in that the NU planning process had been stretched out long enough for Northeast Utilities to recognize that it would be a mistake to install additional peaking capacity in its service territory. This null result was certainly welcomed by those who wanted to keep the Berkshire hills pristine, and anyone familiar with the New England electricity scene in the latter '70s knows that NU ratepayers were probably better off as well. Simply put, slow load growth and the corresponding retardation of nuclear expansion had made the economics of pumped-storage relatively unattractive.

According to any other yardstick for measuring effectiveness in a citizen participation program, however, NU's attempt to involve the public in its site selection process must be said to have failed miserably. In the end it had done nothing to build a relationship of cooperation between the utility and the environmental community, nor even to establish a mutually beneficial dialogue among the concerned parties. Very little in the way of meaningful discussion took place between NU and CEVAL; likewise, the channels of communication between CEVAL and the spectrum of affected interest groups -- especially BLECC -- were clogged more often than not. In sum, that which occurred did not even remotely resemble the sort of constructive interaction one normally hopes will ensue from an open planning process.

Is it, perhaps, somewhat of a 'bum rap' to label Project CEVAL an outright failure, given that the evaluation function originally envisioned as crucial to the outcome of the experiment never actually materialized? Joan Flood, for one, feels there was still a fighting chance for coherent public deliberation on the issues; as she puts it, "we had a good plan (and) we would have done a good job, if NU hadn't dropped the project."* With all due respect to her opinion, however, it seems safe to predict that the process would have continued along the same troubled path it had begun to travel prior to the release of NU's preliminary site study. By that time, CEVAL's organizational 'baggage' had already become so cumbersome as to make it unrealistic to expect the Board could ever improve its capacity to

---

* Personal communication with Joan Flood (August, 1983).

operate in the role of facilitator, a role to which all hopes of success had been pinned.

From the standpoint of industry perceptions of Project CEVAL's outcome, of course, it is largely immaterial whether the course of events might have turned in a more favorable direction. What really mattered was only that which did happen, virtually none of which could be categorized as positive. Without a doubt the project muddied the political waters for NU, insofar as it allowed time for formidable opposition to mobilize, yet didn't produce a supportive constituency of corresponding magnitude or political influence. In addition, to the casual observer the fiasco seemed to confirm all the stereotypes of environmentalist behavior prevalent among utility executives of that era.* While most connected with Project CEVAL had been willing to keep an open mind, more or less, the actions of BLECC had hardly done anything to dispel the notion that intransigent elements would do everything in their power to undermine good faith collaboration. And there was nothing impressive about the way the CEVAL Board had 'spun its wheels', despite having what appeared on the surface as a well-defined task and adequate resources to complete it. In short, the worst case scenario had come true.

Among industry skeptics circa 1972, news of Project CEVAL's woes was understandably greeted with cries of "I told you so" and widely cited as proof that power plant siting and advance citizen participation simply don't mix. This early reaction has since been blunted, of course, by the largely favorable experience other electric companies have had with public involvement in facility planning in the decade after Project CEVAL. Indeed, the best evidence available suggests that for every participatory program that has yielded negative results there are roughly three which have brought satisfaction, on balance, to the sponsoring utility.** Still, the NU case stands as a down-to-earth reminder that processes of informal interaction can turn into a 'can of worms' under certain circumstances. Rather than being dismissed as an exception, then, the episode must be taken as a warning against complacency, and as a golden opportunity to learn something about the forces

---

* See Chapter Two, supra, for a complete listing and discussion of these early perceptions.

** The data to this effect are presented supra, in Chapter Four.

utility-citizen collaboration. No one case can be completely illuminating in this regard, but stories of failure like CEVAL are undoubtedly more likely to be instructive than case studies of success.

Thus, let us proceed now with a sort of "autopsy" on the NU experiment, in search of a cause (or causes) for the misfortune which led to its demise.

The issue of why can be addressed at different levels of complexity, perhaps the most simplistic of which is to try to identify one or more external elements -- unforeseen by and largely beyond the control of the involved parties -- to which the breakdown of the experiment might be which threaten to undermine the prospects for fruitful directly attributable. 'Things would have been different,' such speculative inquiry might have it, 'if only the Ford Foundation had been willing to underwrite the evaluation project . . . . if only the general Berkshire public had been less apathetic . . . . if only the rock conditions at Sheffield had been more suitable. . . . if only the rate of growth in peak power consumption hadn't decreased . . . .' and so forth. One can hardly deny that each of these factors (among others) had a conspicuously negative influence on the process, and it is tempting to say that it was their cumulative impact which spelled the difference between a satisfactory outcome and the disappointment which finally obtained.

Yet, to explain the failure of Project CEVAL as merely the product of a series of 'bad breaks' is somehow unsatisfying. Other endeavors of this kind have since been successful in the presence of adversity, which in some amount must be expected given the intrinsic vagaries of planning and delicacies of conflict resolution. Whether threatening externalities become fatal often depends not so much on the nature of the externalities per se, but on whether the participatory system is resilient enough to withstand occasional setbacks. Obviously, such was not the case with the CEVAL process, which seemed to amplify problems instead of dampen them through constructive adaptation. For some reason, the political equilibrium of the system was highly vulnerable to disturbance.

But what, at root, was the source of this instability? One conceivable hypothesis is that the qualities of leadership and sound judgement were not sufficently abundant within the two groups most directly responsible for program development: the CEVAL Board and its staff; and Hank Foster, along with his associates at the Natural Resources

Center and at the Fund for the Preservation of Wildlife and Natural Areas. An argument could be made, for example, to the effect that the Foster group did not carry out its coordination and liaison functions in an optimal manner, at least if the repeated misunderstandings with the CEVAL Board over spending matters is any indication. Nor did the actions of the Board itself seemingly do much to ensure that the project would get over the rough spots without significant harm. One key mistake was certainly the initial hiring of an executive director without careful consideration of his inclination or ability to perform tasks the group had given priority. Some observers feel, moreover, that the Board made an even more consequential miscalculation in deciding to assume a passive, umpire-like role. Along this line, consider the following reflective statement by Wayne Hanley, then director of publications at the Massachusetts Audubon Society, summarizing his reaction to CEVAL during its formative stages:*

> (I) never was able to understand the position of CEVAL...the structure by which information was to be organized and passed on. Now as near as I could see it, if you have no form and if...as (Lerman and Cate) kept saying to me, "we're entirely neutral in this," you have no interest whatsoever.... the conscious decision nullifies any possibility of doing any thing, which left BLECC out there, BLECCing away to beat the band, because they did have a position.

All this rings true to a certain extent, but on the other hand there is something to be said on behalf of caution when it comes to second-guessing the managerial judgements of the CEVAL community. Take, for example, the Board's subtle (even amorphous) conception of neutrality, which undeniably caused a lot of confusion and misunderstanding on all sides and thus undermined the Board's credibility. But who's to say that its credibility would not have been damaged even more severely had it opted for a more aggressive advisory role vis-a-vis Northeast? Similar arguments might well be made in defense of the way Hank Foster conducted the affairs of his Center in connection

---

* As was the case in previous sections, this quotation is derived from Michael Harwood's unpublished manuscript, Getting Together.

with the project. The point is, without having experienced first-hand the delicate situation confronting the CEVAL leadership, even hindsight fails to provide us with 20-20 vision where the efficacy of their policies and decisions is concerned.

If not mismanagement on the part of Project CEVAL's operators, is its failure perhaps attributable at root to malfeasance on the part of the other two principal actors (and antagonists): Northeast Utilities and BLECC? Of these BLECC is certainly the obvious candidate for the title of scapegoat, insofar as it was obsessed with the goal of blocking the NU project and unwilling to extend to the CEVAL Board -- in its own right a group with blue-ribbon environmental credentials -- the benefit of the doubt for being able to uphold protectionist values and avoid unsavory compromises. But the allegedly unfair tactics of BLECC should not obscure the fact that the actions of NU, albeit somewhat less conspicuous, were no more consistent with the principles of open and cooperative planning. The company's basic style was to hold its cards close to the chest, and in the face of internal pressure from hawkish elements -- which grew steadily throughout the experiment -- it couldn't find the organizational courage to avoid throwing in the towel in the end. Joan Flood would put it even more bluntly; to her, NU was simply guilty of "complete bad faith."*

Such harsh words could easily precipitate an attempt to sort out personal responsibility for the CEVAL imbroglio in more detail, but to do so -- in this author's view -- would be a rather empty exercise, akin to treating symptoms instead of the underlying disease. For one thing, it would be incongruous to point the finger of blame at the circle of people surrounding Lee Sillin, who deserve applause for their willingness to explore a new and risky approach to conflict resolution, not indictment for eventually getting 'cold feet.' Nor is it productive to look for bad guys among the BLECC population, which undoubtedly believed it to be their moral obligation to throw monkey wrenches into NU's plans at every opportunity. This is not to say that the deportment of either the utility personnel or the citizen activists was one of total propriety. But there is no reason to conclude that one group or the other did not act rationally and in good faith, given their respective per-

---

* Personal communication with Joan Flood (July, 1983).

ceptions of the situation and different operational environments. Accepting this proposition, it should be clear that there were really no villains in the CEVAL story -- only victims. All involved parties seem to have been victimized by the fact that their individual actions -- however defensible on a singular basis -- turned out to be quite irrational in sum.

What all this suggests is that the difficulties encountered with Project CEVAL had their foundation not in the ways the various actors dealt with each other per se, but in the context which shaped their collective behavior. The context of reference, of course, was that of a heated controversy which broke out almost immediately after NU's announcement of its public involvement initiative. The formation of BLECC on the heels of that announcement was indicative of how early the battle lines were drawn, as was the fact that BLECC's resistance campaign was well under way even before the CEVAL Board and staff (let alone its informational programs) were in place. As a result the open planning process had to operate in parallel with a full-fledged dispute, a dispute which unduly constrained the range of behavior that would have been open to all parties under less uptight circumstances. As it was, the environment was so permeated by supicion that every relationship had to be forged with extreme caution, every action brought defensive over-reactions, and no communication could escape some measure of distortion.

Needless to say, it is next to impossible for methods of informal cooperation to prove workable in such a crucible. All collaborative approaches to conflict resolution, after all, are based on the premise that the potential for accommodation can only be fully realized before the issues become symbolic, and adversarial behavior (including extremism, which feeds on distrust) begins to predominate. Mediation and other negotiation-based techniques for settling disputes are still available at that point, being designed specifically to operate in a polarized setting. But for advance citizen participation to be effective, conditions must be far less contentious and volatile. Otherwise, the process is deprived of any real chance to succeed.

So the key question is not why the CEVAL experiment foundered amidst an atmosphere of distrust; rather it is why, in the first place, such an atmosphere descended on the project so soon and so permanently. One possible answer is that which the Ford Foundation provided (in advance) when it declined to underwrite Project CEVAL in early 1970. Characterizing the NU sitng process as being "a long way down the road," the foundation staff had expressed serious reservations about the fact that the choice between two sites in the company's service territory was a "very narrow one," and that the alternative of choosing a location outside the region was "clearly beyond the the capability of the citzen's group." Such concerns were derived from what analysts had learned from numerous power plant siting controversies in the '60s, i.e., that political resistance begins when an activist community perceives that the decision has already been made, that their involvement comes too late to make any real difference (except to stop the project altogether). What the Ford people suspected, then, was that to an environmentalist the terms specified for helping NU deal with its siting problem would be only slightly less threatening than a fait accompli.

The Ford Foundation was ultimately proven correct in this prediction, it would seem, insofar as the people associated with BLECC (among others) could see nothing to be gained from participation in any sort of evaluation process. To them NU was a monolithic utility intent on building a pre-ordained type and size of plant at one of two locations, locations which were close enough in space and similar enough in terms of environmental impact to seem indistinguishable in all but the most parochial sense. There simply wasn't enough flex in what the company was proposing for BLECC to conceive of an outcome with which it could live. Indeed, in BLECC's eyes a construction project in the Berkshires looked for all the world like a foregone conclusion, which presented the organization with a very clear choice: acquiescence, or opposition.

Should this be taken to mean that the CEVAL experiment was doomed to fail on account of suffocating groundrules alone? Possibly, but readers should bear in mind that other utilities have since had positive experiences with citizen participation initiated at the point when a specific plant has been proposed and site options have been relatively limited.* Consider, in particular, the first Northern States Power experiment discussed in the preceding chapter, in which the company asked a citizen task force to recommend a site from among only four final candidates. Granted, this slate of options was somewhat larger in number and more geographically diverse than in the NU case;

---

* See Chapter Four, Part B, <u>infra</u>.

but the existence of a tight deadline made it infeasible to conduct an extensive evaluation, let alone expand the site search or delve into still broader issues in long-range planning. Thus in an overall sense the NSP groundrules were comparably circumscribed, yet they did not make it impossible for the company to secure meaningful input from representatives of the environmental community on a voluntary basis.

One must recognize, on the other hand, that there were some important contextual factors in the initial NSP venture that were conducive to a more favorable outcome: the plant in question, being coal-fired instead of nuclear, was not of the same genre that had precipitated bitter confrontation in Minnesota; the participants as a group exhibited a relatively high level of technical expertise and a relatively low level of cynicism toward established institutions; and (most significantly, some would say) none of the sites was really in the 'backyard' of anyone on the task force. It is also important to realize that, notwithstanding such advantages, the NSP interactive process almost did flop at the very beginning, due to citizen resentment over the restrictive premises governing the site evaluation coupled with deep-seated animosity toward the utility and its policies. This close call taught NSP that its initial approach left precious little room for error in establishing its sincerity, and as a result the company provided its next advisory group with much more time and latitude to explore site-related issues.

The sobering message of this digression on the NSP experience is that involving concerned citizens at the 10th hour of planning -- while it is not so late as the 11th -- is a borderline proposition at best when it comes to overcoming the all-important barrier of distrust. And since this is precisely what NU's brand of open planning entailed, there is nothing surprising about the fact that Project CEVAL met with a fate totally different from that of the Minnesota program. Yet, there remains a nagging, intriguing question: if NSP showed it to be at least possible to obtain constructive citizen input on the eve of a decision, why wasn't NU able to "pull it off" as well? Why weren't the company and its cohorts able -- as NSP apparently was -- to dissipate the atmosphere of suspicion and thus prevent a pall of controversy from being cast over the experiment? In search of an explanation, it now behooves us to take a closer look at how the problem of distrust was (and might have been) handled in the CEVAL case.

WHY THE MISMANAGEMENT OF DISTRUST?

As before, this author is inclined to reject deceptively simple answers -- founded on allegations of human failing -- to the question of why public confidence in NU's good intentions never materialized. To begin with, there is no credible evidence that the CEVAL concept was hastily developed or poorly thought out with respect to the issue of distrust. Witness, for example, the considerable length at which Sillin, Stern, Foster and Flood all agonized at the outset over the problem of funding. And if some 'wrong turns' were taken in the course of implementation (as did happen, obviously), it would hardly be equitable to fault the poeple who made them for not knowing any better. After all, theirs was a pioneering and unprecedented effort which, under very trying circumstances no less, had to be played almost entirely by ear. They had no body of knowledge to draw upon for guidance, no seasoned practitioners to turn to for advice -- only their own common sense, intuition, and seat-of-the-pants guesswork.

As an illustration of this point, consider one of the basic assumptions underlying NU's participatory program: that it would be best for a citizen group to have as much functional independence from the company as possible (via association with Foster's Center). Monday-morning quarterbacks might be quick to identify this as a key mistake in light of the Northern States Power experience, where the utility eschewed any form of buffer and took pains to work hand-in-hand with its first advisory task force. But the lessons of the NSP case <u>were not yet available</u> to the principals in the concurrent NU case, to whom it seemed logical that for a citizen evaluation effort to be credible its participants had to be insulated from the forces of advocacy represented by utility staff. How could they foresee that this attempt to keep the process unbiased would actually do more to aggravate the problem of distrust than to alleviate it, i.e., by precluding direct communication between the company and the broader activist community? And how could they foresee that the vehicle for marshalling public input (the CEVAL Board) would encounter such a host of political and organizational complications so acutely distracting as to make the review function itself seem almost secondary?

This is not to say, on the other hand, that NU had no advance indication that the formulation of Project CEVAL left something to be desired. In particular, a red flag had been waved by the Ford Foundation letter rejecting NU's grant application, which clearly warned that citizen activists could suspect the decision had already been made, and thus might regard the public involvement process as an exercise in cooptation instead of cooperation. However prophetic this message turned out to be, though, at the time it must have had a hollow ring to the person who had the final say in the matter -- Lee Sillin. Although it was probably true that the major options were foreclosed in the minds of most other NU executives and planners, there is no reason to believe that Sillin shared in any company predisposition except that toward pumped storage technology. He seemed willing to question the need determination, at least internally; and when it came to siting, it is arguable that he saw nothing imperative about finding a location within the NU service territory.

That the Berkshire region was identified as the prime candidate area was not necessarily inconsistent with this line of argument, because in Sillin's way of thinking it made sense to kick off the discussion with a tangible statement of what looked best to the utility, as long as care was taken to emphasize that the analysis was fluid and all findings would be tentative. He tried to do this, it appears, with a variety of gestures:

* by promising not to acquire any land prior to the conclusion of the evaluation program;
* by naming two prospective sites instead of one;
* by expressing commitment to a full and balanced evaluation of the suitability of those sites, in environmental as well as engineering and economic terms;
* by saying that any NU site deemed suitable after public examination should then be compared to those which might exist elsewhere in New England;
* and, perhaps most importantly, by not imposing a severe time constraint on the public review process.

From Sillin's standpoint, then, there seemed every reason to expect the environmental community to accept on face value his assurance that "a final decision can be reached on a regional basis."

At first, the guardedly positive signals forthcoming from affected interest groups seemed to discount the Ford Foundation's concerns, and probably bolstered Sillin's faith that his offer to share the siting decision with the public would be perceived as genuine. But perception is a mercurial thing, and as word got around that a major utility project was in the making, a dark cloud of suspicion soon appeared on the horizon. This was clearly in evidence at an early point, when NU's proposal to fund the experiment itself (albeit indirectly) made the leadership of two key groups sufficiently nervous to forswear any organizational involvement in -- or even endorsement of -- Project CEVAL. Therein lay a conspicuous danger sign, a sign that attitudes could easily and rapidly change in the negative direction with any false step on the utility's part.

That false step was taken, unfortunately, at Lee Sillin's press conference announcing the CEVAL experiment. Indeed, if the political situation at that point could aptly be described as a powder keg, the event in all likelihood was the spark which caused the explosion of BLECC-led opposition. Everything that occurred there belied the notion that what Sillin really wanted was to facilitate a regional approach to the siting of major peaking facilities. For example, there was no explicit indication that the underlying issue of need for power would be open to review, and the slick presentation about the virtues of the Northfield facility certainly implied that the subject of alternative supply strategies would not be welcome on the agenda, either. Secondly, there was no mention of the many potential sites the company had screened out in targeting the two final candidates (another potential object of scrutiny by citizen evaluators); and nothing was said about how the NU siting effort had been (or would be) coordinated with its counterparts in other utilities. In Sillin's portrayal of the problem to be solved, then, all issues in electricity planning came across as secondary to one fundamental task, the comparison of Canaan to Schenob Brook, a task with which local environmentalists were not the least bit enamored.

As proof of his sincerity, of course, Lee Sillin was probably relying more on his express commitment to a "completely open evaluation" of all relevant factors to be carried out by a local group drawn "from among interested environmental organizations." However, from the perspective of someone inclined to wonder if the participatory program would be meaningful, one can see how this part of the announcement might have been construed as less than believable as well. The basic problem was that since the particulars were still far from being worked out, Sillin had to be silent on all big questions concerning groundrules for the operation of the group. What selection criteria would be used? What relationship would exist with the utility, the Center, and the public at large? What specific input was expected, and how would the company respond? Instead of providing concrete answers, Sillin could only state that Hank Foster was the man in charge of the project and that NU money would be used to finance it. Making matters worse was the fact that Sillin was vague on the actual disbursement mechanism, leaving the impression that direct payments would be made to Foster's Center for Natural Resources, the organization seemingly responsible for orchestrating the local evaluation process.

Thus, ironically, the gestures that Sillin undoubtedly hoped would earn the benefit of the doubt for his proposal actually did just the opposite. The aura of salesmanship that (perhaps unconsciously) surrounded his discussion of the planning problem, coupled with the (arguably false) sense that only superficial attention had been afforded the open planning concept, apparently convinced a lot of people that Sillin's mind wasn't nearly so open as he expected them to believe. What had been created for Project CEVAL was not the image of a process designed to subject the company's tentative ideas to critical scrutiny, but of a campaign to legitimize a point of view which the utility might not otherwise be able to ratify in the regulatory arena. All this would help explain the noticeable erosion in confidence which occurred during the weeks following the press conference, as manifest in the flak Stern and Bragg ran into at the next meeting of the Housatonic Audubon Society; in the suggestion in at least one local paper that the company's financial support gave "the appearance of collusion;" and, above all, in the almost immediate formation of BLECC.

Admittedly, Lee Sillin had no way of knowing that his presentation would send contradictory signals and therefore make people apprehensive. And from all accounts he was honestly taken aback by the subsequent charge that he had paid only lip service to the notion of public participation -- which led him to the disheartening conclusion that his motives had been unfairly pre-judged. Still, it is significant to point out that the negative feedback engendered by the press conference did not precipitate a vigorous attempt by NU to stem the rising tide of suspicion. To the contrary, the company's only additional step was to go ahead with the plan to name the Fund for the Preservation of Wildlife and Natural Areas as the intermediate agent in the funding chain, clinging -- apparently -- to the hope that this would convince people that no strings were attached to the company's money. But by this time it was in the interests of some people to keep the 'taint' idea alive, so in reality the arrangement did little to reduce the discomfiture environmental leaders had previously felt with NU as the sole source of funds. At best it simply called further attention to the (awkward) fact, and at worst it was taken as a sinister 'laundering' scheme to obscure the company's suspected desire to exert a high degree of control over the evaluation process.

Taking a broad perspective, now, it seems abundantly clear that the CEVAL experiment fell victim to mistrust at a very early point because its operators seemed to underestimate how ominous it looked to many activists, and failed to respond accordingly. Despite the fact that the venture had barely reached first base, it was at a crucial juncture not unlike that which Northern States Power had come upon when its first task force started off on a tangent of hostile discussion of electricity growth rates and utility promotional practices. In that case company personnel decided not to aggravate the sensitive situation by attempting to cut off debate, but to 'take the heat' as a gesture of good faith -- even if it meant departing from the original groundrules as well as stretching the process out a bit longer. In the NU case, however, one finds no comparable adaptivity. To the contrary, the company seemed intent on proceeding according to plan, as if oblivious to the time-bomb of distrust which by this point was ticking louder than ever.

What would it have taken to defuse that bomb? In this author's opinion, NU needed to continue the process of face-to-face consultation with affected interest groups (from which it essentially withdrew as Project CEVAL began to take shape), and to go back to the drawing boards on the matter of program design. At a minimum it would seem to have been necessary to reopen the search for an independent source of funding, given that the alternative of using NU money was proving to be such a sore point. And it probably would have been advisable to negotiate a somewhat broader agenda, offering for critical scrutiny not only the logic followed in reducing the list of candidate sites to two, but also the rationale behind the long-range determinations of need for power and choice of plant. Presumably this is what the Ford Foundation staff had in mind when it urged the company to "go back to a more fundamental level" -- an exhortation which the company should have heeded soon after, and preferably before, Lee Sillin formally christened Project CEVAL in his press conference.

When there is a radical fringe bent on opposition under any circumstances, of course, one cannot reasonably expect such a mutual scoping process to satsify everybody and preclude further attempts to discredit the collaborative approach. But goodwill often emanates from repeated elbow-to-elbow contact, and this might have reduced suspicions enough to at least preclude kneejerk attacks on the legitmacy of NU's outreach effort. And in all likelihood a modification in the terms of the evaluation program (along the lines suggested above) would have precipitated a detailed review by Northeast of the defensibility of the planning conclusions it had already reached (to which, as Peter Stern has acknowledged in self-criticism, a much higher standard of engineering proof should indeed have been applied at an earlier point*). In this regard, by the way, one cannot help but imagine how differently Project CEVAL might have turned out had NU agreed to expose the need for power question to active public scrutiny and then arrived, in concert with the environmental community rather than unilaterally, at the conclusion that a slower growth rate allowed for an indefinite deferral of new peaking facilities.

Such is what might have happened, but the real question of course is: why didn't it? Instead of going back to square one before distrust got out of hand, why did Lee Sillin jump the gun with his public announcement, and why didn't he take the time afterwards to make the CEVAL concept more palatable to the activist community? In this author's opinion, the answer lies in the delicacy of his position within the company, where anxiety over facility development was increasing in direct proportion to the rapid pace of annual load growth. With the planning staff consequently in a big hurry to move into the final stages of site selection, the very idea of switching to an open planning mode at that point had raised a lot of eyebrows; indeed, the potential for delay associated just with advance disclosure of the two candidate sites must have sent shudders through certain quarters of the utility. This mood of impatience had already made it difficult to keep the pumped-storage project on hold long enough for Stern and Foster to lay the groundwork for a public involvement program. Think of the intense pressure Sillin would have faced had he suggested not only that the evaluation process should mark time further to allow for additional negotiation with citizen groups, but also that as a result of such negotiation it might be necessary to retrace some of the major steps already taken in defining the size, type, and general location of the plant.

---

* Personal communication with Peter Stern (September, 1983).

Had Sillin been well-established at NU, it might have been possible to resist this inertia if he had really wanted to. But both he and Peter Stern were newcomers to the organization, and in deciding to go ahead with an open planning initiative they had already capitalized on the initial deference which is normally afforded to those assuming a heavy mantle of responsibility. So it would have been painfully obvious that to depart further from the mainstream of thinking within the company might result in a great loss of political capital, perhaps so great as to even jeopardize their reputations as prudent managers and to compromise their ability to provide effective corporate leadership. Weighed against this very serious risk, the benefits of lobbying aggressively for another iteration would have been intangible, uncertain, and maybe even unrecognizable at that point. To this author's knowledge, after all, neither man had a crystal ball in the office which would have shown how bleak was the future for the vaguely-defined Project CEVAL.

The same holds true for Hank Foster and Joan Flood, although one might think that either he -- with extensive academic training in natural resource management -- or she -- with an equally extensive track record in grass roots activism -- would have been more likely to recognize that it was imperative to change the basic framework for participation. To be sure, both worked in their own way to mitigate the most conspicuous problem of funding: in Foster's case this meant helping set up the Fund for the Preservation of Wildlife and Natural Areas as the financial intermediary; and in Flood's case it consisted of a last ditch effort to convince the Ford Foundation to underwrite the project under the auspices of the New England River Basins Commission (instead of Foster's Center). But there is no evidence that either of CEVAL's organizers tried to convince NU to keep looking for a totally independent source of funds, or to negotiate a broader issue agenda upon formation of the CEVAL Board. Indeed, they might well have gone beyond jawboning and insisted on new groundrules as a precondition for their further involvement; without their leadership, after all, there was little hope of mobilizing any sort of participatory effort in the first place.

It was certainly Foster's and Flood's prerogative to employ their leverage on behalf of a better process design, and meeting with success in this regard clearly would have helped reassure the environmental audience that a collaborative approach was worth a try. On the other hand, there is a serious doubt to be raised as to whether such external pressure actually would have evoked the desired response from NU. As noted previously, the internal constraints Sillin and Stern had to operate under were such that it was probably out of the question to undertake any time-consuming reformulation of the CEVAL concept. Thus, any attempt on the part of Foster and/or Flood to 'take the bull by the horns' would have been futile at best; at worst, it even might have backfired by putting Lee Sillin in a no-win situation vis-a-vis the diverse constituencies with which he had to contend. Acceptance of a demand to do more homework on the participatory program would have the company's conservative forces up in arms, yet refusal to do so would further alienate citizen activist groups and make the prospects for constructive interaction even more problemmatic.

Faced with this kind of a choice, it is hard to imagine that Lee Sillin would do other than simply withdraw the offer for public involvement altogether. It is also hard to imagine that either Hank Foster or Joan Flood, both of whom (especially the former) were in close enough contact with the company to appreciate Sillin's predicament, would not realize that an effort to drive a harder bargain could well be the 'straw that breaks the camel's back.' And both, of course, sincerely wanted to encourage the use of participatory methods to resolve conflicts over land use, with Foster perhaps having the additional incentive of wanting to launch his fledgling Center with a project that would impress the broader natural resources community. Like Sillin and Stern, then, it seems highly unlikely that Foster and Flood would have embraced a hard-line position with such great potential to compromise their desire to do something progressive.

The whole point of this exercise in speculative reasoning is to offer a sort of "proof" that the only course of action which might have pointed Project CEVAL in the direction of success -- i.e., an early reformulation of its most basic features -- would have been practically impossible under the circumstance. It matters not whether the need for such a reformulation was known and fully appreciated by the inner circle of people who conceived the experiment; the important point is that any one of them probably would have decided <u>in any event</u> to take their chances with the original approach. To try to correct its flaws would only be to risk missing the "window of opportunity" for innovation, which was already uncomfortably narrow due to the circumscribed planning conditions inside NU. It was a take-it-or-leave-it type of situation, which did not allow for conformity with some of the most important principles of normative citizen participation theory.

Therein lies the basis for the troubles that beset Project CEVAL, whose story is really that of Catch-22. With everyone walking on the eggshells of suspicion from the outset, the project was destined to founder; yet in reality nothing could be done to explicitly and effectively address the critical problem of mistrust. There simply were not enough degrees of freedom in the design function to give the process the second chance it needed to recover from a false start. This has not been widely recognized, and it suggests that the negative outcome of Project CEVAL should not be taken as cause for pessimism about open planning more generally. The NU experiment, to put it succinctly, was simply not a very fair test.

The case does leave us with one fundamentally important legacy, however. It demonstrates that the sine qua non for constructive utility-citizen interaction is that the process be carried out under conditions of truce, which implies a modicum of at least incipient trust as well. Achieving and maintaining such conditions, therefore, should become the first and foremost goal of any public involvement program a power company undertakes in the future.

13

Public Involvement in Corporate Technology Decision-making: The Case of Pennsylvania Power and Light

Eleanor Winsor and
Stephen H. Cutliffe

In recent years technology has increasingly come to be seen as a product of social processes in which abstract economic, cultural, and social values influence the development and implementation of specific artifacts and techniques, through prevailing institutions such as business, government, and the marketplace. Inherent in the social process of technology is a public responsibility for the shaping and control of that process. Thus, active participation in the technological decision-making process is vital if the public is to play an effective role in acquitting any such responsibility.[1]

At present, there exist a number of institutional approaches for dealing with the social impact of technology. These include public hearings, referendums, environmental impact statements, and technology assessments. In addition, a variety of proffered, but as yet untried,

---

Eleanor Winsor is Executive Director of the Pennsylvania Environmental Council and has a broad range of experience in environmental issues and regulation, most recently with respect to hazardous waste facilities planning. Stephen Cutcliffe is Director of the Technology Study Resource Center at Lehigh University, and is co-author of Technology and Human Values in American Civilization. Both Winsor and Cutcliffe have served on the Public Advisory Committee (PAC) to Pennsylvania Power and Light Company. The authors would like to thank the PP&L staff -- particularly Jack Busby, Robert Campbell, Vivian Ross, Gerald Farber, and Bernard Bujnowski -- and Hugh Campbell, past chairman of PAC, for their assistance in the preparation of this article.

approaches have included suggestions for science courts, a technology ombudsman, "adversary forums," and a "technological magistrature."[2] Most of these suggestions, however, are inherently concerned with the discussion or evaluation of technology after the impact upon society has occurred, or at least well after the crucial decisions about implementation have already been made. This is well and good, but something more is needed if public participation is to be truly effective. One possibility that merits consideration is the notion of corporate public advisory committees that have input into the decision-making process right from the beginning of corporate action, not merely after the fact.

In an exploratory article on the application of citizen participation to the public electric utility sector, Dennis Ducsik has discussed the relative merits of such an approach for the siting of large power plants.* Responding to a number of utility concerns regarding open planning, Ducsik concluded that the concept has much more to commend it than the power industry had yet realized. He closed with a call for detailed case studies to further explore the relative merits of the public participation process. Thus, we present here a case history of Pennsylvania Power and Light Company's (PP&L) Public Advisory Committee (PAC), which is useful for what it reveals not only about their particular experiment in public participation, but also about public responsibility for the social process of technology more broadly.[3]

Cognizant of a need to provide wider opportunity for public involvement in the decision-making process, in late 1974 PP&L acted on a suggestion by then Board Chairman and chief executive officer Jack Busby. The company established an in-house Task Force, consisting of a number of middle-level managers under the direction of the Community Services Department, to develop an experimental project to met the following corporate policy or charge[4]:

> Public participation in the decision-making processes of siting PP&L's facilities is not only an appropriate, but an urgent concern. Steps shall be taken, consistent with PP&L's corporate and service responsibilities, to gain such public participation as expeditiously as possible.

---

* The article in question is reprinted herein as Chapter Two.

In pursuing such participation, the corporation sought to enhance public understanding of the siting process (a major area of public concern regarding utility activities), to establish a system for gaining public input, and ultimately to develop a process to involve the public in corporate decision making. However, each of the steps was to be considered very deliberately and, based upon the success or failure of the previous step, the experiment in public participation in corporate decision-making would proceed.

GENESIS OF THE PARTICIPATION PROGRAM

Initially, the PP&L Task Force was interested in establishing a public, ad hoc advisory committee to aid in the siting of one specific, proposed generating facility. If all went well, PP&L management hoped that an appropriately modified process might be utilized system-wide.[5]

The Task Force first established guidelines for the formation and operation of such an ad hoc committee, and then selected a pilot project that offered the widest variety of land use constraints, public interests, and population densities. Institutional, civic, and social groups, as well as non-affiliated interested citizens, began meeting to help PP&L consider the location of a 66-12kV substation and 6500-foot transmission line to serve the citizens of Summerdale, a residential suburb of Harrisburg, Pennsylvania. After four months of many public open meetings, field trips, and analyses, the advisory committee recommended a substation site and transmission route. Both the corporation and the committee were well-satisfied with the results of what was clearly a test case.

PP&L felt justified in its support of the project because of the fact that its System Power and Engineering Staff, in independent studies, chose a similar combination of line route and substation site to that of the ad hoc Summerdale Committee. This fact offered hope that a similar approach to siting projects could be used system-wide. At the same time, the image of public participants as irrational "tennis-sneakered environmentalists" certainly did not materialize. In fact, the PP&L staff developed an increased appreciation and understanding for an objectively motivated committee that was able to work in a cost and time-conscious manner. Similarly, the committee reported feeling "quite encouraged that a utility has involved the community in the decision-making procedures for actions that will affect the residents of the area."[6]

One important specific recognition on the part of PP&L was inherent in its public involvement activities right from the beginning and was probably crucial to their success. The Task Force noted in its recommendation to establish the pilot effort that the project "will require more than a casual philosophical and resource (including manpower) commitment; that many levels of the PP&L corporate structure must not only respond but lead in terms of a PP&L - public community relationship (and that) the manpower requirements while not excessive should not be viewed as insignificant."[7] Similarly, after the project was over, the report noted that "PP&L feels that the most important single factor in public involvement with such projects is a well-chosen liaison, one who is able to devote full time to the project whenever it is necessary."[8] Thus, it became clear at a very early stage that it was vital to have the highest level of corporate support, both intellectually and in terms of personnel and resources, if such an experiment in public participation was to succeed.

Satisfied that public participation could work, PP&L moved to make such participation an integral part of its corporate management process. During the next year, PP&L management discussed the specific ways in which the public participation process should evolve. As a result of many corporate memos and meetings including a public "Delphi" session on power plant siting issues, PP&L developed a clearer sense of direction.[9] In December 1976, its Community Services Department prepared a detailed proposal to incorporate public participation into power plant siting studies permanently.[10] Its report outlined how a Permanent Siting Advisory Committee (P-SAC) could develop, and offered six major reasons why PP&L should involve the public:

(1) Citizen involvement leads to increased understanding and hopefully support for proposed projects;
(2) Increased company understanding of public concerns would improve planning procedures relating to them;
(3) Public dialogue would result in better decisions;
(4) Even where differences remain, public participation would help build mutual respect and understanding;
(5) Public participation should improve the atmosphere of trust and cooperation when public agencies review proposed siting applications; and

(6) Costly delays and legal entanglements may be avoided by having potential problems surface early in the process.

The proposal went on to recommend that PP&L:

> ...establish a Public Siting Advisory Committee to consist of representatives of statewide environmental, economic, and consumer groups, and state agency personnel...This group would be able to advise the company over the years on all siting-related projects under consideration by PP&L. Depending on the relationships developed, this group could also be used as a review and sounding board for such topics as siting legislation, land use policy, etc.

During the first six months of its operation, P-SAC was to have a very specific agenda. Its responsibility was to help PP&L complete a site selection study for coal-fired power plant sites. Thus, out of an amalgamation of corporate responsibility, altruism, and just plain common sense, P-SAC was born.

The make-up of a public advisory committee is extremely important to its ultimate success or failure. The intention of PP&L, which had the final say on membership, was to seek the most balanced committee possible within the constraint that members have a real interest in energy issues. P-SAC's original membership thus consisted of individuals with backgrounds or training in some way related to power plant siting, including environmental, land use, civic, academic, industrial, and governmental experience. PP&L's expectation was (and still is) that the members would not represent the specific constituency to which they belonged, but rather that as an individual citizen they would bring their background and expertise to bear on the committee's deliberations.

Despite the broad range of committee members' experience, there was still some concern on the part of both PP&L and the Committee itself about the total representativeness of P-SAC (e.g., there were no minority representatives, no senior citizens, and no John or Jane Q. Publics). Although the subsequent make-up of the committee evolved in such a way as to include a broader range of participants, the question of who serves on such a committee points to two related problems in involving the public in such a project -- commitment and representativeness. On the one hand, people who are willing or able to give the necessary time tend to be those already deeply involved in public affairs. On the other hand, they are not necessarily representative of the general public. Most P-SAC members agreed to serve for one of several major reasons: they were intrigued by the uniqueness of the idea; they felt the need to participate in the evolution of the public participation process within the corporate structure since they had themselves advocated such a process; or they felt they had something to contribute either personally or as a "representative" of a particular interest group which they believed should be part of the siting process.[11]

PP&L also considered that the undertaking might entail certain risks for the utility. Among the concerns identified prior to convening P-SAC were that: (1) the committee might not accept the need for a facility or the planned size to the point where revisions in the work already done by a consultant would be required, or the scope of the rest of the work might be changed; (2) premature site disclosure might raise undue concern or speculation and could affect acquisition to the point where PP&L would have to accept a less desirable site or use condemnation powers; (3) P-SAC and the consultant might not be compatible; (4) P-SAC might not be able to reach consensus on a site; (5) there might be local opposition to a statewide group evaluating the site; and finally, (6) the recommendations which came out of P-SAC might not be in line with corporate policies or priorities (which would necessitate providing an acceptable explanation to the committee).

All of these potential problem areas, PP&L knew, might involve increased costs or time delays.[12] However, if the process of public participation was valuable enough to be pursued and was to have meaning, these risks had to be accepted. That PP&L chose to advance the level of public involvement at this point indicated their belief that the process was a viable one.

THE P-SAC EXPERIENCE WITH SITING

The formation of P-SAC began early in 1977. The contact letter to potential members indicated that the committee's initial task would be the evaluation of a specific coal-fired generating facility, but also made it clear that in the future P-SAC would advise PP&L on other projects and issues.[13] The first meeting of P-SAC was held on March 29 in Allentown. Jack Busby, PP&L Board Chairman, opened the meeting. His presence, along with that of Leon Nonemaker,

Vice President of Division Operations, and John Kaufman, then Vice President of System Power and Engineering, in addition to members of PP&L's middle management, made it clear to P-SAC members that PP&L's management was deeply committed to public involvement. This strong sense of corporate commitment went a long way toward dispelling any fears that P-SAC members had that they might be "used as window-dressing." In interviews at the conclusion of the siting phase of P-SAC's work, members repeatedly reiterated the importance they attached to this high level support. As one member said: "Without the clear commitment of support from the top down at PP&L, P-SAC would have found itself functioning in a vacuum. This support was a strong motivating force, for it made us realize that the utility would use our findings."[14] PP&L staff, as well, considered this support critical. Vivian Ross, the Coordinator of P-SAC (an employee of PP&L's Community Services Department), noted that the support of Mr. Busby and other top management officials helped middle managers support the project when P-SAC asked them to provide information.

The March 28 meeting was primarily an organizational one. PP&L staff explained that, although it was envisioned that in the long run the agenda for P-SAC would be open-ended, the group was being asked to deal with a specific issue during the first six months. This issue was the general siting study then underway and the resultant selection of several plant sites. P-SAC also received at this time a tentative set of guidelines, drafted by Community Services, to serve as the basis for the committee's relationship with the company. Included in those guidelines were the following statements:[15]

Guidelines for P-SAC

Long-Range

* P-SAC understands that PP&L is undertaking this program in order to benefit from the informed evaluation of this advisory group but also understands that final responsibility for siting-related decisions remains with PP&L management.
* P-SAC members understand that some information provided to them will be considered confidential and they agree to respect PP&L's judgement in this.
* P-SAC members expect and deserve full cooperation and careful consideration of all recommendations

and reports. P-SAC members will work in the same spirit of cooperation with PP&L.

Short-Range

* For the initial study phase, P-SAC members accept the necessity of PP&L to complete the study for a fossil-fueled power plant in the 800 MW range as the next power generation facility for the 1980s.
* P-SAC members will attempt to meet the target dates for new site recommendations which have been mutually established with PP&L.

Guidelines for PP&L

* PP&L will not make decisions relating to siting facilities being studied by P-SAC outside of the framework of this program.
* PP&L will be completely open in its relations with P-SAC and will make available all relevant information.
* PP&L will commit sufficient resources of time and personnel to be responsive to all reasonable requests for information, programming, etc.
* Although PP&L will retain the responsibility for all final decisions, management will seriously consider and explore all P-SAC recommendations. If judgements continue to differ, PP&L will respond with substantiated reasons for these disagreements.

Although clearly an experimental effort, it was important to establish a jointly reached and mutually understood set of "groundrules" to provide direction for the program. A majority of the committee members said PP&L's guidelines were so specific that P-SAC did not need to clarify them further; however, the committee did make it clear that they expected from PP&L genuine consideration of P-SAC deliberations. As one member put it, "we wanted to be certain that our considerations and thoughts were taken into serious account. PP&L may not follow them, but they had to consider them seriously." This feeling was conveyed not in confrontation, but as a mutual understanding of the value of the endeavor. PP&L repeatedly gave P-SAC assurance that it respected this condition.[16] The importance of this last guideline for the success of P-SAC cannot be overemphasized, and PP&L has maintained its serious consideration of the committee's recommendations through its present deliberations. At the same time, P-SAC's willingness to participate in a rational and constructive manner assured PP&L of the sensibility of the project.

In public participation projects there is frequently a concern as to whether lay persons can effectively take part in a complex planning process without necessarily having technical training or backgrounds. Although the process for selecting members emphasized that they have some direct interest in energy matters, P-SAC quickly realized that it needed additional information. While a few members had substantial knowledge of how a utility operates, most did not. Therefore, in order to evaluate the siting process, P-SAC members asked for and received briefings by both PP&L's staff members and some outside speakers on the rationale for building a new large generating facility, and on other aspects of the utility's operation. The committee was readily able to assimilate the technical material as it continued its evaluation process, thus laying to rest any concern about competency. Also relevant here is the fact that decisions in certain areas (such as land use and community impact) are based on economic and societal considerations on which the public citizen is very capable of passing intuitive value judgements and, in fact, may bring to the decision-making process a broader range of perspective than that contained within the utility.

Another concern related to the proprietary nature of some of the material with which the committee dealt. While P-SAC viewed itself as a public body, it recognized that premature publicity or discussion could prove detrimental to the entire siting process. Therefore, the committee agreed not to send out any press release until it had completed its initial review of the siting study. P-SAC furthermore agreed to consider all the information in the draft study dealing with specific sites as proprietary because of its tentative nature and its possible disruptive effects both on the project and the people living in those areas.[17]

Although the early stages of the project were clearly organizational and educational, PP&L made sure that the committee had a clear understanding of what was expected of it, the time frame in which it had to operate, and the attitude of management toward the undertaking. Consequently, as P-SAC members began to evaluate the siting process, they did not have to waste time defining where they were going and how they would get there. The door was always open so that P-SAC could get additional information. The company made no effort to judge whether or not this material was actually needed, believing that only P-SAC members could determine what they needed to know. This freedom enabled the committee to grow and, ultimately, to branch out into other areas upon the completion of the siting study.

After getting organized, P-SAC began in April of 1979 to evaluate the process used by PP&L and its consultant to establish a preferred site, and several back-ups, for the construction of a one or two unit, 800 megawatt, coal-burning power plant. As with the Summerdale project, there was a strict deadline for completion; in this case, six months. The committee's approach was to read each chapter of the draft study thoroughly, discuss their comments, develop a consensus, and submit their findings to PP&L. The rationale behind reaching a consensus rather than providing a range of views was to provide PP&L with a clear hierarchy of sites, a necessity in the siting process.[18] If the company and consultant, based on their expertise, agreed, these modifications were incorporated into the report. When they were not, PP&L made sure that P-SAC members understood why.

As P-SAC reviewed the study, they had many criticisms, most of which fell into two categories. In the first category were editorial comments designed to make the study more understandable to the general reader by clarifying and amplifying the information presented. The second area of criticism was more serious. There were instances where P-SAC members felt that the utility and its consultant failed to review data sufficiently to ascertain whether or not a site was satisfactory. In these instances P-SAC asked for, and received, additional information about the site. Although study costs increased, PP&L felt the criticisms were valid. P-SAC completed its review in December 1977 and endorsed the study the following February.[19]

From the outset, PP&L had expected P-SAC to recommend a "prime site" for the power plant. As the review process evolved, committee members debated this issue extensively. There were some who believed that P-SAC should approve the process, not a particular site. If the process was adequate, then any site that was identified as a result of such a review procedure should also be sound. Also, because they were associated with specific constituencies, some P-SAC members felt they should avoid saying, "This is it." Other committee members felt just as strongly that the committee was abrogating its responsibility by failing to select a specific site, and that to approve only the process was to weasel out of the committee's original charge of agreeing upon a site. Ultimately, those P-SAC members who supported the process, but not a site, won out. The committee did not select a particular site. The PP&L staff members who worked with P-SAC were disappointed, but believed that this decision did not detract from the process. According to the PP&L staff closest to the project, the important thing was that the group had reached a consensus about the siting process. Upon the conclusion of its review, P-SAC spent some time discussing how it felt PP&L should present the report to the public. Many of its ideas were the same as those of PP&L's management. P-SAC felt that a popular condensation of the review process, in addition to the formal report and an executive summary, would be valuable. This report would discuss the siting process and P-SAC's efforts. P-SAC members felt that a popular version would allow the casual reader to learn more about the public participation aspects of PP&L's approach to power plant siting than he or she could learn from the study itself.

Finally, in December 1977, P-SAC made a presentation to PP&L management recapping the first phase of its work and looking ahead to its future. Mr. Busby, on behalf of the utility, reiterated his enthusiastic support of P-SAC and gave the committee PP&L's full commitment as it considered new topics in the future. P-SAC's first phase had come to an end.

How did the review process compare to expectations? For the most part, PP&L accepted P-SAC's study. It felt that P-SAC's recommendations would "result in a greatly improved study which will be more usable ultimately as a public document."[20]

P-SAC's members felt also that the process was worthwhile. It was a pioneer effort in which people with differing interests sat down to discuss their viewpoints and work out a common conclusion. Some members saw the effort as "a model to the region and to other corporations on how one can use interested citizens to assist in the corporate management process." One individual went on to say that the committee "changed the way PP&L dealt with the consultant. By questioning and probing, P-SAC got PP&L to look differently at the consultant's work and helped it review the study more critically." Other members felt that their involvement helped make PP&L realize "public participation doesn't have to be negative. There is a great deal to be gained from a continual evaluation of policies and issues and open decision-making." Also, members noted the P-SAC process stimulated PP&L to rethink its fundamental positions. It helped the utility to acknowledge the need for balance among the issues involved. Gradually, it helped the utility modify its thinking and adopt different postures in terms of energy issues. This change may have been evolving before PP&L started P-SAC, but the committee acted as a stimulant to bring the change about more quickly.[21]

PP&L was not the only party to gain from involvement in P-SAC; the committee members benefitted as well. They perceived the advantages to themselves in many ways. Members were almost unanimous in their feeling that they had gained a greater knowledge and understanding of the complexity of siting and energy issues, personal satisfaction in trying to evaluate a complex issue objectively, a greater awareness of corporate problems, and an opportunity to listen to different viewpoints and learn from them all. What's more, the mutual respect participants gained for one another enabled them to work together, later, to solve other problems. As an industry-oriented P-SAC member noted, and not facetiously, "I came to realize that environmentalists are not altogether unreasonable." At the same time, the environmentalists learned that all businessmen are not out "to rape the environment."

Finally, the project had shown clearly that the formation of a successful committee depends upon the integration of different interests into a group with a clear goal toward which it may work.[22] While the specific make-up of such a committee might vary, other utilities, industries, or government agencies that decide to emulate the process could benefit from several salient points that emerged from the PP&L experience. First, no representative of the utility should serve on the committee. The corporate staff members assigned to assist must aid the committee, but not participate in its deliberations. They should provide information, set up meetings, and coordinate committee and corporate activities. Second, it is important that the committee represent diverse interests, especially interests which normally do not interact with one another, and therein lies one of the great strengths of the committee. The members not only review new material that the utility provides, but they also debate and discuss matters with one another. One of the primary difficulties facing a group with such diverse interests is that of allowing the viewpoint of each member to surface, be expressed, and be understood, while at the same time allowing a consensus to emerge. Furthermore, it is important that this process occur within a reasonable time frame.

P-SAC GIVES WAY TO PAC

As the evaluation of the power plant siting study drew to a close, P-SAC became involved in other aspects of the utility's policy-making process. Committee members and PP&L management reviewed P-SAC's involvement with the corporation and simultaneously reached the conclusion that the committee had a broader role in PP&L's future. Consequently, in August 1978, the name of the committee was changed to the Public Advisory Committee (PAC),[23] reflecting a transition in the nature and scope of the committee's activities.

In the field of facility siting, PAC wished to revise and modify the power plant siting process so that other utilities could adapt PP&L's experience to meet their needs. During the siting study review some PAC members indicated to the committee, and to PP&L, that they would like to revise the siting process to reflect PAC's criticisms and prepare a report describing a model siting process. PAC set up a subcommittee to prepare a draft report. After reviewing the consultant's work again, discussing how PAC was involved in the siting study, and agreeing on how they would modify the procedure, the subcommittee presented a draft report to PAC for its review. The report, modified by the entire committee's critique and review, was published by PP&L in October 1978, as <u>A Model Process for Site Selection of Power Facilities</u>, and it placed emphasis on considering environmental and social concerns first and only then the engineering considerations. It was a concrete expression of the committee's commitment to the process and their feeling that they had a responsibility to help other utilities learn from their experience.[24]

More recently, PAC has been involved in reviewing the feasibility and siting of two, very different, potential projects -- a hydro/pumped-storage facility and a mine-mouth anthracite generation plant. Both projects were very clearly exploratory in nature and while they may never be built for technical and economic reasons, it is nonetheless instructive that PAC input has been requested by PP&L from the outset.

The first of these projects, the hydro/pumped-storage study, provides a very specific example of how PAC's input has altered the decision-making process. Based on the model siting process, which reverses the traditional utility approach of evaluating potential power plant sites primarily upon engineering considerations, PP&L first examined critical land use constraints and environmental criteria for several potential sites in the Lower Susquehanna River Basin. PAC's review of this phase of the study recommended that PP&L expand its study area to include all of the main stem of the Susquehanna River Basin within the utility's service territory. PAC's rationale was the lack of diversity in the potential sites, which were all in the same vicinity and offered no real alternative if local opposition to the chosen site should subsequently emerge. Although this meant a significant increase in cost and an extension of the time frame for the study, PP&L agreed to the committee's recommendation. Having a sense that there really is a range of alternatives in a given siting situation, or that the company is willing to consider other options, is an important part of public involvement in planning.

In the revised subsequent phase of the study, the environmental and community impact evaluation of twelve potential sites was done by an independent consulting firm. At a workshop held in November 1980, the consulting firm presented their findings to PAC, asking the committee to participate in a comparative site analysis based upon the assignment of importance weights to evaluation criteria previously agreed upon. During the workshop PAC raised a number of significant concerns about several of the assumptions the consultant used to evaluate the sites. Perhaps most important was the consultant's assumption that placing the facility far away from people was more important than preserving a previously unspoiled, remote site; many members of PAC did not agree. Among other considerations, PAC chose to downplay the relative importance of the number of viewers likely to be exposed to a pumped-storage facility, emphasizing instead the quality of an area independent of the number of people who might be exposed to it.

Following the workshop, the consultant produced a series of sensitivity analyses with revised importance weights to reflect PAC's concerns. Despite the initial differences between PAC and the consulting firm, two sites were recognized as clearly inferior and were dropped from further consideration. Subsequent engineering and economic reviews showed an additional six sites to be unacceptable. Eventually the study was put on hold when it proved economically unattractive to pursue the project, given present levels of peak demand for electricity. Should changing demand warrant it, however, the hydro-pumped-storage alternative to new generating capacity can be reexamined from this stage onward without having to replicate earlier phases. Most important, in terms of decision-making, PP&L agreed to incorporate PAC's concerns as noted above into the subsequent site-selection phase and worked closely with the committee in reviewing the "siting criteria, data collection, decision process and final results." This clearly indicates the committee's ability to provide direct and meaningful input into the decision-making process in more than an abstract manner.[25]

Questions related to facility siting, it should be noted, have comprised only a portion of the PAC agenda since the Committee's purview was broadened in mid-1978. By early 1979, for example, PAC had conducted a review and analysis of corporate energy planning so that they could better understand the interlocking, but seemingly separate, factors upon which the utility must make its decisions. This activity culminated in the preparation of a report entitled An Initial Report on PP&L's Energy Management/Demand Management Program.[26] to achieve the broad goals of using less electricity and achieving greater efficiency in that which is used, PAC made a number of specific recommendations such as encouraging retrofit insulation and experiments with off-peak rates. PAC continues to monitor PP&L activities in the Energy Management/Demand Management area.

During 1980 PAC devoted a portion of its time to reviewing and commenting on PP&L's Corporate Mission Statement, which resulted in PAC drafting a revised mission statement. In its explanatory comments, PAC noted that it believed the statement's scope should be expanded to encompass PP&L's more active role in such areas as energy research and development, energy conservation, and energy education -- areas which go beyond the traditional requirement of meeting electricity demands. PAC's review has been incorporated into PP&L's continually updated Corporate Objectives.[27]

During the second half of 1982, PAC was asked to review and comment on a PP&L position paper on acid rain. In its response to the position paper PAC members were quite critical and characterized the draft text as "unconvincing, undocumented, and self-serving." In the committee's view, the draft attempted to avoid environmental issues and was not consistent with the company's stated business philosophy, nor did it do justice to PP&L's demonstrated history of environmental concern. In response to this criticism, PP&L's acid rain task force redrafted the position paper to take into account PAC's comments. In particular, the new draft reflected a much clearer recognition of the existence of a problem and proposed positive ways to seek solutions, specifically an accelerated research program by PP&L. Although for perceived economic reasons the new draft did not go so far as PAC had suggested in terms of immediately increasing abatement controls, PP&L did express a willingness to take the necessary steps to decrease the emissions "if it is reasonably shown that coal burning in general and PP&L's power plants in particular are contributing to the acid rain problem." As with other projects, PAC's involvement in the redrafting of PP&L's position on acid rain has had a clearly discernable effect on corporate policy at a very basic level. PP&L has acknowledged the need to maintain a certain flexibility in its position on acid rain as new evidence emerges and events in Washington affect developments, and the company has expressed its willingness to continue PAC's involvement in its ongoing deliberations.[28]

Brief mention of several other projects and reviews that PAC has considered provides further evidence of the wide range of the committee's interest. Such reviews have included PP&L's capacity planning strategy, the Governor's Energy Council draft of a Pennsylvania state energy plan, and the corporation's public education activities. In early 1982, moreover, PAC developed a plan for becoming involved in PP&L's long-range planning process, particularly its corporate sales and load forecasting activities. This plan, to be discussed further <u>infra</u>, represents perhaps the most significant opportunity to date for PAC to influence the ways in which the company approaches its basic mission of producing and marketing electricity.

Before proceeding with that discussion, however, it is necessary to indicate that the switch from a committee that looked at a narrow utility problem (i.e., siting) within a limited time frame, to one that looks at broad corporate issues, was not entirely smooth. For example, a tactical error was probably made when PP&L asked for PAC's input on the corporation's 1983 goals, when in reality they had already been printed for tentative distribution. In effect, at least some members of the committee were affronted by what appeared to be a request for input on something that had already been finalized. In point of fact, it subsequently became clear that there was still some room for last minute adjustment, but this knowledge did little to allay the distaste the poorly-timed request left in the mouths of some members of the committee. In its response the committee made it clear that it was willing to help the company achieve its goals during the coming year and would be pleased to be involved in the process of setting the 1984 goals, but that it saw no need to collectively respond to the 1983 goals themselves. If such an ill-timed request had been made when the committee was initially formed, the consequences might have been much more damaging to its successful operation.

Fortunately, this situation was a rare exception and not typical of the utility's relationship with PAC.[29] Nevertheless, it did add somewhat to the concerns of some members of the committee who were already uncomfortable with their sense of the committee's role within the corporate decision-making process. In interviews following the P-SAC stage, a number of committee members indicated their feeling that PAC is most useful when it has a specific issue to evaluate, a definite time frame within which to operate, and a product to produce. As the committee's advisory role broadened, this specificity decreased. As a result, the committee felt the need for a clearer sense of overall direction from the corporation.

The basic question of who should provide direction, PP&L or PAC, has not been completely clarified, but to minimize this concern the PP&L liaison staff has tried when possible to provide the committee with specific issues with definite time frames for their evaluation and recommendations (for example, the review and comment approach to the acid rain position). Of course, PAC may also make suggestions as to topics it would like to investigate. For example, in the summer of 1982 PAC formed a subcommittee on corporate communications to look into the ways in which the company communicated its activities and concerns to both its employees and its customers. The subcommittee has been focusing on PP&L's communications as they relate to bringing on line the new Susquehanna nuclear power plant facility.

The end result has been an evolution of a two-fold role for PAC. On the one hand PAC has clearly taken the lead in suggesting a long-term agenda for itself, while on the other hand it has retained a short-term role of responding on a review-and-comment basis to topics brought to it by PP&L. PP&L has evidenced a desire to remain flexible in its dealings with PAC and, perhaps because of this desire for flexibility, has left an impression of a lack of direction on the part of at least some committee members who feel uncomfortable with the dual function for PAC.

As a result, PP&L has tried to provide a sense of administrative direction by assigning responsibility for PAC to the Vice President for Customer and Community Service, who attends all meetings. While this administrative change may not answer the question of direction completely, it should help to rectify some of the committee's misgivings.

RECENT EVOLUTION OF THE PAC ROLE

In an attempt to further clarify PAC's emerging role relative to the corporate decision-making structure, in early 1982 the Committee initiated and agreed in cooperation with PP&L to undertake a review of the company's long-range planning activities. More specifically, PAC decided to take an in-depth look at PP&L's corporate sales and load forecasting activities, on the assumption that they held the key and in some ways drove company policy. Ultimately PAC hoped to identify emerging social or economic issues that might affect the corporate forecasting process, which in turn would have an impact on company operations.

In June 1982 both PAC and PP&L agreed to a specific plan of action proposed by a special Long-Range Planning Subcommittee. The plan included four steps or stages:

1) an electric industry overview focusing on the regulatory and rate request processes in Pennsylvania;
2) a review and analysis of sales and load forecasting;
3) a review of emerging issues; and
4) an examination of the impact of such emerging issues on the corporate forecasting process and of the options open to the company.

The Committee embarked on steps one and two, after agreeing to defer the third and fourth stages pending the outcome of the first two. The time schedule for completion of this part of the work agenda was early 1983. During the next several months PAC undertook several site visits and listened to numerous presentations on utility operations, particularly in the targeted area of sales forecasting, in an attempt to reach an understanding of how forecasting related to long-range planning.[30]

During March and April of 1983, it came to the Committee's attention that PP&L had undertaken to develop a computerized Utility Strategic Implications Model (USIM). The USIM was designed as a short time-frame tool through which the utility could alter certain of the inputs into its forecasting process in order to see how such changes might affect the projected outcome. A number of commitee members felt USIM was duplicative of PAC efforts and expressed disappointment that PP&L had not kept the committee better informed of its actions in this area. PP&L's Director of Corporate Planning Edward Reis disagreed with PAC, noting that it did not perceive the development of USIM and PAC's activities to be the same, although it agreed they were obviously related.

After discussing a variety of possible PAC options on how to proceed from this point, PAC ultimately agreed to accept the Long-Range Planning Sub-committee's report of April 20th. After relating the project's findings, this report recommended that the current work program be terminated because of the duplicative nature of PP&L's USIM work, which PAC could not hope to match given the limited time and resources available to the Committee. PAC also endorsed the sub-committee's final recommendation that PP&L and PAC re-examine PAC's role as an advisory committee, in terms of how agenda items and priorities were set and in terms of communications procedures. In a letter to Corporate President Robert Campbell, Committee Chairman Stephen Cutcliffe noted PAC's continued interest in long-range planning but requested specific guidance from PP&L as to how PAC might best contribute toa joint effort in the long-range planning area. In a related letter he also asked for a more formalized agenda-setting mechanism and regularized communication between the PAC chair and the corporate CEO.[31]

President Campbell responded postively to PAC's request by proposing that PAC, working closely with PP&L's Corporate Planning staff, develop a range of alternative operating scenarios which the company might have to face in the future. "The purpose of developing alternative scenarios was to provide a test of PP&L's corporate strategy to determine whether it is sufficiently flexible to respond to a range of plausible futures." It was expected that the end product would be discussed at PP&L's June 1984 Executive Planning Conference.[32]

Working with the assistance of a public involvement consultant, the Committee began its task by reviewing PP&L's strategy and planning assumptions as contained in its <u>1984 Corporate Planning Assumptions</u> document. Following this review PAC developed a set of additional assumptions that the Committee believed might affect electricity demand during the 1984-2000 time frame. After developing its own assumptions, the Committee moved on to estimate the likelihood of occurrence and the impact that such occurrence might have on electricity supply and demand. During early 1984 PAC drew upon this background data to develop a set of alternative operating scenarios. This work was completed in March and submitted to Corporate Planning.

In a previously agreed upon parallel effort to that of PAC, Corporate Planning was also developing alternative scenarios for consideration at PP&L's executive planning conference. Each of the PAC as well as the corporate planning scenarios was quantified to as great an extent as possible, using the USIM model, and reviewed in detail at the Committee's May meeting. Following that meeting, PAC's alternative scenarios along with a set of issues for management consideration were then discussed at a separate meeting, attended by the vice presidents responsible for presenting the scenarios at the June PP&L Executive Planning Conference.[33]

At PAC's June 1984 meeting, Vice President for Marketing and Customer Service Grayson McNair and Director of Corporate Planning Edward Reis reported on the Executive Planning Conference's discussion of the alternative future scenarios. Following this the Committee engaged in a retrospective analysis of its involvement. In a letter to President Campbell, recently-elected PAC Chairman Oliver Smith and past-Chairman Stephen Cutcliffe summarized the Committee's reaction to the project as a whole. In brief, the Committee was very pleased with the way the process actually chosen evolved over the course of its approximately eight month lifetime, believing it was an intriguing and worthwhile endeavor for both PP&L and PAC. A number of members felt that the principal value to the company was the manner in which PAC's work with company staff shaped or influenced PP&L's own evolving thinking about its corporate plan. PAC closed its response by noting that PP&L might want to consider having the Corporate Planning staff evaluate the process to determine PAC's helpfulness from their perspective. The Committee felt this might also help it understand its past role relative to the company more completely, and provide a better sense of possible future involvement in planning activities.[34]

President Campbell's response confirmed the Committee's sense that this had been a mutually beneficial and rewarding experience.[35] Indirectly it also pointed to the value of having strong direction from the highest corporete levels in determining the direction of PAC activities. The Committee retained a great deal of flexibility in designing its own work program and setting an appropriate schedule, but the basic request and project had been laid out by President Campbell. Coincidentally, the constant interaction between Corporate Planning and PAC avoided the perception on the part of either group that there might be any sort of hidden agenda. The ensuing positive results of this project were in sharp contrast to those of the Committee's first attempt at involvement with long-range planning through its analysis of corporate forecasting. The differences in the way in which the two projects evolved, both in terms of how the initial directions were set and in terms of end results, continues to point toward the need for strong corporate guidance in clarifying the Committee's role relative to the company by helping to set the Committee's agenda and goals.

This point was perhaps further confirmed by two other projects with which PAC became involved during this same two year period (1982-1984). Coincident with the corporate forecasting evaluation, PAC was also involved in an evaluation of PP&L's model siting process for transmission lines, a process somewhat similar to the company's generation facilities siting process. During this project PAC conducted a trial siting experiment to see whether use of the model would yield a route similar to that tentatively identified by PP&L. Interestingly, there were many basic similarities between the routes chosen, which suggested the validity of the model siting process. From the standpoint of particpatory program design, however, the exercise served to demonstrate the value of concrete projects for advisory committees to sink their teeth into. As in the early P-SAC years, siting issues provided just such concrete issues.

Similarly, in the late spring of 1984 PAC became involved at a general level in a system-wide ash disposal siting project for which PP&L sought the Committee's advice. Although the Committee has not carried out specific site evluations, it has suggested that PP&L investigate alternatives to simply expanding disposal facilities, e.g., increased use of ash by-proudcts or the return of waste proudcts to the original mine sites. The company has agreed to do this. Again, the enthusiasm and ease with which PAC focused its attention on the specifics of this project continue to suggest, if not confirm, the need for strong corporate leadership in establishing appropriate roles for public involvement in corporate decision-making.

FURTHER REFLECTIONS ON THE PP&L EXPERIENCE

A Public Advisory Committee's effectiveness is very dependent upon its make-up. Membership on the committee should be open in the sense of being non-exclusionary and include a wide diversity of interests. In addition to members of the general citizenry, it helps to insure credibility for such a committee by involving personnel from state agencies or public interest groups whenever possible. However, these people often have extensive demands upon their time, making it difficult for them to make the commitment necessary for such a committee's success. Problems of attendance and turnover have not been extensive for PAC but have occurred. At present PP&L has adopted a three-year rotating term of office for PAC members to insure that new blood and ideas will infuse the committee in its deliberations while at the same time retaining a sense of continuity. This system seems to be working well, although it does mean that new members have to be "brought up to speed" with respect to past committee decisions and general utility information. Although special orientation sessions are held periodically, PP&L is presently considering ways by which to enhance the process.

As noted with P-SAC's experience, both PP&L and the committee have found that a strong support staff is necessary to PAC's success. The staff provides contact with corporate management, organizes meetings, records minutes, distributes materials, and collects information and obtains speakers when necessary. These activities essentially take the full time of a professional staff member as well as periodic assistance from other Community Services personnel. PAC minutes and requests for materials also keep one secretary busy about one-half time. Other PP&L staff have frequently provided assistance by providing answers to specific questions or by making presentations in their particular areas of expertise.

In addition to providing strong administrative support, it is crucial that the corporation be open and receptive to the input requested from an advisory group, and be well-informed about what it is doing. Prior to establishing PAC, PP&L formed an in-house steering committee to clarify what input it wanted and to be sure that PAC got the information it wanted. The steering committee did not interfere with PAC's activities or attend meetings, but it did enable management to keep abreast of PAC's activities at all times. Comprised of different departmental managers, the steering committee identified areas where the managers felt PAC could help them. In this way, PP&L established one more link between the corporation and the advisory committee.

PP&L subsequently reformulated and streamlined the administrative link with PAC by assigning the responsibility to provide support to two Senior Vice Presidents -- System Power and Engineering (who is responsible for the hardware of power plants and distribution systems), and Division Operations (who is responsible for service). Most recently, a further refinement has taken place (as noted above) with the assignment of PAC responsibility to a single Vice President, that of Customer and Community Services, who attends all PAC meetings. The result has been a fairly tight administrative structure providing visible forms of high level corporate support.

While the financial cost of operating PAC has not been insignificant, it has not been unreasonable. Direct costs have run from $12-18,000 per year. This figure includes expenditures for travel, meals, and rooms (which have been the primary expenses), plus minimal duplicating and printing costs. It should be noted that PAC members are not paid for their time. Larger indirect costs to the corporation in the form of salaries for the liaison staff and its office space are unavailable and are not reflected in cost estimates, nor are other indirect costs such as the presentations made by corporate employees. PP&L has estimated that each presentation takes approximately three days of staff time including preparation, review with the liaison staff, and the actual presentation.

A final category of costs that PP&L has incurred includes revisions and additions to a given study or program. These costs are difficult to measure and not available for every PAC study, but PP&L estimated that for the 800 MW coal plant siting project they amounted to approximately $86,000 out of a total study cost of $450,000. For the hydro/pumped storage project the corporation originally budgeted $145-178,000. After accepting PAC's recommendation that the study area be enlarged, this figure was revised upward to $244-263,000 including costs for both the outside consultants and PP&L staff. While these may at first glance appear to be large figures, PP&L believes the extra money to be well spent in the long run, principally in insuring the best possible study by using the PAC process to identify otherwise hidden issues. Not to be overlooked, in addition, is the benefit of avoiding greater unexpected costs due to public reaction at a later point in a project, for example, at the licensing stage where delays are very costly, or by discovering that they might have had to take certain steps anyway. It should also be remembered that such study dollars are ultimately only a very small percentage of a project's total costs.

In addition to PAC itself, and primarily as a result of its success, there has been a proliferation of other public participation activities at PP&L in the past several years, as indicated in Table 13.1. At this time there is no common point of coordination, with individual departments undertaking citizen involvement activities independently of each other. There is no standard set of administrative practices, although the use of an advisory committee is the most prevalent technique. However, efforts to coordinate the various public participation activities are beginning to evolve, in particular through an evaluation of the Plant Advisory Committees.

For obvious reasons, the ultimate influence of public input on management policy decisions is difficult to document, let alone measure. Clearly PAC has affected specific siting studies, e.g., the original coal-fired plant and the more recent hydro/pumped-storage facility, and at the same time has had an impact on much broader questions such as the mission statement and long-range planning. Most recently the revision of the company position on acid rain has made it clear that public involvement in corporate decision-making can have an important impact, but this still leaves open the questions of with which corporate issues should the public be involved, at what stage, and in what manner? These questions are not easily answered, and PP&L is clearly still wrestling with them. A recognition that the public cannot and probably should not be involved in every utility decision is important, just as is a recognition on the part of the utility that limiting public involvement where it naturally has a role to play in the social context of technological development would be counter-productive. Presumably, PP&L's desire to remain flexible in its utilization of public participation activities is a result both of the difficulty of answering the question as well as of a recognition of the changing nature of technological decision-making, especially now in the utility industry.

In conclusion, what can be learned from this case about the worth of public participation? In an assessment of PAC's effectiveness, executives at PP&L noted the committee's value for improving utility operation and programs by serving as a sounding board and by raising perceptive questions that might otherwise be overlooked. The corpora-

Table 13.1
Summary of PP&L citizen participation programs

PROGRAM	FUNCTION	TYPE/FORMAT	DATE ESTABLISHED	NUMBER OF PARTICIPANTS	FREQUENCY OF MEETINGS	COMMENT
Public Advisory Committee (PAC)	Review & Comment on Various Programs	Permanent Advisory Committee	April 1977	20–25	10/yr.	Intensive involvement in siting activities. PAC reports directly to senior management.
AD HOC Groups: Transmission Lines & Sub-stations	Review & Comment on Transmission Line Siting Proposals	AD HOC Committee	October 1975	15	Varies	Four committees used since 1975 at: — Summerdale — Bartonsville — Benton — Rheos
Mitigation Committees* (PHRAC)	Committee with Management on Project Impacts	Advisory Committee	November 1977	20	Varies	Recently integrated with Susquehanna Plant Committee to consolidate public participation activities at the site.
Plant Committees	Contact Point with Public on Plant Issues	Advisory Committee		Varies 20–30	Approx. Quarterly	Five plant committees currently organized at: -- Montour -- Sunbury -- Susquehanna -- Brunnar Island -- Holtwood
PP&L Consumer Advisory Panel	Provide Input to PP&L Consumer Policies	Permanent Advisory Committee	August 1978	16	Approx. Monthly	PP&L employees on panel, non-PP&L members receive compensation for time.
Division Consumer Panels	Provide Input to PP&L Consumer Practices	Permanent Advisory Committee Divison Level	1977	9–11	Varies	Advises on consumer activities at the division level.
Energy Education Advisory Panels	Develop Education Materials for Classroom Use	Advisory Committee	March 1979		Monthly, When Established	Prepares conservation education materials for classroom use; membership consists of professional educators, activated on as-needed basis; two committees operational.
Emergency Plan Advisory Committee	Review & Comment on Development of Susquehanna SES Emergency Plan	Advisory Committee	June 1979	20	Weekly-12 Held	Local community leaders assembled working committee to review development of emergency plan for Susquehanna SES — disbanded in December 1979, upon submission of report to PP&L.
Employee Consumer Panel	Employee Input for Improved Consumer Practices	Advisory Committee	1980	9–11		Oriented towards practices & procedures.

* Pond Hill Reservoir Advisory Committee.
Source: Community Services Department, March 1982.

tion perceives an added value of public involvement in the increase of credibility and goodwill that the utility gains in the general public's eye. By virtue of the fact that PP&L has moved from a single site-specific experiment through a permanent siting committee to a public participation element in all transmission and substation siting projects, as well as a wide-ranging advisory committee, it should be obvious that the utility has found what it considers to be, if not the complete answer, at least a vital component of public participation in the technological decision-making process.

But what of other corporations in fields outside the utility industry? Could they too find a public advisory committee useful? While other industries may not have a mandate to serve the public in as direct a way as the utility industry, it still seems feasible that on economic grounds alone, corporations might find the concept of use. By involving the public at a very early stage it may well prove possible not only to provide better products and services, but also to avoid public reaction that can prove costly, even fatal, to a project's outcome. Here one need only think of such varied examples as the SST, the Aswan Dam, DDT, the Alaska Pipeline, automobile safety and recalls, the Green Revolution, and the ban on cyclamates. On economic grounds alone, the message seems obvious, but when one considers the addition of social considerations as well, the answer appears abundantly clear.[36]

A public advisory committee is worthwhile because it creates one more mechanism through which different interest groups can work together to solve problems which have a direct impact on all of us. This is perhaps the greatest value and the one that is the most intangible to assess. No doubt there are other forms of public participation which other corporations and government agencies can use. Presumably no two companies or corporations would proceed in exactly the same manner, but the crucial point is the recognition that public participation is important, even necessary, to the well-being of our society. If we are concerned about the quality of life and the nature of society, we should give every such avenue and possibility a real chance to prove itself. We must learn to incorporate more fully public participation and responsibility because of the very social nature of our technology.

NOTES

1. An expanded version of this view is contained in Steven L. Goldman and Stephen H. Cutcliffe, "Responsibility and The Technological Process," Technology in Society 1 (No. 4, 1979): 275-86.

2. The following articles discuss one or more of these suggestions in some depth: Allan Mazur, "Science Court," Minerva 15 (Spring 1977): 1-14; Arthur Kantrowitz, "Controlling Science Democratically," American Scientist 63 (September/October 1975): 505-509; Allan Mazur, Andrew A. Marino, and Robert O. Becker, "Separating Factual Disputes from Value Disputes in Controversies Over Technology," Technology in Society 1 (Autumn 1979): 229-237; Barry M. Casper, "Technology Policy and Democracy," Science 194 (1 October 1976) 29-35; George Bugliarelli, "A Technological Magistrature," Bulletin of the Atomic Scientists 34 (January 1978): 34-37; and Goldman and Cutcliffe, "Responsibility and The Technological Process."

3. Dennis W. Ducsik, "Citizen Participation in Power Plant Siting: Aladdin's Lamp or Pandora's Box?," Journal of the American Planning Association 47 (April 1981): 154-66.

4. Pennsylvania Power and Light Co. Report of PP&L Task Force on Research Project in Public Participation. May 20, 1975, p. 1.

5. Ibid, pp. 1, 5.

6. Pennsylvania Power and Light Co. Summerdale: A Pilot Project in Public Participation, no date, pp. 2, 21, 25.

7. PP&L, Task Force on Public Participation, p. 4.

8. PP&L, Summerdale, p. 2.

9. As part of the plan to increase public involvement and "to give greater emphasis to the environmental, land use and social aspects of power plant siting as compared to the technical considerations," PP&L's Community Services Department organized a two-day "Delphi" session in June 1976 to "thoroughly examine power plant siting issues to insure that a broad range of concerns are appropriately reflected in our decision-making process." Participant Workbook: Site Evaluation Delphi, June 17-18, 1976.

10. PP&L, A Proposal for Public Participation in Power Plant Siting in Community Affairs, (December 1976, revised Jan. 11, 1977): 2-3.

11. Interviews with P-SAC members, Fall 1978.

12. PP&L Dept. of Community Affairs, A Proposal for Public Participation in Power Plant Siting (January 11, 1977): 15-17.

13. Letter from John T. Kaufman, Vice President, or Leon L. Nonemaker, Vice President, to participants, February 1977.

14. Interviews with P-SAC members, Fall 1978.

15. Minutes, P-SAC meeting, March 29, 1977; PP&L Guidelines for Discussion: Implementation of P-SAC Program April 1977.

16. Interviews with P-SAC members, Fall 1978. Since that time PAC has revised its working guidelines but on the important points of PP&L responsiveness to PAC, they remain in essence the same. Guidelines for PP&L's Public Advisory Committee. September 1, 1979.

17. "Policy Guidelines -- Proprietary Information," March 27, 1977, appended to P-SAC minutes, April 27-28, 1977.

18. When the committee later moved to considerations of broader PP&L policy issues, as discussed below, there was less emphasis placed on reaching a consensus. In general, the committee has felt that a consensus was desirable both from the point of view of value to the company, but also in terms of the committee's sense of need for having made some concrete recommendation. At no time, however, have minority views been repressed or discouraged, especially in areas where the company is truly unsure of its future direction. Rather, the company has taken the view that a range of opinion as well as a consensus, should one emerge, are both valuable input from the advisory committee.

19. Dames and Moore, Draft Site Selection Study for Coal Fired Power Plants, prepared for Pennsylvania Power & Light Company (April 1977). It should be noted that because of a leveling off of the projected energy growth rate, this plant has not yet been built. As a result, the siting study is presently on hold and has not been forwarded to the local public for their input and review.

20. Gerald S. Farber, PP&L Community Planning supervisor, Remarks on P-SAC to the PEA Land and Land Rights Committee Meeting, February 2, 1978.

21. Interviews with P-SAC members, Fall 1978.

22. Ibid.

23. Minutes, P-SAC meeting, August 24, 25, 1978.

24. Public Advisory Committee, A Model Process for Site Selection of Power Facilities (September 1978). Copies of this report were sent to several industry-oriented periodicals where they were advertised as publications available to the utility industry. Through this mechanism, numerous copies of the Model Siting Process have

been disseminated. Letter, Gerald Farber, PP&L Community Planning Supervisor to Editors, Electric Light and Power, Electrical World Week, Executive Perspective, Public Utilities Fortnightly, Utility Spotlight, January 8, 1979.

25. Minutes, PAC meetings March 19, 20, November 19, 20, 1980, January 21, 33, March 18, 19, April 22, 23, 1981; PP&L Pumped Storage Siting Study: Environmental and Community Impacts, Draft Report, Site Screening Phase (January 1981) and Revisions Memorandum, Scott Cragle to PAC, April 16, 1981.

26. Public Advisory Committee, An Initial Report on PP&L's Energy Management/Demand Management Program (March 1979).

27. PAC Review of Corporate Mission Statement (July 1980); Mark McClellan, PAC Chairman to Robert K. Campbell, July 28, 1980; Robert K. Campbell to Mark McClellan, September 9, 1980; March 6, 1981; Minutes, PAC meeting, November 1980.

28. PAC response to PP&L's "Position on Acid Rain," August 9, 1982; PAC comments on Final Draft 9/14/82 of PP&L "Position on Acid Rain," Oct. 13, 1982; Minutes, PAC meetings June 23, 24, Sept. 16, 17, Nov. 17, 18, 1982; Letters from Gene Gockley, Manager-Environmental Management division to Stephen H. Cutcliffe, Chairperson, PAC, Sept. 8 and Nov. 1, 1982.

29. Minutes, PAC meetings Oct. 20, 21 and Nov. 17, 18, 1982; Letter Stephen H. Cutcliffe, PAC Chairman to Robert K. Campbell, Nov. 22, 1982.

30. Sub-Committee on Long Range Planning, Proposed Plan of Action, June, 1982; Minutes, PAC meetings, June 23, 24, 1982, September 16, 17, November 17, 18, 1982; January 26, 27, February 23, 24, March 23, 24, 1983.

31. Ibid, April 20, 21, May 18, 19, 1983; Sub-Committee on Long-Range Planning, Report, in letter from Sub-Committee Chairman Frederick G. Carlson to PAC Chairperson Stephen H. Cutcliffe, April 20, 1983; Stepehen H. Cutcliffe to President Robert K. Campbell, June 7, 1983.

32. Robert K. Campbell to Stephen H. Cutcliffe, July 13, 1983; Minutes, PAC Meetings, September 21, 22, October 19, 20, 1983.

33. PP&L, 1984 Corporate Planning Assumptions, PAC Summary Report, Alternative Future Operating Scenarios, June 1984; Minutes, PAC meetings, November 16, 17, December 14, 15, 1983; February 15, 16, March 28, 29, May 16, 17, 1984; PAC Alternative Futures Operating Scenarios Sub-Committee to Vice Presidents William F. Hecht, Grayson E. McNair, and Charles E. Russoli, May 23, 1984.

34. Minutes, PAC meeting, June 27, 28, 1984; PAC Chairman Oliver F. Smith and Stephen H. Cutcliffe to Robert K. Campbell, August 15, 1984.

35. Robert K. Campbell to Oliver F. Smith and Stephen H. Cutcliffe, September 21, 1984.

36. Edward W. Lawless, Technology and Social Shock (New Brunswick, N.J.: Rutgers University Press, 1977) contains forty-five brief studies illustrating the effect of technology on contemporary lives and institutions. Although primarily written to suggest the importance of better technology assessments, the volume also implies that such assessment might contribute to reduced economic loss. See especially pages 512 and 514.

Appendixes

A
Prominent Endorsements of Open Power Plant Siting

1) "The Nuclear Regulatory Commission shall take such action as it deems necessary or desirable....to encourage or require persons proposing to construct and/or operate nuclear power reactors and persons proposing to obtain site permits for such facilities....to engage in open and advance planning for the addition of such generating capacity and the selection of such sites, with opportunity for the Commission and interested persons to participate in such open and advance planning.... Open and advance planning under this subsection is intended to promote the early identification and resolution of issues which may exist with respect to a potential site or facility."

   - Excerpt from the proposed Nuclear Siting and Licensing Act of 1978, S. 2995, introduced in the 95th Congress, 2nd Session (March 12, 1978).

2) "Optimizing the opportunity for participation of all potentially affected parties and providing for open planning at the site selection phase assures the best possibility of public acceptance of the validity of the (siting) process."

   - Southern States Energy Board, Energy Facility Siting in the United States, Volume I (executive summary), pp. xiii-xiv, Atlanta, Georgia (February, 1978).

3) "On the average, the earlier you get people involved the better.....Part of the reason for open and public participation in energy matters is the opportunity given

the public to find out what's going on, and to ask questions. Some think that, by opening things up, a lot of fears of just not knowing what's out there are reduced. It's the difference between a black box and a glass box. If you find a black box you might call the bomb squad. If you find a glass box you can look into, you can easily see there's nothing threatening. What we want to do is to put everything into that glass box so people can look."

- Excerpt from interview with Robert I. Hanfling, Deputy Undersecretary, U.S. Department of Energy, in Electrical World (December 1, 1977).

4) "Adequate opportunity for public participation in facility site planning and site analysis at an early stage must be further developed. Citizens should not have to attempt to influence site decisions long after all important decisions have been made. Therefore, utilities should disclose facility plans at the earliest possible time, and an improved planning process at the state and regional levels should provide, throughout the process, expanded ways in which individual and group views and opinions can be expressed. With improved citizen access throughout the process, relevant issues can be identified and dealt with on a timely basis."

- National Governors' Association, "Policy Statement on Energy Facility Siting," item no. 4, issued at Detroit, Michigan (September 9, 1977).

5) "Greater public participation in energy siting processes is necessary and desirable, particularly because of the large scale, financial cost, and impact of decisions involving modern power plants. The greatest public participation in planning efforts should occur at the earliest possible time, including energy growth and load forecasting and the earliest possible consideration of alternative future sites."

- National Governors' Conference, "State Perspectives on Energy Facility Siting: A Workshop Summary," preliminary finding no. 11 of the Subcommittee on Energy Facility Siting, Washington, D.C. (August, 1977).

6) "There is much concern expressed that the public does not have timely opportunity to see or understand the

long term system plans of utilities and that the identification of specific sites takes place too late to avoid a clash of commitments...To remedy these deficiencies, we have concluded that there must be a higher degree of public participation earlier in the process. There must be early disclosure of system plans and sites."

- Office of State Programs, U.S. Nuclear Regulatory Commission, Improving Regulatory Effectiveness in Federal/State Siting Actions, preliminary staff report, NUREG-0195, at 1-1, 1-4, Washington, D.C. (May, 1977).

7) "(When a) power company is evasive about how it estimates future electricity requirements, the environmentalists will suspect the worst and work to prevent construction of any plant. It is psychologically impossible to eliminate the pattern completely, but it could be altered sufficiently to give the informed public a reasonable access to and understanding of corporate and governmental decision-making processes. A public relations presentation and a pro forma public hearing are not enough. One specific reason why the courts are overused is that they are among the few places where the agency or corporation can be cross-examined and where the decision-maker is expected to listen."

- J.H. Barton, "Behind the Legal Explosion," 27 Stanford Law Review 567, at 575 (February, 1975).

8) "Early and continuous public involvement in power system planning is essential. The utilities should be required to disclose the results of their early planning to the public. Individuals and groups should be given the opportunity to make their views known before utilities are committed to particular sites, and throughout the site selection process."

- R.H. Ball, et al., California's Electricity Quandary: II. Planning for Power Plant Siting, Rand Corp., report no. R-1115-RF/CSA, at xi (September, 1972).

9) "The electrical power producing industry should continue to expand its processes of public information and, to the extent possible, open discussion. The utilities should also expand their practices of working informally with the regulatory bodies and interested persons to

include a substantial amount of discussion about factors relevant to a new proposed power plant site prior to the formal application and hearing."

- National Academy of Engineering, Committee on Power Plant Siting, Engineering for Resolution of the Energy-Environment Dilemma, at 12, Washington, D.C. (1972).

10) "The Committee recommends that all stages of the decision-making process be opened up to effective public participation. That group of recommendations flows from two conclusions: first, that a more open decision-making process is more likely to arrive at sound decisions in the public interest, and, second, that the credibility of such processes cannot withstand the cynicism that inevitably flows from the exclusion of persons or interest groups who reasonably believe themselves to be affected by the results of such proceedings."

- Association of the Bar of New York City, Special Committee on Electric Power and the Environment, Electricity and the Environment: The Reform of Legal Institutions, at 8-9, West Publishing Co. (1972).

11) "As the pattern of public resistance discussed in this paper shows, opposition starts when a company first initiates long-range planning without consulting the public on the acceptability of that planning. Consultation with state and public groups should begin, therefore, long before any public resistance to the reactor proposal exists. The utility company should pull the public into the planning and decision process for nuclear plants, including the design, siting and construction activities. The company should abandon any traditional public relations policy that emphasizes corporate anonymity and should establish a wide range of unobstructed and flexible communication channels with the general public."

- D. Jopling and S. Gage, "The Pattern of Public Political Resistance," Nuclear News, at 35 (March, 1971).

12) "Minimizing adverse effects on the environment and providing for the energy needs of society must now be regarded as co-equal objectives right from the begin-

ning. Public acceptability and related political considerations are no longer 'external' factors. Although the engineer may sometimes regard such considerations as 'irrational,' they are often powerful and even decisive, and they need to be included in any generalized systems analysis that is supposed to deal with the total problem.

In arriving at a balanced solution the systems engineer may find it imperative to involve working groups of volunteer environmentalists and other public interest groups in the planning operation from the outset...Of course, if this process is to be meaningful the systems engineer needs to keep open a wide range of possible choices of fuels, plant sites and waste heat management schemes, and not foreclose certain alternatives in advance because of relatively narrow cost considerations."

- P. Borelli, L. Lees et al., People, Power, and Pollution, Environmental Quality Laboratory Report No. 1, California Institute of Technology (September, 1971).

13) "Public knowledge of utility expansion plans and an opportunity for participation in the decision-making process for site selection is a vital element in any program for resolving conflicts over power plants and extra high voltage (EHV) transmission lines."

- Executive Office of the President, Office of Science and Technology, Electric Power and the Environment, at vi, prepared by the Energy Policy Staff (August, 1970).

14) "The planning process must to a large extent be worked out in a fish bowl where the agencies who are responsible for the quality of our air and water and for the wise use of our land can examine all the alternatives that are considered. These government agencies and the public must be satisfied that the sites selected are the optimum from the environmental point of view long before it is necessary to begin construction."

- Freeman and Weinhold, "Policy Alternatives for Resolving the Power Plant Siting Problem," in Environmental Aspects of Nuclear Power Stations, International Atomic Energy Agency Symposium, at 762 (August, 1970).

B
The Federal Government Should Encourage Early Public, Regulatory, and Industry Cooperation in Siting Energy Facilities

Planning can be improved and costly, time-consuming licensing conflicts can be minimized if energy facility sponsors effectively consult with regulators and the public about their concerns early in project plans, while plans are still flexible.

The Chairman, House Committee on Energy and Commerce, asked GAO to find out whether such "open site planning" can help balance energy and environmental concerns, and what role, if any, the Federal Government should play in increasing the use of open site planning processes.

Before preserving the Nation's environment became a major concern, utilities based their decisions for selecting and developing sites for major energy facilities on economic and technical considerations.

The traditional industry approach of deciding privately on a site, then announcing the siting commitment and defending it, complements the typical regulatory process which is normally structured to focus on a single industry proposal. However, the result of this "decide-announce-defend" siting process has often been extended conflict and controversy. Reasons cited include:

-- Project sponsors are reluctant to revise plans after applying for licenses because of their time, money, and psychological commitments.

Reprinted from Comptroller General's Report to the Chairman, Committee on Energy and Commerce, U.S. House of Representatives, EMD-82-18 (November 13, 1981).

-- Misunderstandings occur between industry and regulators about topics to be covered and techniques to be used in supporting applications.
-- Eleventh-hour public hearings raise valid issues requiring additional time and money to address.
-- The adversarial nature of regulatory proceedings promotes conflict and polarization, not negotiation and compromise.
-- Opposing sides may continue conflicts through time consuming and costly administrative and judicial appeals.

A NEW APPROACH

Opening the site planning process to regulators and the public can potentially save time and money and result in more acceptable energy facility planning. Although open site planning is not yet common practice, elements of it have been used in many situations.

"Early" and "open" are the two most important characteristics of this still evolving planning process. "Open" means comprehensively identifying, understanding, and addressing during site planning the concerns of not only industry but also regulatory officials and the public who will be called on to accept proposed energy facilities. "Early" means timing these consultations among interested parties before the project sponsor's invested time, money, and detailed planning commit it to one facility site and development plan.

In several open site planning processes GAO reviewed, industry initiatives included regulators and the public as early advisors rather than just reactive reviewers and often adversaries. In other instances, regulators and the public took major initiatives in finding sites for energy facilities rather than just reacting to private project sponsors' site planning. Facilities concerned range from power plants to synfuel facilities. Numerous organizations and individuals were contacted in reviewing these siting processes.

The relative newness of open siting and Federal regulations requiring earlier, more open Environmental Impact Statement (EIS) processes limit the information available. While available experience and evidence are insufficient to fully assess the effects of open site planning, GAO found that most participants were satisfied that it improved the siting process.

However, open site planning is not yet common practice. While its newness partly accounts for its limited use, more important reasons range from lack of knowledge about managing effective public involvement processes, to distrust between participants and reluctance to change traditional practices.

POSSIBLE BENEFITS

Early consultations with regulators and the public can potentially improve industry's facility siting plans and reduce uncertainty regarding the acceptability of industry proposals for energy facilities. Many observers see more open site planning as saving time and money for their companies. Such processes also can improve industry's credibility and image. By participating during early stages of site planning, the public potentially can better influence the nature and extent of energy facilities' environmental and socioeconomic impacts. Regulators can pursue their goals more efficiently and effectively because they are given more information and time for planning their licensing and EIS workloads, and because they have early opportunities to advise on criteria and methodologies for siting studies and environmental analyses.

While the potential benefits of open site planning are substantial, they are not assured. These approaches are new and evolving, and designing and managing such planning processes to suit varying siting environments is a formidable challenge. All must be concerned with making public involvement substantive and not superficial as has often been the case.

Open site planning is, in GAO's view, a flexible, voluntary, and cooperative supplement to later regulatory processes. Both these processes can help in balancing domestic energy development with environmental protection and public participation values. Backed by a real commitment and careful organization, open site planning can benefit industry, the public, and the Government.

A MODEST FEDERAL ROLE

Although much of the initiative in siting matters rests with the energy industry and the States, the Federal Government also has a role and can assist in promoting more open, rational, and efficient siting processes. Existing

Federal environmental policy is already basically in agreement with the open site planning concept and provides a basis for promoting its expanded use. To achieve such expansion will, however, require the cooperation of industry, the public, and the States. Given better cooperation between siting process participants, open site planning can address many objectives of recently considered Energy Mobilization Board legislation -- legislation intended to expedite the regulatory review process for energy facilities.

There are at least three situations where Federal agencies could, if requested, actively assist in promoting more open site planning:

- -- Many open site planning processes operate independently of, and sometimes begin well before, initiation of the EIS process. Supporting these industry and State initiatives is, in GAO's opinion, an appropriate Federal activity. For example, one State's site "banking" process searches for acceptable powerplant sites well in advance of specific development proposals. Such processes can raise questions about how Federal regulations might affect potential sites' acceptability. The resulting uncertainty may be minimized by early consultation with Federal officials to clarify the regulations' likely effect.
- -- Through early and open consultations, the EIS process can be, but usually is not, used as a forum to open up site planning processes. This is due partly to widespread reluctance to involve Federal agencies at an early stage of site planning, and partly to the uneven performance of Federal agencies in initiating EISs before siting commitments are made. This situation can, however, be improved if Federal agencies actively encourage project sponsors and States to voluntarily support starting the EIS scoping process at an early stage of project planning, while siting options are still open. Also, industry and States can use the EIS regulations as a basis for involving Federal agencies early in clarifying potential sites' acceptability.
- -- Information on different approaches and their results might encourage more traditionally inclined energy facility sponsors and States to try open site planning. Providing such information is, in GAO's opinion, an appropriate Federal activity.

RECOMMENDATIONS

GAO recommends that the Federal Government actively encourage the voluntary growth of open site planning processes for energy facilities. These siting processes should supplement, not supplant, traditional Federal roles in balancing energy development with environmental protection and public participation values.

The Secretaries of Energy and the Interior, the Administrator of the Environmental Protection Agency, and the Chairman of the Council on Environmental Quality should, where appropriate:

-- Cooperate with established open site planning processes where later Federal involvement is likely. Some industry and State processes that operate independently of, and begin well before, the EIS process of permitting process may want early input from Federal agencies.

-- Encourage an early open EIS process, as conceived under the 1978 regulations implementing the National Environmental Policy Act (NEPA), that facilitates more open site planning for energy facilities. Specifically, early scoping that identifies regulatory and public concerns about alternative facility sites can help all interested parties clarify sites' acceptability and plan early to minimize conflicts.

-- Advise siting process participants who are unfamiliar with it about experiences with open site planning so they can assess its usefulness, and cooperate with efforts to begin using such processes. This should be done in connection with agencies' existing NEPA responsibilities to consult with project sponsors during early planning.

AGENCY COMMENTS

The four agencies that provided formal comments on this report generally considered it accurate and agreed with its recommendations. The comments came from the Council on Environmental Quality, the Environmental Protection Agency, and the Departments of Energy and the Interior.

C
Annotated Bibliography on Citizen Participation in Power Facility Planning

This annotated bibliography is the result of an extensive literature search on "open planning," conducted in support of the Edison Electric Institute's Workshop on Utility Experience with Advance Public Participation in Planning (March 4-5, 1982; Washington, D.C.). For purposes of that project the term "open planning" was taken to mean some form of citizen participation in the process of evaluating alternative courses of action, as it occurs before decisions are reached. The concept is thus one which does not normally embrace adversarial forms of public involvement in decision-making, as manifest for example in legal intervention in various types of regulatory proceedings.

In the first section of this bibliography, entries focus for the most part on generic discussions of principles, issues, methods and practices -- not empirical work. The second and third sections, on the other hand, cover case study material that is either accessible through standard publishing sources or is available in report form only. The fourth and last section comprises a listing of other bibliographic literature in the general field of energy facility siting.

The information provided herein was gathered using conventional library research techniques supplemented by direct contact with a variety of professional personnel. The general literature review -- conducted at six different libraries on the Clark University and MIT campuses -- covered central card catalogs and indices, some twenty-two

---

* Research Assistant, Energy Studies Group, Center for Technology, Environment and Development, Clark University, Worcester, MA (Fall, 1981).

journals, several newsletters and other periodicals, and three major computer data banks (via the Dialog Search System). The supplemental sources, tapped via personal communication, included approximately 20 utility companies (many of whom had been identified in a pre-workshop questionnaire distributed by EEI staff), together with the private collections of several researchers and consultants working in the citizen participation field.

Throughout this undertaking, our chief concern was to be as comprehensive as possible in locating documentation on public involvement in both power plant siting and transmission line routing. This is not to say that the listing is an exhaustive one, which is unlikely because utility-generated literature tends to appear in diffuse and relatively obscure outlets. We do feel confident, however, that whatever stones remained unturned at this point are probably minor ones.

GENERAL ARTICLES AND REPORTS

Alpert, Geraldine. "Consumer Advisory Boards and Investor-Owned Utilities: Rhetoric and Reality." Public Utilities Fortnightly, V. 108, N. 5 (August 27, 1981): 19-22. (See also Chapter Three, herein.)

The establishment of consumer advisory boards or consumer councils has been one response by utility companies to mounting demands, during the past decade or more, for greater citizen participation in all processes which affect citizens' lives. The author of this article cautions that it would be well to examine the full implications of such citizen participation in utilities' decision-making processes prior to its adoption on any wider scale.

A critique of this article, in the form of letters from readers, may be found in the December 3, 1981 issue of Public Utilities Fortnightly, pp. 65-68. Much of the response favored the concept of consumer advisory boards and came from persons associated with utility companies which have formed advisory boards.

American Bar Association Panel, "Symposium on Public Participation in the Location of Facilities Dedicated to Public Use." Administrative Law Review, (Winter, 1972).

The central problem addressed by the Symposium is the conflict between the need to enable the public to enjoy the greatest possible degree of participation in planning processes and the need to reach a swift and certain executive decision. In a statement by John N. Nassikas, it is stressed that rapid technological changes and environmental restraints call for greater planning and accelerated adaptation to change. There is no orderly provision for public participation in siting decisions to conform with these various standards of different regulatory authorities.

Bauman, Eric. "The Role of Socioeconomic Assessment in Energy and Industrial Facilities Planning." Tempe, Arizona: 4th Annual Applied Geography Conference, Arizona State University, October 22-25, 1981.

Socioeconomic assessment has, at the minimum, two basic roles in the process of energy and industrial facilities planning, and environmental planning. First, socioeconomic assessment has a defined place in day-to-day energy facilities project planning, impact assessment, and impact mitigation. Second, socioeconomic assessment serves as a pathway for interjecting individual and community values into the traditional planning and decision-making processes. This paper briefly discusses each of these two complementary roles of socioeconomic assessment.

Borelli, Peter et al. People, Power and Pollution: Environmental and Public Interest Aspects of Electric Power Plant Siting. Pasadena, CA: California Institute of Technology, Environmental Quality Laboratory, Report No. 1, September, 1971.

This report involves a discussion of all aspects of power plant siting, particularly the public interest factors. Case studies are provided which help to illustrate the variety of procedures that have been developed by the utility to best suit its needs in dealing with conflict resolution. The case studies include: Northern States Power Company (Minnesota); Consumers Power Company (Michigan) and Yankee Atomic Power (Vermont); Commonwealth Edison (Illinois); Southern California Edison Company (Huntington Beach); and Alternate Sites of Electric Generating Capacity for California. In addition, the Report offers conclusions and recommendations in support of open planning, increased provision for inclusion of public interests, and increased research and development in the energy effort.

Borko, B. et al. Legal and Institutional Problems of Electrical Facility Siting. A summary of workshop presentations held by MITRE Corporation and sponsored by National Science Foundation, May, 1975.

This document summarizes the presentations and discussions of a workshop on Legal and Institutional Problems of Electrical Facility Siting held at the MITRE Corporation in McLean, Virginia on February 27 and 28, 1975. The purpose of the National Science Foundation sponsored workshop was to prioritize issues and identify areas for further research and development. The presentations address the current structures and problems of electrical facility siting, alternative solutions, and the impacts of utility problems on the consumer. NTIS: PB-245-220. MITRE Technical Report MTR-6917.

Cavanaugh, H.A. "The Management Report: Does Open Planning Work?" Electrical World (February 15, 1979): 37-40.

This Management Report provides a preview of a book by David Myhra concerning the role of the public in power plant siting. The book describes guidelines to be followed in establishing an open-planning and public participation program in energy planning and site selection. Specific case studies describing conflict between the public and the utility over new construction of power plants are given. These cases include Seabrook, Kaiparowits, and Kosh Konig. Myhra states that "particular attention is given to the attitudes and positions taken by project management and its interaction with site area residents and various government agencies." This Management Report also includes a thirteen-project scorecard describing the utility, project and location, key issues, and overall results with public involvement.

Creighton, James L. Public Participation: A Manual for EEI Member Companies. Report prepared for Public Participation Subcommittee, Consumer Affairs Committee, and Public Participation Task Force, Energy and Environment Committee, Edison Electric Institute, Washington, D.C.: 1984.

The product of a joint effort between two standing committees of the Edison Electric Institute, this manual is written for practitioners -- the individuals responsible for carrying out any type of public participation program. The first section describes the philosophy and rationale behind electric utility public participation programs. The second part lists and explains the various techniques that can be used to meet a utility's public participation needs. The final part consists of a series of case studies on actual utility experience with public participation programs.

D'Amore, L.J. and Associates Limited. United States Utility Experience in Public Participation. Report prepared for Ontario Hydro, Route and Site Selection Division, Toronto, Ontario, November, 1978.

This report reviews the American utility experience, based on information from twenty-nine utilities and other relevant agencies, in incorporating public participation in the context of facility siting. The background and evolution of public participation as it relates to the utility industry are reviewed, as well as the dimensions, scope and mandate given to public participation programs. Illustrative models based on approaches used in the U.S. are provided along with techniques and mechanisms to develop participation. Participation in a regulatory context is discussed.

Ducsik, Dennis W. "Integrating Coastal Zone and Electric Facility Planning: Weak Links in the Institutional Chain." Coastal Zone Management Journal, V. 8, N. 4 (1980): 263-288.

It is widely acknowledged that planning for future energy-related development along the coast must pay special attention to a variety of societal objectives regarding the wise use of the nation's shoreline. After a background discussion of the difficulties coastal agencies may face in relating to the specialized units of government that license power plants, this paper takes a critical look at both utility siting practices and government regulation thereof. Arguments are presented to the effect that important coastal considerations may not be sufficiently taken into account by utility planners and that, moreover, state and federal licensing authorities are not in any position to ensure that tradeoffs made throughout the siting process are consistent with public vis-a-vis purely private interests. In light of these findings, coastal zone management agencies are advised to strengthen their interface with the utilities and not to limit their attention to problems of interagency coordination alone. The most productive way to accomplish this is seen as a process of direct interaction between coastal planners and the electric companies, with opportunity for citizen participation as well, in which the coastal vs. inland issue and other siting matters are analyzed in an open and collaborative manner.

Ducsik, Dennis W. "Power Plants and People: A Profile of Electric Utility Initiatives in Cooperative Planning." Journal of the American Planning Association, V. 50, N. 2 (Spring, 1984): 162-174.

Many empirical studies have been done on citizen participation in planning major public works like highways and water projects, but relatively little has been written about planner-public interaction in developing power plants. This article reviews the evolution of attitudes toward public involvement within the electricity industry, then analyzes the characteristics of fifty-eight episodes of participatory planning by individual companies. A preliminary evaluation of outcomes indicates that a high proportion of cooperative ventures have been mostly positive experiences for the sponsoring utilities. The bad news is that case study information on most pilot projects is sorely deficient, and so, therefore, is our understanding of key issues in process design and implementation.

Heilman, C. Bruce; Graybill, Donald; and Schicker, Leonard Gilbert. "Information Program Gets Public Involved in New Plant's Siting and Development." Electric Light and Power. V. 56, N. 5 (May, 1978): 19.

Conflicts about where to locate electric utility facilities dictate that rational, systematic methods and analyses be used to select sites that will minimize environmental and societal effects. Each element of power facility development provides an opportunity to involve or inform the public. A utility may want to implement a policy program to: create public awareness of complex issues; instill public confidence and establish credibility; minimize intervention by preventing misunderstanding; reduce litigation and regulatory delays by addressing public concerns before commitments are made; and guide regulatory and legislative bodies with factual data about public interests.

Hendrickson, P.; Bahl, R.; Gray, B.; and Maynard, W. Measuring the Social Attitudes and Aesthetic and Economic Considerations Which Influence Transmission Line Routing. Richland, WA: Battelle Pacific Northwest Laboratories, July, 1974.

A methodology previously developed by Batelle Pacific Northwest Laboratories for siting nuclear power plants has been adapted to the problem of selecting an optimum electric power transmission line route from among selected alternatives. The unique feature of this methodology is the systematic inclusion of public values into the decision-making process.

Kloman, Erasmus H., Coordinator. "The Federal System and Energy Facility Siting," A Symposium of the National Academy of Public Administration, May, 1977.

A report based on a Symposium of the National Academy of Public Administration held in conjunction with the Spring meeting of the Academy on May 19 and 20, 1977. The report includes papers concerning these issues: forces for change in the Federal-State relationship; the state perspective; the regional perspective; the federal perspective; institutional stalemate; and commentary. The commentary included a call for early disclosure by the utilities to the public concerning long-range planning, and timely public participation.

Link, L.E. "Power Plant Siting on Lake Michigan." _Proceedings of the Symposium on Social and Economic Aspects of Water Resources Development._ Ithaca, New York: Cornell University, June, 1971.

Opinions relating to siting of electrical generating plants on Lake Michigan are summarized. A forum is needed to help achieve compromise solutions and the public must participate in the process to determine least obnoxious alternatives. The Lake Michigan situation is typical. However, the impasse can be downgraded by more sensible forum approaches, where all parties of the controversy may participate in the arbitration.

Metz, William C. "The Mitigation of Socioeconomic Impacts by Electric Utilities." _Public Utilities Fortnightly_, V. 106, N. 6 (September 11, 1980): 34-42.

This article provides a thorough compilation of measures taken by over thirty electric utilities for the mitigation of social and economic impacts of major project construction. It describes current utility involvement in mitigation efforts, changing attitudes toward utility responsibility, and potential future public policies regarding such responsibility. Issues involve: transportation; housing; education; public utilities; public health, safety and recreation; and utility and community interactions.

MITRE Corporation. _Public Participation in Energy-Related Decision-Making: Workshop Proceedings._ Sponsored by Division of Policy Research and Analysis, National Science Foundation, Washington, D.C.: January, 1977.

This document contains edited transcripts of presentations and discussion at plenary sessions of a workshop on Public Participation in Energy-Related Decision-Making sponsored by the National Science Foundation and held at the MITRE Corporation in McLean, Virginia on September 21 and 22, 1976. The Emergency Core Cooling System rulemaking, the consideration of Energy Parks in Pennsylvania, and the Seabrook, New Hampshire Nuclear Station decisions are summarized, and the process of public participation in each decision is analyzed by actual participants in the respective cases. Also summarized are the North

Anna decision, the Sears Island decision and the Big Rock Point decision. The conclusions and recommendations from working group discussions on the role and process of public participation are presented. An overall summary is provided, along with the final report of the National Academy of Public Administration Panel which was convened to assist in the design and conduct of the workshop.

Myhra, David. "The Management Report: Utility Decisions: What Voice for the Public? I-II" Electrical World, (December 1, 15, 1977).

In this two-part series, David Myhra, an expert on public participation in the siting of energy projects, conducts an interview with Robert I. Hanfling, a deputy undersecretary in the Department of Energy under the Carter Administration. The focus of the interview concerns the Nuclear Reform Act of 1977, increased public participation, the energy crisis that is not a crisis, and how utility executives can fit into the public participation role.

Myhra, David. "Will the Public Have More Say in Power Plant Siting?" Planning, V. 44, N. 6 (July, 1978): 23.

Robert I. Hanfling, Department of Energy, was interviewed concerning a nuclear power plant siting bill reintroduced in Congress in 1978. Sections of the Act include provisions for early site selection, early public notice and participation, and increased state responsibility for setting power needs and environmental standards. These changes are expected to reduce construction lag time from ten years to six. The need for the bill, the relationship between public participation and the utilities, and potential problems of implementing the bill are discussed.

Myhra, David. Energy Plant Sites: Community Planning for Large Projects. Atlanta, GA: Conway Publications, Inc., 1980.

The information provided in this book is geared toward helping local officials, planners, and private citizens to minimize the adverse socioeconomic impacts that accompany major energy facility development and construction. The author suggests a well-structured and open-planning approach. Also included are: guidelines describing the energy impact planning process; the nature of socioeconomic impacts; energy facility site selection pnergy impact management.

A major case study considers nuclear power plants and socioeconomic change. A typical planning sequence is described and it is shown that socioeconomic changes vary between project sites. Public participation issues are considered including: issue definition; past experiences by both government agencies and utilities; who is the public; public information/education as an objective; and selecting an appropriate public participation strategy.

National Governors' Conference. State Perspectives on Energy Facility Siting: A Workshop Summary. Washington, D.C.: National Governors' Conference Subcommittee on Energy Facility Siting, August, 1977.

This publication summarizes workshop discussions and preliminary findings and conclusions concerning state insights in examining existing difficulties and means by which the siting of nuclear electrical generating facilities could be expedited, including a greater need for public involvement in the siting process. The workshop was attended by represented, including a greater need for public involvement in the siting process. The workshop was attended by representatives of state government, NRC and other federal agencies, and individuals from environmental, utility, and industrial organizations.

Office of Economic and Community Development. The Siting of Major Energy Facilities. Washington, D.C., 1979.

The needs of OECD regions for new energy facilities to meet energy demand by 1985 are projected. Current siting policies and procedures are described. The need for new siting policies and procedures, including public participation, is discussed. A framework for a major energy facility siting procedure is presented and public participation in that siting procedure is recommended.

Ontario Hydro. *North American Electrical Utilities Public Participation in Route and Site Planning*. Preliminary report. Toronto, Ontario: Ontario Hydro, Community Relations Department, November, 1980.

The purpose of this study was to gain insight into the public participation experience of North American electrical utilities. Twenty-one electrical utilities responded to a survey. The study documents how and during what stage utilities involve the public in route and site planning; the types of social profiles some utilities undertake in the study area; techniques that are used to involve the public, particularly public meetings; how and what areas of public participation are assessed; the varied roles assumed by the public during a site study; and recommendations for future public participation program development. These include expanding the public participation program and increased public communication through in-house seminars with senior management; maintenance of an on-going liaison program with the public; and staff training in effective public participation techniques.

Randle, Russel V. "Coastal Energy Siting Dilemmas." *Natural Resources Journal*, V. 21, N. 1 (January, 1981): 125-159.

This article examines the rural and urban patterns of opposition which have brought coastal energy siting to a near standstill. It explores the possibility of using a tradeoff principle as a means to compromise these disputes and expedite siting. In order to understand the obstacles to using a tradeoff approach, disputes over the siting of a liquid propane gas terminal and an oil refinery in coastal North Carolina are examined.

Robertson, Jack B. "Electric Utilities and Future Shock: Choice or Chance?" *Public Power*, V. 34, N. 1 (January/February, 1976): 20.

The utility industry can expect some major social changes in the future. Effective utility executives should make something positive of these changes instead of meeting them passively. Six areas of future shock are described: increasing competition for decreasing amounts of U.S. capital; control of electricity demand; a possible nuclear moratorium; community impact of utility developments; the public's involvement in decision-making; and comprehensive land use planning.

Royal Commission on Electric Power Planning, The Decision-Making Framework and Public Participation. Issue Paper N. 8, Toronto, Ontario, Canada: May, 1977.

This paper describes the existing decision-making framework related to power system planning with respect to Ontario and Ontario Hydro, the Government of Ontario, elected officials, citizen groups and the general public, as well as the structure of the decision-making process. Issues related to public participation are discussed, including Ontario Hydro's experience with public involvement, consultation and joint planning, and the public hearing. A concentrated initial educational effort with continuing communication is called for to ensure informed participation in the decision-making process.

Stevens, David W., and Helminski, Edward L. State Perspectives on Energy Facility Siting: Improving Regulatory Effectiveness in Federal/State Siting Actions. Washington, D.C.: Prepared for the Nuclear Regulatory Commission by the Energy and Natural Resources Program, National Governors Association, March, 1978.

This publication is a composite of views expressed by states participating in NGA Energy Facility Committee workshops and in other NGA sponsored meetings. Negative features of current siting practices were identified and remedial action recommended, concerning establishment of timely planning processes, thorough analyses of site suitability, and necessary public participation mechanisms.

The Tosco Foundation. Siting Major Energy Facilities: A Process in Transition: Executive Summary. Prepared for Rockefeller Foundation, State of Delaware, State of New Mexico, Tosco Corporation. Boulder, CO/Washington, D.C.: The Tosco Foundation, October, 1979.

This report summarizes the major findings of a research project on energy facility siting, sponsored by the Tosco Corporation. These major findings involve: key siting issues; federal/state interaction issues; state energy facility siting program; and regulatory reform and streamlining the siting process. Public participation in siting processes is a commonly noted benefit of state level siting programs, though the operation of public participation process is often criticized. A variety of public participation techniques are presented and their use by states described.

U.S. General Accounting Office. The Federal Government Should Encourage Early Public, Regulatory, and Industry Cooperation in Siting Energy Facilities. A report prepared by the Comptroller General for the Chairman, Committee on Energy and Commerce, U.S. House of Representatives, Washington, D.C.: U.S. General Accounting Office, November 13, 1981. (See also Appendix B, herein.)

This report describes the state-of-the-art in open planning for energy facility sites, analyzes why the open site planning concept is not used more frequently, and recommends actions the Federal government can take to encourage its use. The report resulted from direct contact with participants in energy facility siting: industry; the public; state and local government; and the Federal government. Illustrations to reflect the variations in processes used to open up site planning are given. These include proposed power plant case histories from Northern States Power Company, Pennsylvania Power and Light Company, Utah's Interagency Task Force, and the Eastern Kentucky Power Cooperative.

Warren, Frederick, H., and Howard, E. David, eds. Land Use and Nuclear Facility Siting: Current Issues. Program report from Conference on Land Use and Nuclear Facility Siting: Current Issues. Washington, D.C.: Atomic Industrial Forum, November, 1976.

This program report outlines and summarizes the conference held on Land Use and Nuclear Facility Siting: Current Issues, held in Denver, Colorado in July, 1976. The Report includes introductory material concerning progress trends, current practice, and major policy gaps; a program report providing information on siting practices, socio-economic impacts, cooperative planning, and regulatory issues; and conclusions and recommendations. In addition, a summary of the Kaiparowits Power Project is provided. One portion of the conference dealt with current trends in nuclear facility siting in terms of public participation in that siting process. Topics included constraints in planning nuclear facilities, siting of nuclear power plants in New York State, impact on NRC licensing, and public participation and the "site bank" concept.

Williams, Edward; Massa, Alison. Siting of Major Facilities. New York, NY: McGraw-Hill Book Company, 1983.

This book offers a comprehensive, dependable, and mutually acceptable method for dealing with many issues involved in siting a major industrial or public facility -- a method that both handles the enormous quantities of complex information that go into a siting decision and promotes positive public involvement in the siting process. It thoroughly explains and illustrates the various phases of the site selection process, and includes a full chapter on data needs and analysis. Finally, a special section of the book provides a series of actual site selection case studies, each chosen to demonstrate unique solutions to specific problems.

Willrich, Mason. "The Electric Utility and the Energy Crisis, Parts I and II." Public Utilities Fortnightly, V. 95, N. 1 (January 2, 1975): 22-28; and V. 95, N. 1 (January 16, 1975): 25-34.

Part I of the series states that until a new political consensus emerges, electric utilities will have to operate in a political environment which is ambivalent at best. For utilities the issues are: how to operate in an ambivalent, volatile, and potentially hostile environment; and whether to take an active or passive role in efforts to form a new consensus.

Based upon these key factors in Part I, Part II addresses the issue of how should utility managements decide what to do. A general approach to strategic corporate planning and policy decisions is offered. The author suggests that utilities consider new approaches to decisions concerning new power plant development. First, utilities should consider ignoring present landholdings unless deemed suitable for consideration by unbiased research. Second, a new methodology for conducting alternative site searches should be developed, thus providing for public participation at the earliest stages of the site selection process. Third, an open-ended search for potential sites would begin followed by the utility formally proposing and seeking government approval for construction on any alternative site. Not only would the utility have established a site selection methodology, but also have continuing public involvement and improved public confidence in utility managers.

PUBLISHED CASE STUDIES

Anonymous. "The Cancellation of the Willamette Falls Hydro Project." P.R. Casebook: A Monthly Compendium of Selected Public Relations Programs, V. 2, N. 11 (November, 1981): 8-10. (See also Chapter Seven, herein.)

This article provides a case study involving public participation in the siting of the proposed Willamette Falls hydro project by Portland General Electric Company in Oregon. The case is unique in that the community to be affected was placed in an advisory rather than an adversary role. Although PGE already held a license for the site, a twenty-member Community Advisory Committee had responsibility for reviewing technical material, soliciting community reaction and concern, and making a final recommendation to PGE. The Committee recommendations indicated that PGE had community support from the energy standpoint but not from an environmental standpoint, so the project was cancelled and the "voice of the community" was heard.

Blackmon, Donald B. "Utility Industry Experience With Early Site Review." Paper presented at Conference on Current Issues on Environmental Regulation of Nuclear Power Facilities. Washington, D.C.: Atomic Industrial Forum, October 12, 1977.

The author describes Duke Power Company's experience and concerns with early site review processes. He details the Company's involvement in the development of a facility siting procedure demonstration project. A power plant siting process in three phases is outlined. Phase I develops site-plant alternatives, through a logical, subjective and objective identification of potential sites. Phase II concerns the evaluation by state agencies, federal agencies, and the general public of the site-plant alternatives with the inputs being used to select the two best fossil and nuclear site-plant combinations. Phase III of the process is the ultimate decision, by Duke concerning the site for the next thermal baseload power station. Based upon the positive demonstration project experience, it is believed that legislative and procedural changes in the siting process are necessary.

Comstock, Roland. "Critical Precedent." Water Spectrum Magazine. V. 4, N. 4 (Winter, 1972): 1-6. (See also Chapter Eleven, herein.)

NSP's Director of Environmental Affairs recounts how the company began its ongoing affair with open planning by laying a major plant siting decision before a panel containing some of its most skeptical detractors. By every standard -- including some that were not anticipated -- this initial effort was a success both from the corporate point of view and from the community point of view. At the same time, it was learned that the major obstacle to such an effort is distrust, and the company must work long and hard and patiently to overcome this obstacle.

Chaney, Henry W. "Public Participation From A Utility's Viewpoint: Benefits and Problems." Paper presented at Engineering Foundation Conference on Energy Complexes, August 3-8, 1975, New England College, Henniker, New Hampshire; New York, NY: American Society of Civil Engineers, 1976.

A Duke Power Company representative describes a number of siting projects in the utility's history that involved public participation to some degree in the siting process. In the early '60's, the Company used a press release and a Governor's Speech to inform the public about a new hydro plant near Charlotte, North Carolina. The Company built the Keowee-Toxaway Visitor's Center to help explain nuclear generation to the public. The Company's first attempt at getting the public involved in siting took place with the Bad Creek Project. A series of meetings with the public, environmental groups, and academicians provided valuable input and convinced the Company that public participation can indeed work.

Dietz, Joseph F. "Public Participation in Power Plant Siting: A Case History." Paper presented at Conference on Land Use and Nuclear Facility Siting: Current Issues. Washington, D.C.: Atomic Industrial Forum, November, 1976.

In trying to resolve the problem involved with ranking candidate siting areas for power plants, San Diego Gas and Electric Company formed an Ad Hoc Environmental Advisory Committee to assist in the environmental evaluation of new power plant sites. A process was developed in which the public could voice an effective opinion. Methods for effecting meaningful participation of the public in the site selection process are discussed along with the public response to various invitations, interviews and public hearings.

Foster, Charles H.W. The Northfield Mountain Pumped Storage Project: Counter-point to Con Ed. Petersham, MA: Harvard University, Harvard Forest, Paper N. 19, April, 1970.

This analysis of the Northfield Mountain pumped storage project concentrates on the process used to obtain an acceptable balancing of economic and environmental interests as well as the conservation ethic. The success of the Northfield Project is contrasted with the ill-fated Storm King proposal in New York. The utility involved in the Northfield Project chose to be entirely candid with state officials from the outset; recognized the technological aspects of the project; and provided positive responses and initiating actions in relation to natural resources.

Holdren, John; Herrera, Philip. Energy: A Crisis in Power. San Francisco, CA: Sierra Club Books, 1971.

This book explores the energy/environment dilemma on two levels. First, a scientist examines the sources and consumption of energy in the United States and describes how the production of power inevitably leaves its mark upon the environment. Then a journalist takes over, reporting on the efforts of concerned citizens to avoid air, water, and landscape pollution. The next-to-last chapter, "Planning with People," is a brief and cursory account of open planning initiatives taken by Northern States Power Company and Northeast Utilities.

Jopling, David G. "Plant Siting at the Florida Power and Light Company." Public Utilities Fortnightly. (June 21, 1973): 27-31.

The new Florida Power and Light Company's power plant siting program is centered in the Environmental Planning and Research Department, which is responsible for coordinating and managing all company environmental research. The Department is required to prepare environmental impact statements for new power plants, and aims for greater public participation and more efficient planning approaches. The planning process for the Company's Martin County fossil-fueled plant is described.

Jopling, David G. "Plant Site Evaluation Using Numerical Ratings." Power Engineering. (March, 1974): 56-59.

Florida Power and Light Company's new site evaluation technique, capable of comparing 100 alternative sites at one time, has been used in siting three of the Company's power plants. Use of this technique, which estimates and compares variables such as engineering, cost, and environmental and political tradeoffs, is aimed at evaluation of site alternatives to be included in a ten year plan required by state law. The site rating system is considered to be a highly effective segment of the environmental planning process, particularly when supplemented by other programs.

McGinnis, R.S. Power Plant Siting - Social/Political Considerations. Reprint No. 1528. American Society of Civil Engineers, ASCE Annual and National Environmental Engineering Meeting. St. Louis, October 18-22, 1971.

Major facility siting involves local and community consideration. Since social considerations vary among regions, Northern States Power decided to go directly to the public for precise problem definitions. In 1970, leading environmental and conservation groups along with Minnesota's environmental cabinet were invited to attend meetings to discuss future siting and development of transmission lines and power plants in the area. A plant siting task force, which is still being used, was formed and assisted in making a recommendation on the location of a generating station.

MITRE Corporation. Public Participation in Energy-Related Decision-Making: Six Case Studies. Washington, D.C., 1976.

This document contains case studies of public participation in six decisions on or related to energy facility siting. These studies document the events of each case and then analyze and evaluate the role and mechanisms of public participation in the decision-making process. The six cases studied are the Emergency Core Cooling System; Energy Parks in Pennsylvania; the Seabrook, New Hampshire nuclear power plant; the North Anna, Virginia nuclear power plant; Big Rock Point, Michigan; and the Searsport, Maine oil refinery.

Myhra, David. "One Nuke Gets a Warm Welcome." Planning, V. 41, N. 8 (September, 1975): 13-18.

The article describes a precedent-setting process through which a municipal body and a power company came to an agreement involving the siting of a nuclear power plant. The case in point involves the Puget Sound Power and Light Company of Bellevue, Washington, and Skagit County, Washington. The power company agreed to provide impact payments to schools and law enforcement agencies in the county as a condition for obtaining a rezoning permit.

Roberts, Marc J., and Bluhm, Jeremy S. The Choices of Power: Utilities Face the Environmental Challenge.

Cambridge, MA and London: Harvard University Press, 1981.

Six case studies of electric utilities, three public and three private, are used to illustrate different ways of dealing with regulations to protect the environment and to offer a framework for understanding the decision-making process in large firms. Based on in-depth interviews with managers, the authors describe the formulation process for each utility's environmental policy and describe a number of factors motivating organizational behavior. The six case studies involve the following: the Tennessee Valley Authority; the Pacific Gas and Electric Company; Ontario Hydro; the Southern Company; the Los Angeles Department of Water and Power; and Boston Edison.

"San Diego G & E Tries Sweet Reason." Electrical World (April 15, 1972): 25-26.

This article describes an experimental two-day open meeting held by San Diego Gas and Electric Company involving the public in the study of a proposed new power plant. It was the first meeting of its kind on the West Coast, and its purpose was to gather information on public concerns. The meting was conducted quietly, was broadly advertised beforehand, and resulted in a number of valid recommendations.

Seeley, Wiliam S. "Open Planning for Corporate Responsibility. Public Utilities Fortnightly, June 21, 1973. (See also Chapter Eleven, herein.)

The Northern States Power Company has persuaded Minnesota's Environmental Quality Council that open power plant siting -- involving public and governmental participation -- can save substantial amounts of money in minimizing plant permit delays, legal fees, and management time spent in prolonged public confrontation. The article describes the successful use of a citizens advisory task force that included members from the environmental community to participate in the power plant siting decision.

Seiple, W. Ray. "New Approach to Site Selection Studies." Journal of the Power Division, American Society of Civil Engineers (July, 1974): 63-69.

Site selection and environmental impact studies for conventional and nuclear power plants historically have been performed under security restrictions. San Diego Gas and Electric Company used a new approach, hiring the consulting firm, Dames and Moore, to complete a site evaluation study for a combined cycle power plant and to provide full public disclosure. An ad hoc committee of representatives from environmental and planning groups was formed early in the project. A number of suggestions for improved communications and relations with committee members are offered including optimal length, style and content of presentation, and meeting length and frequency.

Smock, R.W. "The Greening of Northern States Power." Electric Light and Power. (February, 1973): 37-39. (See also Chapter Eleven, herein.)

Northern States Power Company officials realized during the 1960s that new approaches to plant siting that involved the public were necessary. This article describes the company's efforts to meet its corporate responsibility by creating a siting task force and beginning education programs for its employees. With respect to the task force, which was made up of state government and environmental group representatives, members were asked to choose one from among four possible sites for a coal-fired plant. The task force submitted two separate final reports and the Company ultimately agreed with the environmentalists' site choice. The company feels that the task force is of great value in paving the way for new site acceptability.

Stroube, Hal. "Fishbowl." Public Utilities Fortnightly, V. 93, N. 13 (June 20, 1974): 31-37.

It is commonly said that regulated utilities must live in a fishbowl because of the continuous surveillance of their fiscal and other operations. For electric utilities, a new tier of regulatory inspection and supervision has been imposed by Atomic Energy Commission regulations for those seeking nuclear plant licenses. The resulting difficulties, in terms of additional cost and delay, are discussed. The article describes how and why Potomac Electric established its Office of Public Responsibility, its new effort to emphasize the intention of using open planning with respect to nuclear power development.

U.S. Department of the Interior. Resource and Land Investigations (RALI)Program: Considerations in Evaluating Utility Line Proposals. Prepared by MITRE Corporation for U.S. Department of the Interior, Bureau of Land Management, September, 1975.

Considerations involved in evaluating the location of a proposed utility line and its probable environmental impact are presented. The principal environmental factors are analyzed, together with the technical variables involved in constructing electric transmission lines and pipelines. The legal requirements that generally must be met at the various governmental levels are summarized. Considerations affecting shared or multiple use of a utility corridor are addressed. A set of case studies illustrates methodologies available to utilities in locating lines. The case studies include: Virginia Electric and Power Company concerning transmission lines serving Northern Virginia; Ontario Hydro concerning a transmission corridor; Environmental Decision Alignment Process (EDAP) and Processing of Energy Routing Minimizing Impact from Transmission Systems (PERMITS) methodology used in Southern Wisconsin; and the Northern Border Pipeline Company concerning a natural gas pipeline through Canada and the United States. Each case study focuses on public participation techniques used as well as other criteria used in locating utility lines.

U.S. Department of the Interior. Scoping Analysis and Public Involvement: Summary for the Proposed New Mexico Generating Station Environmental Impact Statement and the San Juan Basin Action Plan Cumulative Overview. Santa Fe, New Mexico: U.S. Department of the Interior, Bureau of Land Management, May, 1981.

This document has been prepared to provide agencies and interested parties with the results of the Bureau of Land Management's scoping and public involvement efforts relating to the New Mexico Generating Station Environmental Impact Statement and the Cumulative Overview Portions of the San Juan Basin Action Plan. The information contained in this summary document was obtained through fifteen public meetings, and represents views of hundreds of participants. The scoping process is federally required and is meant to provide public involvement to the environmental impact statement process. The Public Service Company of New Mexico is indirectly involved in this process.

Whitlatch, E. Earl, and Aldrich, John A. Energy Facility Siting Procedures, Criteria, and Public Participation in the Ohio River Basin Energy Study Region. Prepared for the Ohio River Basin Energy Study. Washington, D.C.: U.S. Environmental Protection Agency, Office of Research and Development, August, 1980.

Findings are presented on the adequacy of current review procedures, criteria, and public participation in energy facility siting for nuclear and coal-fired plants. Three case studies of energy facility siting are presented in detail for coal-fired power plants, all including specific accounts of public participation activity. The case studies are: East Bend in Kentucky; Killen in Ohio; and Mountaineer in West Virginia. Although there has been no experience with an open planning approach to EFS in the ORBES region, recommendations are made to this effect. These include: public involvement during early site screening stages; regional siting reports to be made available to the public; to include more of the public's concerns at an earlier stage in the sitng review process; and to initiate an experimental open planning process.

UNPUBLISHED CASE STUDIES

Clemente, Frank, and Ferrar, Terry A. "The Energy Park Experience in Pennsylvania." Paper prepared at Center for the Study of Environmental Policy, the Pennsylvania State University, 1976.

This paper presents a history and an analysis of the events surrounding the Energy Park controversy in Pennsylvania, including a description of the public participation process involved and an evaluation of the role of key groups in the controversy. Public involvement was essentially confined to public meetings in each candidate county. Suggestions are made for more effective public participation efforts in the future.

Harwood, Michael. Getting Together: Can the Electric Utilities and the Environmentalists Learn to Communicate Sensibly Thus Saving Us All A Lot of Trouble? Unpublished manuscript, Washington, Connecticut: (October, 1974). (See also Chapter Twelve, herein.)

This manuscript provides an in-depth study of Project CEVAL (Citizens' Evaluation), a unique and privately-funded experiment conducted by Northeast Utilities in its 1971-1972 planning for a large pumped storage facility on the Connecticut/Massachusetts border. The author evaluates the experiment in book format, includes impacts that all major parties to the controversy had on the experiments, and compares that to a similar experiment in Minnesota.

Ontario Hydro. North Channel Site Selection Program. Report submitted by West Citizens Committee (Blind River) to Ontario Hydro, January 16, 1978.

This report represents the views and opinions of the Western Citizens Group involved in the selection of a site for an energy center on the North Channel of Lake Huron. The Committee Report outlines the project; personnel involved; study activity; observations, conclusions; recommendations; and, a postscript suggesting that Group members felt the objective of their participation, involvement and input, was achieved.

Ontario Hydro. Public Participation Workshop: Canadian Electrical Utilities. Final Summary Report of Workshop held in Toronto, November 4-6, 1981. (See also Chapter Five, herein.)

This Report provides a detailed summary of the three day public participation workshop, the goals of which were: to share experiences and corporate philosophies; to identify and discuss some common problems as practitioners in the public participation field; and to discuss techniques. Representatives of participating corporations provided an overview of their company's approach to and problems encountered with public participation programs. Three case studies from Ontario Hydro, British Columbia Hydro, and Saskatchewan Power Corporation provide specific details involving public involvement techniques and areas of concern. The two most important areas for concern dealt with internal problems: lack of commitment to and understanding of public participation; and attitude and involvement of senior management.

Ontario Hydro. Public Participation in Route and Site Planning. Report No. 80384. Ontario Hydro, Community Relations Department, October, 1980.

Major strides were taken during the seventies to include citizens, interest groups and communities in the planning of transmission corridors and power plant sites. The process to involve the public in locational planning of new electrical facilities is continually evolving in Ontario Hydro. This document describes the public participation process as it evolved up to 1980, and reviews the reasons for involving the public. It also discusses the role of the public in Ontario Hydro's planning studies, methods of participation, and the means of evaluating the effectiveness of the public participation program.

Pennsylvania Environmental Research Foundation. Public Involvement in Corporate Policy Decisions: A Report on the Experience of the Public Advisory Committee of Pennsylvania Power and Light Company, Philadelphia, PA (March, 1979). (See also Chapter Thirteen, herein.)

Pennsylvania Power and Light Company began a public participation program in the spring of 1977 in an effort to involve the public in the decision-making activity associated with power plant siting. This report explains how the Permanent Siting Advisory Committee (P-SAC) was developed and expanded to become the Public Advisory Committee (PAC) of the utility. One of the Committee's first tasks was to evaluate the site selection process for an 800 megawatt coal-fired generating facility within six months. The study was evaluated, and substantive as well as editorial changes were recommended and in most instances accepted by the utility. PAC is also involved with discussing energy management and demand management issues and has prepared a model siting process. Organization, faith, and trust are critical to successful public participation.

Pennsylvania Power and Light Company. Summerdale: A Pilot Project in Public Participation (undated).

This report describes a Pennsylvania Power and Light Company (PP&L) pilot project to permit citizens of Summerdale, Pennsylvania to participate in the location of a substation and transmission line. The project was deemed a productive public participation experience by PP&L in that the public can actually work with a utility in facility location. The project involved the formation and utilization of an ad hoc committee made up of twenty-six people representing a balance of political, institutional, civic and social groups, as well as non-affiliated individuals.

Pennsylvania Power and Light Co. Benton: Public Participation in Electrical Facility Siting: A Pilot Project (undated).

Pennsylvania Power and Light Company (PP&L) undertook a pilot project in the primarily rural area of Benton, Pennsylvania, to gain further experience in public participation. The objectives of the project were twofold: to obtain meaningful participation in the site and line selection process, and to develop public participation policies that are consistent with PP&L's corporate and public service responsibilities. This report describes the experiences of the Benton project which demonstrated that public participation can be achieved in a primarily rural area.

Pennsylvania Power and Light Co. A Model Process for Site Selection of Power Facilities. Report of the Permanent Siting Advisory Committee to Pennsylvania Power and Light Co., September, 1978.

This report describes a model process for identification of sites for power facilities that is designed to be flexible; to function as part of the larger process of facility planning, siting, licensing, and construction; and to help shorten the time necessary to build a power plant. The process was developed by the Permanent Siting Advisory Committee (P-SAC), a committee created by Pennsylvania Power and Light Company (PP&L) because it believes that early public involvement in successful power facility siting is crucial.

Pennsylvania Power and Light Co. Report of PP&L Task Force on Research Project in Public Participation, May 20, 1975.

The Pennsylvania Power and Light Task Force recommended the creation of a public ad hoc facility siting advisory committee for the purpose of routing or siting a specific proposed facility. This report outlines the steps necessary for establishing such an ad hoc committee and for testing it on a pilot basis.

Portland General Electric Company. Final Report on Public Information/ Involvement Process for Proposed Willamette Falls Hydroelectric Project. Prepared by Don Barney and Associates, Inc. (undated).

This Final Report traces the progress of and provides an evaluation of a public education/public involvement program designed to give Portland General Electric Company recommendations on its proposed hydro plant in Oregon City. The utility management had made a commitment that the community's response would play an important role in the final building decision. The process involved distinct public information and public involvement components. The information component provided information on the public involvement process and the involvement component centered on the Citizen Advisory Committee. The evaluation outlines the positive features of the process and suggests areas for improving public participation.

Portland General Electric Company. Report to the General Electric Company From Willamette Falls Citizen Advisory Committee. Portland, Oregon: Portland General Electric Company, February 11, 1981.

This report provides a detailed overview of a four-month decision process, performed by the twenty-member Willamette Falls Citizen Advisory Committee, in connection with Portland Gas and Electric Company's proposed hydro project. The report includes subcommittee reports on energy needs, community impact, and fish and recreation. A list of recommendations to PGE from the Committee advised that the plant not be built unless certain conditions were met in order to protect fish runs and the scenic and historic beauty of the falls.

Short, Robert H. "Public Participation in the Willamette Falls Hydro Project (or) How PGE Finally Got a Hearing Aid." Remarks before the Tri-Cities Chamber of Commerce, Oregon, December 9, 1980.

The speaker describes the process by which Portland General Electric Company involves the public in the decision process concerning the proposed Willamette Falls Hydro project. This public participation process included the formation of a twenty-member Citizen Advisory Committee, meetings, discussions, hearings, and questionnaires. The utility stated that for a "go" decision on the project, it must be cost-effective and have community support. The speaker also describes the utility's commitment to listening to the voice of the community.

Solandt Commission. Report of the Solandt Commission: A Public Inquiry into the Transmission of Power Between Nanticoke and Pickering. Ontario: Solandt Commission, 1974.

The report details a study conducted for Ontario Hydro, by Bhi Limited, concerning the need for a major new east-west transmission line. Based upon preliminary yet extensive research by Ontario Hydro for such a transmission line, a more thorough and systematic study of potential routes as well as public involvement from the outset were called for before any route decisions were made. As a result of this experience that included citizens in the siting process, a number of considerations for public participation in future planning activities are presented.

Tennessee Valley Authority. Consumer Roundtable Series on the Statement of Corporate Purpose and Direction. From Citizen Action Office Files, TVA, Knoxville, TN, April 2, 1981.

The Citizen Action Office, in cooperation with the Office of Planning and Budget, developed a public participation program to integrate consumer opinion into the development stage of policy and program planning by soliciting public response to TVA's Statement of Corporate Purpose and Direction. This paper outlines the criteria addressed in developing the public participation process, including not only objectives and identification of participants, but also the how to's of setting up and facilitating roundtable discussions.

Tennessee Valley Authority. 1981 TVA Annual Public Participation Report. Prepared by TVA Citizen Action Office, September 15, 1981.

This report, published by the Citizen Action Office, is designed to be used as background information for both office situation assessments and the annual TVA policy conference. Public interests are quantified based upon calls received over the Citizen Action Line (90% of the calls originated in Alabama and Tennessee) over a one year period from September 1, 1980 to September 1, 1981. Trends in public interest include water resources, energy conservation, and employment. An increase in calls cconcerning power rates was noted. Also in the report is a list of public meetings held during the year on a variety of issues.

OTHER RELEVANT LITERATURE REVIEWS

Cluett, Christopher; Greene, Marjorie; Redford, Linda. "Individual and Community Response to Energy Facility Siting: A Review of the Literature." Seattle, WA: Batelle Human Affairs Research Center, November 30, 1979.

This study reviews selected literature areas, including the boomtown studies, that focus on the social, economic, and political effects of energy development, and primarily on the impacts of coal extraction and processing on rural communities. Abstracts of selected studies within this literature are appended to the report as an annotated bibliography. The abstracts summarize research findings that pertain to the process of community development, social change, and public reaction to rapid energy development in rural areas. The primary objective is to describe and explain who the actors are, how they respond to prospective development, and why they either encourage, resist, or remain indifferent to facility siting and development.

Freedman, A., ed. Energy Facility Siting: Selected Abstracts. McLean, VA: MITRE Corporation, August, 1976.

This bibliography contains energy facility siting information from Federal, state, and private agencies. The material is divided into thirteen technology and policy sections as follows: coal; other fossil fuels; nuclear; geothermal; ocean thermal; solar; other energy forms; economics; finance; environmental; sociological; energy centers; policy, planning constraints; and resource availability. Energy facility siting was defined to include all issues that affect the time, cost or acceptability of bringing new energy facilities on line.

Hamilton, Michael S. "Power Plant Siting: A Literature Review." Natural Resources Journal, V. 19 (1979): 75.

The author reviews literature concerning electric utility planning and operations with regard to issues such as interconnection and interdependence, interconnections in public policy, coordination, public/-private ownership, size of planning units, federal power agencies, the Northeast Power Failure of 1965, reliability, regional coordinating councils, increasing scale of technology, energy centers, regional perspective, state siting controls, coastal zone siting, location and investment analysis, power systems engineering, and siting methodologies.

Hundemann, Audrey S., ed. Power Plant Siting: A Bibliography With Abstracts. NTIS/PS-78/1007, Washington, D.C.: National Technical Information Service, September, 1978.

Abstracts pertain to the siting of nuclear power plants, solar thermal power plants, ocean thermal gradient plants, and coal-fired utilities. Emphasis is on nuclear power plant siting, including the siting of underground nuclear facilities. Subject areas covered include geology, seismology, safety, social effects, economics, and environmental impacts. abstracts dealing with siting such facilities in countries other than the U.S. are included. 310 abstracts. Period covered: 1964 -- September, 1978.

Klema, Ernest D., and West, Robert L. Public Regulation of Site Selection for Nuclear Power Plants: Present Procedures and Reform Proposals -- An Annotated Bibliography. Research Paper R-2. Washington, D.C.: Resources for the Future, Inc., 1977.

This bibliography was prepared as a first stage of a project to identify and evaluate feasible alternatives to the present procedure for the siting of nuclear power plants. The bibliography is divided into these four sections: the process of power plant siting; practices of enterprises, including siting procedures and analytical tools for site assessment and a range of siting considerations; public regulation of power plant siting; the contemporary critique of regulatory experience; and reform proposals.

Nakamura, Sandra L., ed. An Annotated Bibliography: Energy Facility Siting and Licensing. Washington, D.C.: Atomic Industrial Forum, Inc., September, 1978.

This annotated bibliography covers siting and licensing of power plants and associated energy facilities. Material was prepared from the Atomic Industrial Forum's INFORUM data base which includes abstracts from environmental reports, environmental impact statements, and generic studies. The bibliography is divided into five sections. The general siting and licensing section includes abstracts of generic studies prepared primarily by state and federal regulatory agencies. The methodology section includes analytical assessment techniques used by industry. The third section focuses on population, land use, and environmental variables as factors in siting decisions. The fourth section identifies candidate areas for potential plants, alternate sites, and cost-benefit analyses and early site review studies. Section five is a corporate author and geographic index.

Made in the USA
Monee, IL
12 November 2026

955cc816-3e85-4807-a1c5-c1a6d105e944R01